Sensor Engineering Systems: From Theory to Applications

Sensor Engineering Systems: From Theory to Applications

Edited by **Marvin Heather**

WILLFORD PRESS
New York

Published by Willford Press,
118-35 Queens Blvd., Suite 400,
Forest Hills, NY 11375, USA
www.willfordpress.com

Sensor Engineering Systems: From Theory to Applications
Edited by Marvin Heather

International Standard Book Number: 978-1-68285-094-7 (Hardback)

Printed in the United States of America.

Contents

Preface

Every book is a source of knowledge and this one is no exception. The idea that led to the conceptualization of this book was the fact that the world is advancing rapidly; which makes it crucial to document the progress in every field. I am aware that a lot of data is already available, yet, there is a lot more to learn. Hence, I accepted the responsibility of editing this book and contributing my knowledge to the community.

Sensors form a crucial part of measurement systems. This book elucidates the concepts and innovative models revolving around present as well as prospective developments of sensor engineering. Included in this book are coherent discussions on topics such as sensor principles, measuring systems, sensor materials, MEMS technology, etc. The objective of the text is to give a general view of the different aspects of sensor engineering. It will be an excellent resource guide for students as well as academicians pursuing sensor engineering and associated fields.

While editing this book, I had multiple visions for it. Then I finally narrowed down to make every chapter a sole standing text explaining a particular topic, so that they can be used independently. However, the umbrella subject sinews them into a common theme. This makes the book a unique platform of knowledge.

I would like to give the major credit of this book to the experts from every corner of the world, who took the time to share their expertise with us. Also, I owe the completion of this book to the never-ending support of my family, who supported me throughout the project.

Editor

Work area monitoring in dynamic environments using multiple auto-aligning 3-D sensors

Y. Wang, D. Ewert, T. Meisen, D. Schilberg, and S. Jeschke

Institute Cluster IMA/ZLW & IfU Institute, RWTH Aachen University, Germany

Correspondence to: Y. Wang (ying.wang@ima-zlw-ifu.rwth-aachen.de)

Abstract. Compared to current industry standards future production systems will be more flexible and robust and will adapt to unforeseen states and events. Industrial robots will interact with each other as well as with human coworkers. To be able to act in such a dynamic environment, each acting entity ideally needs complete knowledge of its surroundings, concerning working materials as well as other working entities. Therefore new monitoring methods providing complete coverage for complex and changing working areas are needed. While single 3-D sensors already provide detailed information within their field of view, complete coverage of a complete work area can only be achieved by relying on a multitude of these sensors.

However, to provide useful information all data of each sensor must be aligned to each other and fused into an overall world picture. To be able to align the data correctly, the position and orientation of each sensor must be known with sufficient exactness. In a quickly changing dynamic environment, the positions of sensors are not fixed, but must be adjusted to maintain optimal coverage. Therefore, the sensors need to autonomously align themselves in real time. This can be achieved by adding defined markers with given geometrical patterns to the environment which can be used for calibration and localization of each sensor. As soon as two sensors detect the same markers, their relative position to each other can be calculated. Additional anchor markers at fixed positions serve as global reference points for the base coordinate system.

In this paper we present a prototype for a self-aligning monitoring system based on a robot operating system (ROS) and Microsoft Kinect. This system is capable of autonomous real-time calibration relative to and with respect to a global coordinate system as well as to detect and track defined objects within the working area.

1 Introduction

The ability to autonomously acquire new knowledge through interaction with the environment has been in the focus of significant research in the field of dynamic work area. Challenging research topics arise in pose estimation, sensor alignment and object recognition. In order to accurately manipulate the objects in a dynamic work area, a reliable and precise vision system is required in a robotic system to detect and track workpieces and to monitor the operation of the robots to accomplish manufacturing tasks such as assembly planning (Ewert et al., 2012). Such a vision system not only has to be aware of the presence and location information in the working site, but also needs to have the information of its own real-time position and orientation as sensors.

Rather than being fixed, the vision system has to be able to move accordingly to provide a complete coverage in a dynamic scenario. To meet the above-mentioned requirements, we present a prototype for a self-aligning monitoring system based on an ROS and Microsoft Kinect. The main tasks of the vision system are autonomous self-calibration both relatively and with respect to the global coordinate system and target detection and tracking within the working area.

The proposed 3-D monitoring system, comprised of multiple Microsoft Kinects, is capable of self-alignment through calibrating Kinect both individually and as a stereo camera with reference to markers to obtain the relative location information between each other, as well as their pose in the global coordinate system. Two Kinects placed with a certain angle and distance with regards to each other can enable a

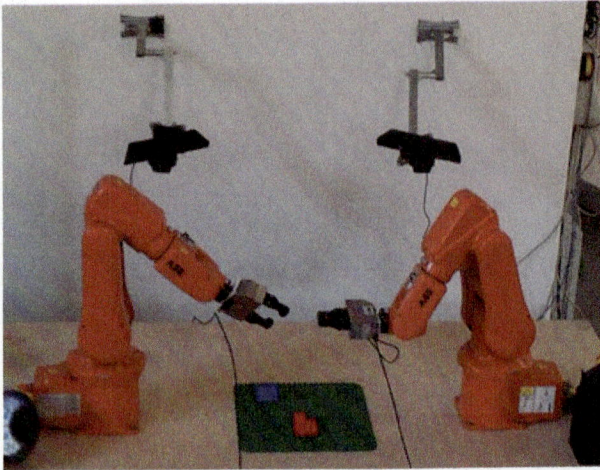

Figure 1. The test platform of the monitoring system.

full view of the working site if their image data are correctly aligned and fused. Experimental studies are carried out in the test platform which uses two Kinects and two ABB robots to represent the general case of multiple sensors and robots as Fig. 1 shows. While single 3-D sensors already provide detailed information within their field of view, complete coverage of a complete work area can only be achieved by relying on a multitude of these sensors. However, to provide useful information all data of each sensor must be aligned to each other, integrated and fused into an overall world picture. Therefore, it is of vital importance for sensors to be aware of not only its real-time pose in the real world but also their relative position and orientation to each other, so as to reconstruct a 3-D view of the working site.

To be able to align the data correctly, the position and orientation of each sensor must be known with sufficient exactness. To address this problem, a fixed marker is introduced into the system as an anchor. With the marker in sight, the Kinect matches the marker's location in the 2-D image with that in the real-world coordinate system to get the transformation from real-world coordinate system to the camera system. As individual Kinects are not fixed in the dynamic work area, there are circumstances where these Kinects do not detect the same geometrical marker for direct estimation of relative pose between each other or where one or both Kinects do not detect the anchor marker for self-positioning in the real world. Different relationships between Kinects and markers are considered and classified and corresponding solutions are presented in the following section.

Other than being able to be aware of its sensing element's pose relative to each other and with regards to the world coordinates, a vision based monitoring system is required to interpret a scene, which is defined as an instance of the real world consisting of one or more 3-D objects, to a determination of which 3-D objects are where in the scene. Therefore two main problems are involved: the first is object recognition,

in which a label must be assigned to an object in the scene, indicating the category to which it belongs. The second involves the position and orientation estimation of the recognized object with respect to some global coordinate system attached to the scene. We adopt the viewpoint feature histogram (VFH) method to deal with the object recognition and six-degrees-of-freedom (6DoF) pose estimation will be discussed. It uses a two-dimensional Cartesian histogram grid as a world model, which is updated continuously and in real time with range data sampled by Kinect thus enabling real-time performance of the vision system.

The remainder of the paper is organized as follows: Sect. 2 presents a brief review of recent literature on object recognition approaches in industrial vision that are relevant to our proposed vision system. The architecture and workflow of industrial vision monitoring systems are discussed in Sect. 3. Software and hardware tools, sensor alignment and object recognition approaches that are used in assisting the development of the proposed vision systems are presented in Sect. 4. Section 5 summarizes the contribution of this work and plans for future work.

2   Related work

Much research attention has been drawn to workpiece position and orientation estimation in the industrial robot area, which is the primary requirement of industrial robot monitoring. A good variety of approaches have been proposed to solve object pose detection and their categorization. Literature differentiates between model- and view-based approaches (Bennamoun and Mamic, 2002; Bicego et al., 2005), feature- and appearance-based approaches or introduces several classes (Belongie et al., 2002). Among all other methods, the model of the object and the image data are represented by local geometric features. Geometric feature matching is used to interpret images through matching the model of object-to-data feature and estimating 3-D pose of the model. The shape, texture or the appearance of the object is always the center of attention. Because the object identification depends on this information to make reliable judgments by matching the model and scene data. We apply the model-based pose estimation approach in our research, which is done by matching geometric representations of a model of the object to those of the image data.

Besides object pose estimation, sensor self-positioning is another topic that researchers have been interested in and many efforts have been made in using and comparing marker and markerless pose estimation. Quite a few vision-based applications: camera calibration, augmented reality, etc., have benefited from the use of artificial 2-D and 3-D markers. These markers are designed to be easily detected and require very simple image processing operations. As to geometry, some applications are specially designed to avoid the trouble of estimating object pose. Typically, markerless object

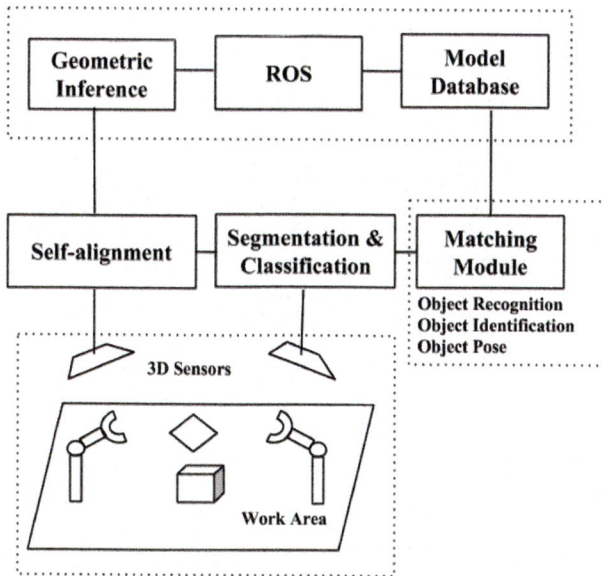

Figure 2. System architecture.

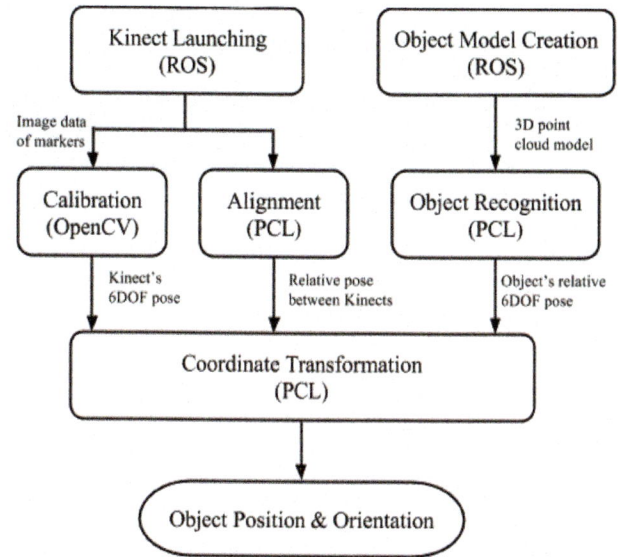

Figure 3. Workflow of the vision monitoring system.

detection and pose estimation start with feature extraction (Canny, 1986; Forstner, 1994; Harris and Stephens, 1988; Smith and Brady, 1997). Other methods based on affine invariant regions determined around feature points were proposed (Kadir et al., 2004; Matas et al., 2002; Mikolajczyk and Schmid, 2004; Tuytelaars and Gool, 2004) in order to obtain invariance to out-of-plane rotations and translations. However, these algorithms are too time-consuming to meet the requirement of real-time computing speed. A registration method was proposed by State et al., 1996, using stereo images and a magnetic tracker. Vision techniques, multiple fiducial markers and square markers were used respectively for identifying 2-D matrix, markers robust tracking and fast tracking (Neumann et al., 1999; Rekimoto, 1998; Klinker et al., 1999). In our research, markers with distinct and simple geometrical patterns are used to attach on objects for recognizing and tracking, as they are easy to detect and recognize, thus achieving both robust and fast tracking.

We are proposing a real-time self-aligning multi-sensor vision monitoring system for a dynamic work area. Model-based pose estimation approach and VFH method are applied for object recognition and 6DoF estimation; anchor markers are used for sensor self-alignment and simple geometrical markers are attached on objects to distinguish and track them, which enables the monitoring system to be aware of the real-time position and pose status of sensing elements, robots and objects in it.

3 System overview

3.1 System architecture

The monitoring mechanism of the proposed test system is shown in Fig. 2. The markers form geometric inference, which is used on a robot software development platform, ROS, to implement self-alignment of multi-sensors. An ROS is also used to create 3-D object point cloud models which compose a model database for object recognition and pose estimation by matching module after the overall scene image is processed by segmentation and classification module.

3.2 Workflow

Figure 3 is the workflow of monitoring object's movement in the work area.

- Kinect launching: An ROS camera driver launches Kinect and outputs 2-D/3-D image data.

- Calibration: calibrate a single Kinect with an anchor marker in work area. From the calibration the location of the points on the marker and its counterparts in the image, the transformation between marker and camera coordinates can be obtained. The location in the world coordinate is already known, thus, Kinect implements self-positioning.

- Alignment: align every Kinect pair as a stereo camera. As two Kinects detect the same marker, they register their captured images at the corresponding points and compute the relative position and orientation between the Kinects, thus align the image from the two Kinects to visualize the work area.

- Object model creation: create 3-D point-cloud model of object for later recognition and alignment.

- Object recognition: recognize and position the object from the scene. Object relative position and orientation will be obtained through aligning the object model to the point cloud of current scene.

- Coordinate transformation: transform object pose which is relative to scene in camera coordinate system to global coordinate system.

4 Tools and methods

4.1 Tools

4.1.1 Kinect

The robot has to rely on its sensory feedback to build a model of its surroundings. The 3-D sensor used in our research is Microsoft Kinect. It is able to capture the surrounding world in 3-D by combining the information from depth sensors and a standard RGB camera as shown in Fig. 4. The result of this combination is an RGB-D image with 640×480 resolution, where each pixel is assigned color information and depth information. In ideal conditions the resolution of the depth information can be as high as 3 mm, using 11 bit resolution. Kinect works with 30 Hz frequency for both RGB and depth cameras. On the left side of the Kinect is a laser infrared light source that generates electromagnetic waves with the wavelength of 830 nm. Information is encoded in light patterns that are deformed as the light reflects from objects in front of the Kinect. Based on these deformations captured by the sensor on the right side of RGB camera, a depth map is created. According to the light coding technology PrimeSense, this is not the time-of-flight method used in other 3-D cameras (Tolgyessy and Hubinsky, 2011). The interaction space is defined by the field of view of the Kinect cameras. To increase the possible interaction space, the built-in tilt motor supports an additional $+27$ and $-27°$, which also allows for the dynamic interaction in front of the sensor.

4.1.2 ROS

Robot operating system (ROS) (http://www.ros.org) is a software framework for robot software development, providing standard operating system services such as hardware abstraction, low-level device control implementation of commonly used functionality, message passing between processes, and package management. It is based on a graph architecture where nodes that receive, post and process messages from sensors, control, state, planning and actuactor.

An ROS is composed of two main parts: the operating system ROS as described above and ROS-pkg, a suite of user contributed packages that implement functionality such as simultaneous localization and mapping, planning, perception,

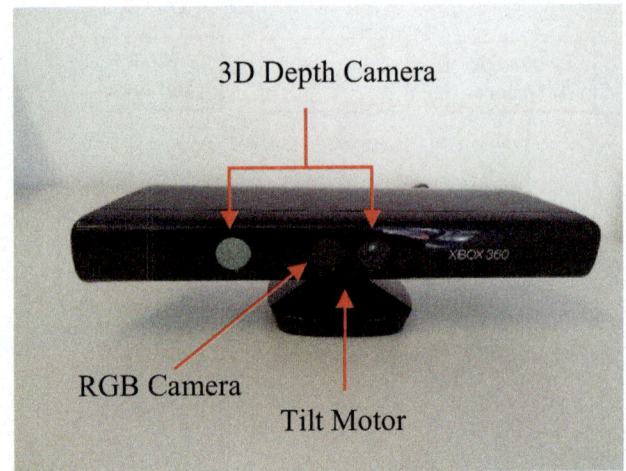

Figure 4. Kinect.

simulation, etc. The Kinect node package provides a driver for using the Kinect RGB-D sensor with an ROS, which launches an OpenNI device and loads all nodelets to convert raw depth/RGB/IR streams to depth image, disparity image and registered point clouds. So it outputs point clouds, RGB image messages and its associated camera information for calibration, object recognition and alignment.

4.1.3 PCL

The point cloud library (PCL) (http://pointclouds.org) is a large-scale, open project for 2-D/3-D image and point cloud processing. The PCL framework contains numerous state-of-the-art algorithms including filtering, feature estimation, surface reconstruction, registration, model fitting and segmentation. These algorithms can be used, for example, to filter outliers from noisy data, stitch 3-D point clouds together, segment relevant parts of a scene, extract key points and compute descriptors to recognize objects in the world based on their geometric appearance, and create surfaces from point clouds and visualize.

4.1.4 OpenCV

OpenCV (Open Source Computer Vision Library) (OpenCV.org) is an open source computer vision and machine learning software library. OpenCV was built to provide a common infrastructure for computer vision applications and to accelerate the use of machine perception in the commercial products. The library has a comprehensive set of both classic and state-of-the-art computer vision and machine learning algorithms.

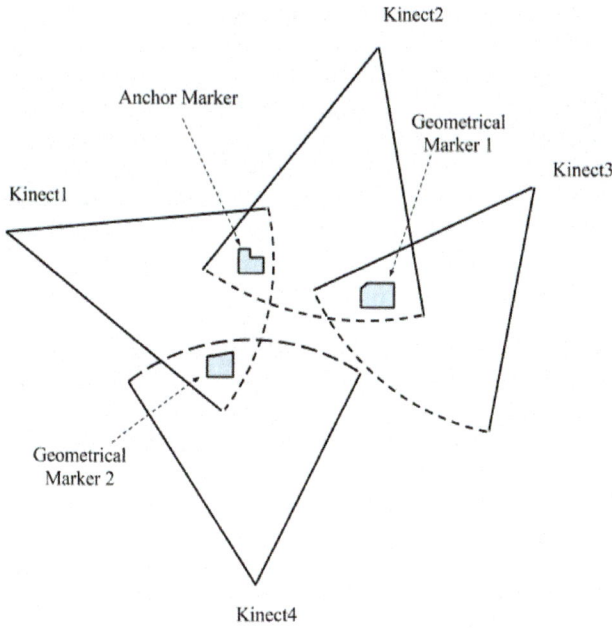

Figure 5. Four possible relative cases of multiple cameras.

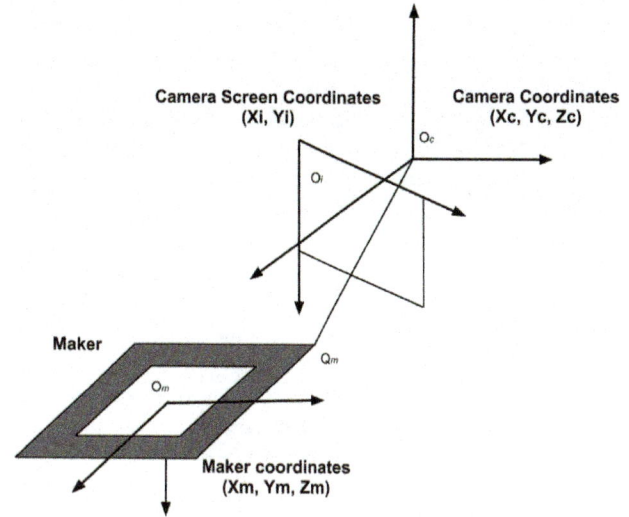

Figure 6. The relationship between marker and camera coordinates.

4.2 Methods and mechanism

4.2.1 Sensor alignment

In the proposed vision monitoring system, multiple sensors are used and must be aligned to each other and fused into an overall world picture. In order to align the sensing data accurately, the position and orientation of each sensor are of priority for aligning the sensing data accurately. Our work employs two types of 2-D markers respectively fixed on the work area as landmarks for camera self-positioning and attached on the objects as name tags for object identification, namely anchor markers and geometrical markers. The introduction of anchor markers and geometrical markers ensures the reconstruction of the whole scene of the work area. Instead of being fixed in the work area, Kinect moves up and down, left and right on its base to obtain visual information of the work scene from different viewpoints. Therefore, the spatial relationships of anchor markers, geometrical markers and Kinects vary from time to time. The relative pose of Kinect can be generally summarized and classified into four cases, as shown in Fig. 5:

1. Kinect 1 and Kinect 2 have at least one anchor marker in their intersected vision area.

2. Kinect 2 and Kinect 3 have at least one distinguishing marker and no anchor marker in their intersected vision area.

3. Two Kinects have no common marker in their intersected vision area:

a. Kinect 1 can position itself by an anchor marker; Kinect 3 has no anchor marker in its sight but a geometrical marker.

b. Both Kinect 3 and 4 detect no anchor marker but geometrical markers.

For case 1, Kinect 1 and 2 can use the anchor marker in sight for their own 3-D pose estimation by relating camera measurements with measurements in the real, three-dimensional world. In this model, a marker scene view is formed by projecting 3-D points of the marker into the image plane using a perspective transformation as Fig. 6 shows.

Projective transform maps the points Q_m in the global world coordinate system (X_m, Y_m, Z_m) to the points on the image plane with coordinates (x_i, y_i) and to the points on camera plane with coordinates (X_c, Y_c, Z_c). The projection from global world coordinate system to camera image coordinate system can be summarized as in Eq. (1):

$$
\begin{bmatrix} X_c \\ Y_c \\ Z_c \\ 1 \end{bmatrix} = \begin{bmatrix} r_{11} & r_{12} & r_{13} & t_1 \\ r_{21} & r_{22} & r_{23} & t_2 \\ r_{31} & r_{32} & r_{33} & t_3 \\ 0 & 0 & 0 & 1 \end{bmatrix} \begin{bmatrix} X_m \\ Y_m \\ Z_m \\ 1 \end{bmatrix}. \tag{1}
$$

For case 2, Kinect 2 and 3 capture the same non-anchor marker. For any given 3-D point P in object coordinates, we can put P in the camera coordinates $P_l = \mathbf{R}_l P + T_l$ and $P_r = \mathbf{R}_r P + T_r$ for the left and right cameras, respectively. It is also evident that the two views of P (from the two cameras) are related to $P_l = \mathbf{R}^T (P_r - T)$, where $\mathbf{R}$ and T are, respectively, the rotation matrix and translation vector between the cameras. Taking these three equations and solving for the rotation and translation separately yields the following

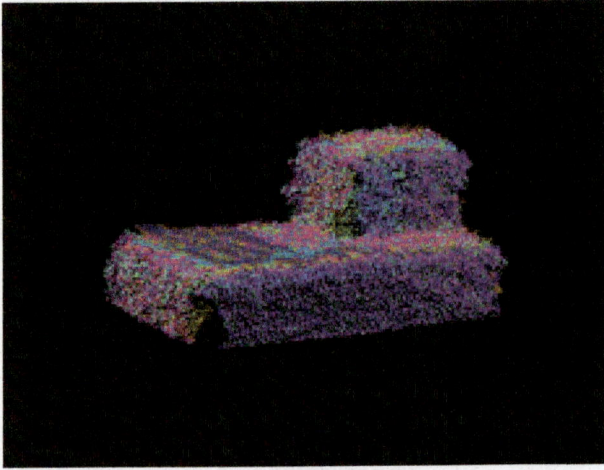

Figure 7. Overlapping of 7 point clouds at different viewpoints.

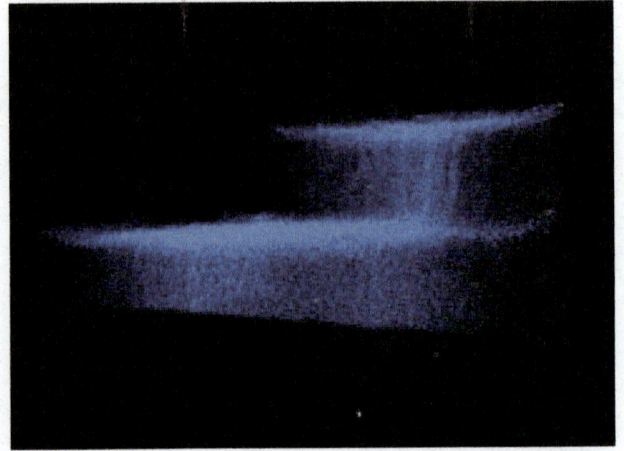

Figure 8. Merging of 34 point clouds at 34 different viewpoints.

simple relations (OpenCV.org):

$$\mathbf{R} = \mathbf{R}_r (\mathbf{R}_l)^T, \tag{2}$$

$$T = T_r - \mathbf{R}T_l. \tag{3}$$

Then the relative rotation and translation from Kinect 1 to Kinect 2 can be obtained, and in the chain of Kinects that detect the same marker with Kinect 2 directly or indirectly, there must be one that has an anchor marker in vision range. Therefore, the second case can be solved in the same way as case 1, only with the corresponding transformations.

For case 3, Kinect 1 and Kinect 3 do not have the same marker in their vision ranges. We apply a similar strategy here by searching for an anchor marker in the chain composed of overlapped Kinects to estimate the 6DoF pose of at least one Kinect and then to make pose estimation of others through coordinate transformation.

4.2.2   Object Recognition and 6DoF pose estimation

Object recognition is the process of automatic identification and localization of objects from the sensed images of scenes in the real world. For object recognition in this system, scene point clouds with the object's presence are downsampled by corresponding sampling algorithm from PCL for analysis and computation. To obtain the surface normals of the specified input point clouds, Kd-Tree (http://ros.org) is used to search for neighboring point and the radius that defines each point's neighborhood. The VFH (http://pointclouds.org) descriptor is employed as a representation for point cluster recognition and its 6DoF pose estimation. The computation of VFH descriptors is implemented from the input point cloud and its surface normals. The resulted features are invariant to image scaling, translation, rotation and partially invariant to illumination changes and affine or 3-D projection. With the normals and local feature descriptors, the object point cloud model is aligned into the current scene cloud

to get final transformation and a fitness score to evaluate the aligning results.

Object recognition is achieved by matching features derived from the scene with stored object model representations. One of the most common ways to create the object model for recognition is to extract the target as a cluster from the point cloud. However, in this way only a partial model is created out of the object, which provides very limited information for object identification.

Model creation

In this research, the approach adopted to create a 3-D point cloud model from an object is to use an object recording API of the package RoboEarth from an ROS along with a Kinect camera and a marker pattern. The target object is placed in the middle of the marker template and either the camera or marker pattern and object are moved to record a complete pose. It is always a better idea to move the object, otherwise the illumination might not be constant and therefore color effects might arise.

Figure 7 shows the overlapping of point clouds of the object captured at seven different viewpoints and all the point clouds are created at 34 different viewpoints and are finally processed and merged into one 3-D point cloud model as Fig. 8 shows.

Normal estimation

Given a geometric surface, it is usually trivial to infer the direction of the normal at a certain point on the surface as the vector perpendicular to the surface in that point. The problem of determining the normal to a point on the surface is approximated by the problem of estimating the normal of a plane tangent to the surface, which in turn becomes a least-square plane fitting estimation problem. The solution for estimating the surface normal is therefore reduced to an analysis of the eigenvectors and eigenvalues of a covariance matrix created from the nearest neighbors of the query point. More specifically, for each point p_i, we assemble the covariance matrix $\mathbf{C}$ as follows:

$$\mathbf{C} = \frac{1}{k} \sum_{i=1}^{k} (p_i - \bar{p}) \cdot (p_i - \bar{p})^T, \tag{4}$$

$$\mathbf{C} \cdot \bar{v}_j = \lambda_j \cdot \bar{v}_j, j \in \{0, 1, 2\}, \tag{5}$$

where k is the number of point neighbors considered in the neighborhood of p_j, p represents the 3-D centroid of the nearest neighbors, λ_j is the jth eigenvalue of the covariance matrix, and v_j the jth eigenvector (Bradski and Kaehler, 2008).

Feature description

Features define individual components of an image and can be categorized into two major groups: global features and local features. Global features are defined as properties of an image based on the whole image. Local features are defined as properties of an image based on a component of the image and these will be used for object recognition. Therefore, we need a way to describe the features of an image.

VFH descriptor is a novel representation for point cluster recognition and its 6DoF pose estimation. VFH has its roots in FPFH (Fast Point Feature Histograms) descriptor and add in viewpoint variance while retaining invariance to scale. The main idea of object recognition through VFH descriptors is to formulate the recognition problem as a nearest neighborhood estimation problem. Let p_c and n_c be the centroids of all surface points and their normals of a given object partial view in the camera coordinate system (with $||n_c|| = 1$). Then (u_i, v_j, w_i) defines a Darboux coordinate frame for each point p_i:

$$u_i = n_c, \tag{6}$$

$$v_i = \frac{p_i - p_c}{||p_i - p_c||} \times u_i, \tag{7}$$

$$w_i = u_i \times v_i. \tag{8}$$

The normal angular deviations $\cos(\alpha_i)$, $\cos(\beta_i)$ and $\cos(\varphi_i)$ for each point p_i and its normal n_i given by

$$\cos(\alpha_i) = v_i \cdot n_i, \tag{9}$$

$$\cos(\beta_i) = n_i \cdot \frac{p_c}{||p_c||}, \tag{10}$$

$$\cos(\varphi_i) = u_i \cdot \frac{p_i - p_c}{||p_i - p_c||}, \tag{11}$$

$$\theta_i = a\tan 2(w_i \cdot n_i, u_i \cdot n_i). \tag{12}$$

Note that $\cos(\alpha_i)$, $\cos(\beta_i)$ and θ_i are invariant to viewpoint changes, given that the set of visible points does not change.

For $\cos(\alpha_i)$, $\cos(\beta_i)$ and θ_i histograms with 45 bins each are computed and a histogram of 128 bins for $\cos(\beta_i)$, thus the VFH descriptor has 263 dimensions (Aldoma and Vincze, 2011).

Pose estimation

As the point cloud data of an object model is stored and the corresponding Kd-tree representation is built up, objects are extracted from the given scene as clusters and for each of them, an individual cluster; for each cluster, their VFH descriptor from the current camera position is computed for searching for candidates in the trained Kd-tree. After find the best candidate for recognition, the position and orientation of the object that the model represents can be determined by registering the model to the scene point cloud.

5 Conclusions

In this paper, we have introduced a new approach for work area monitoring in a dynamic environment using multiple 3-D self-aligning Kinects. The anchor marker is used to calibrate Kinect to correct for the main deviations from the pinhole model that Kinect uses, to obtain the transformations from a global coordinate system to a camera coordinate system and relative position and orientation between the Kinects. In this way, Kinect is able to have an awareness of its own positions and 6DoF poses as well as the object's location in the working scenario at any moment, enabling robots to accommodate changes in the workpiece position/orientation and to perform complex operations like automated assembling and sorting. Simple geometrical markers are used to distinguish objects, which achieves robust and fast tracking of objects in dynamic work sites. In conclusion, addressing the requirements of real-time monitoring of a dynamic industrial production area, the proposed vision monitoring system is able to provide overall vision of the work area and estimate 6DoF pose of multiple objects with defined geometrical markers and anchor markers.

To evaluate and optimize the performance of our proposed approaches in this vision system, we will involve the following aspects as future research topics. Firstly, adopt color information for object recognition and extraction; secondly, implement boundary analysis using the combination of a photogrammetric processing algorithm and point cloud spatial information; thirdly, compare the results of using different models to align to scene image: 3-D CAD model, model generated based on both digital image and point cloud obtained by depth camera, scanned object 3-D point cloud model and object model extracted from the scene image.

Edited by: R. Tutsch
Reviewed by: two anonymous referees

References

Aldoma, A. and Vincze, M.: CAD-Model Recognition and 6DOF Pose Estimation Using 3D Cues, IEEE International Conference on Computer Vision Workshops, 585–592, 2011.

Belongie, S., Malik, J., and Puzicha, J.: Shape matching and object recognition using shape contexts, IEEE T. Pattern Anal., 24, 509–522, 2002.

Bennamoun, M. and Mamic, G. J.: Object recognition: fundamentals and case studies, London, Springer, 2002.

Bicego, M., Castellani, U., and Murino, V.: A hidden Markov model approach for appearance-based 3D object recognition, Pattern Recognition, 26, 2588–2599, 2005.

Bradski, G. and Kaehler, A.: Learning OpenCV: Computer Vision with the OpenCV Library, O'Reilly, 378–386, 2008.

Canny, J. F.: A computational approach to edge detection, IEEE T. Pattern Anal., 8, 679–698, 1986.

Ewert, D., Mayer, M., Schilberg, D., and Jeschke, S.: Adaptive assembly planning for a nondeterministic domain, Conference Proceedings of the 4th International Conference on Applied Human Factors and Ergonomics, 1253–1262, 2012.

Forstner, W.: A framework for low-level feature extraction, European Conf. on Computer Vision, 383–394, 1994.

Harris, C. and Stephens, M.: A combined corner and edge detector, Proc. of the 4th Alvey Vision Conf., 147–151, 1988.

Kadir, T., Zisserman, A., and Brady, M.: An affine invariant salient region detector, European Conf. on Computer Vision, 2004.

Klinker, G., Stricker, D., and Reiners, D., Augmented Reality: A Balancing Act between High Quality and Real-Time Constraints, Proceedings of ISMR '99, 325–346, 1999.

Matas, J., Chum, O., Urban, M., and Pajdla, T.: Robust wide baseline stereo from maximally stable extremal regions, British Machine Vision Conf., 2002.

Mikolajczyk, K. and Schmid, C.: Scale & affine invariant interest point detectors, Int. J. Comput. Vision, 60, 63–86, 2004.

Neumann, U., You, S., Cho, Y., and Lee, J.: Augmented Reality Tracking in Natural Environments, Mixed Reality – Merging Real and Virtual Worlds, Ohmsha and Springer-Verlag, 101–130, 1999.

OpenCV.org: http://docs.opencv.org/modules/calib3d/doc/camera_calibration_and_3d_reconstruction.html (last access: 9 April 2013).

PCL: http://www.pointclouds.org/documentation/tutorials/ (last access: 9 April 2013), http://ros.org, http://www.ros.org/wiki/ (last access: 9 April 2013).

Rekimoto, J.: Matrix: A Realtime Object Identification and Registration Method for Augmented Reality, Proceedings of Asia Pacific Computer Human Interaction, 15–17, 1998.

Smith, S. M. and Brady, J. M.: A New Approach to Low Level Image Processing, Int. J. Comput. Vision, 23, 45–78, 1997.

State, A., Hirota, G., Chen, D. T., Garrett, W. F., and Livingston, M. A.: Superior Augmented Reality Registration by Integrating Landmark Tracking and magnetic Tracking, Proceedings of SIGGRAPH96, 429–446, 1996.

Tolgyessy, M. and Hubinsky, P.: The Kinect Sensor in Robotics Education, In Proceedings of 2nd International Conference on Robotics in Education. Vienna, Austria, 143–146, 2011.

Tuytelaars, T. and Gool, L. Vn.: Matching Widely Separated Views Based on Affine Invariant Regions, Kluwer Academic Publishers, 2004.

A catalytic combustion-type CO gas sensor incorporating aluminum nitride as an intermediate heat transfer layer for accelerated response time

A. Hosoya, S. Tamura, and N. Imanaka

Department of Applied Chemistry, Faculty of Engineering, Osaka University,
2-1 Yamadaoka, Suita, Osaka 565-0871, Japan

Correspondence to: N. Imanaka (imanaka@chem.eng.osaka-u.ac.jp)

Abstract. A catalytic combustion-type carbon monoxide gas sensor exhibiting good sensing performance even at moderate temperatures was previously developed by employing a Pt loaded CeO_2–ZrO_2–SnO_2 solid solution as the CO oxidizing catalyst. The addition of aluminum nitride as an intermediate heat transfer layer between the Pt coil and the CO oxidizing catalyst drastically accelerated the response of this device to CO at temperatures as low as 70 °C.

1 Introduction

Carbon monoxide (CO) is well known as a highly toxic gas which can act as a severe health hazard if inhaled, even at relatively low concentrations. Because the ability of the CO molecule to bind to hemoglobin is approximately 250 times as high as that of O_2, exposure to elevated levels of CO gas (especially over 0.15 %) can be fatal. Since CO gas is both colorless and odorless, detection equipment is typically required to prevent accidental exposure to CO in situations where high levels of the gas may be generated. For these reasons, it is desirable to install compact, inexpensive CO sensors at sites with the potential for elevated CO concentrations in the ambient atmosphere.

To date, various types of compact CO gas monitoring devices have been developed such as those based on semiconductors (for example, Göpel and Schierbaum, 1995; Korotcenkov, 2007), potentiostats (for example, Blurton and Stetter, 1978) and catalytic combustion (for example, Sakaguchi et al., 2009; Ozawa et al., 2005). Semiconductor-type CO gas sensors exhibit stable performance due to their simple construction and a sensing mechanism based on the electrical resistance change caused by CO adsorption on the semiconductor surface. These monitors, however, have a basic deficit in that gases other than CO can also be adsorbed on the semiconductor surface and produce a signal, meaning that the selectivity of this type of sensor is poor. Potentiostat-type CO sensors allow selective CO detection but cannot operate over prolonged time periods because of the eventual evaporation of liquid electrolyte in the device. In contrast, catalytic combustion-type CO gas sensors have a simple detection system consisting of a Pt coil combined with a CO oxidation catalyst, resulting in both stable and rapid sensing performance over long periods of time. Although catalytic combustion-type sensors detect CO gas via changes in the resistance of the Pt coil, similar to the semiconductor-type sensors, selective CO gas detection can be realized by employing a catalyst that oxidizes only CO gas at a given temperature, since in such cases resistance changes are caused solely by the heat generated from combustion of CO gas on the catalyst loaded on the Pt coil. Moreover, the sensor signal is completely proportional to the resistance change of the Pt coil during CO combustion and this resistance change is precisely correlated with the amount of CO gas oxidized by the catalyst. Therefore, the sensor signal is directly proportional to the CO gas concentration. Unfortunately, a significant problem associated with conventional catalytic combustion-type CO gas sensors remains; the catalysts (Pt/Al_2O_3 or Pd/Al_2O_3) require temperatures over 400 °C for the complete oxidation of CO gas. Because other gases such as methane and volatile organic compounds (VOCs) also combust at such elevated temperatures, these sensors will not be entirely selective for CO gas. To obtain reliable selectivity, therefore, it is necessary to find novel catalysts which can oxidize CO at lower temperatures at which gases other than CO are not oxidized. Until now, some kinds of low-temperature operating sensors were reported such as the semiconductor-type sensor based on a SnO_2 thin film combined with a Pd/Al_2O_3 thick film (Tabata et al., 2005) and the catalytic combustion-type sensor applied a Ce-doped cobalt oxide employed as a catalyst (Xu et al., 2008). For the sensor based on a SnO_2 thin film, the low temperature operation at $80\,^\circ$C was successfully realized, there still remains a possibility of interference of other gases. However, the catalytic combustion-type sensor with Ce-doped cobalt oxide could detect CO selectively at $92\,^\circ$C in dry atmosphere, but the CO oxidation activity of the catalyst employing a Co-based solid is deteriorated by water vapor. For the practical use of the sensor, the resistance toward water vapor is a concerning problem. Therefore, water-durability is also requested for the catalyst.

Based on this demand, we have succeeded in developing a low-temperature catalytic combustion-type CO gas sensor, by employing $10\,\text{wt\%}$ $Pt/Ce_{0.68}Zr_{0.17}Sn_{0.15}O_{2.0}$ as the catalyst (Hosoya et al., 2013). Since this catalyst is capable of completely oxidizing CO at $65\,^\circ$C, we have achieved sensitive CO detection at temperatures as low as $70\,^\circ$C. Moreover, since the present catalyst exhibited an excellent CO oxidizing activity even under humid conditions, the sensor also showed responses to CO even in a humid atmosphere. The response time of the device, however, is still sluggish at 180–240 s, due to insufficient heat transfer from the catalyst to the Pt coil, and thus increasing the response time is vitally important in terms of allowing the practical application of this sensor.

In the study reported herein, we improved the response time of a catalytic combustion-type CO gas sensor incorporating a $10\,\text{wt\%}$ $Pt/Ce_{0.68}Zr_{0.17}Sn_{0.15}O_{2.0}$ catalyst by adding the thermoelectric material aluminum nitride (AlN) as an intermediate heat transfer layer. We subsequently investigated the CO sensing performance of this improved device at $70\,^\circ$C.

2   Experimental

The $10\,\text{wt\%}$ Pt-loaded $Ce_{0.68}Zr_{0.17}Sn_{0.15}O_{2.0}$ solid solution was prepared via the sol–gel method, as described in our previous report (Hosoya et al., 2013), and the sample composition and its CO oxidation activity were confirmed by X-ray fluorescence (Rigaku, ZSX100e) and X-ray powder diffraction (XRD) (Rigaku, SmartLab) analyses as well as by employing a conventional fixed-bed flow reactor.

The CO gas sensor element was fabricated using a Pt coil whose size is 1 mm in length with a $30\,\mu$m diameter Pt wire; a schematic illustration of the element is provided in Fig. 1. AlN (Toyo Aluminum K. K.) was dispersed in ethylene glycol to form a slurry which was painted over the Pt coil, and then ethylene glycol was drove off at

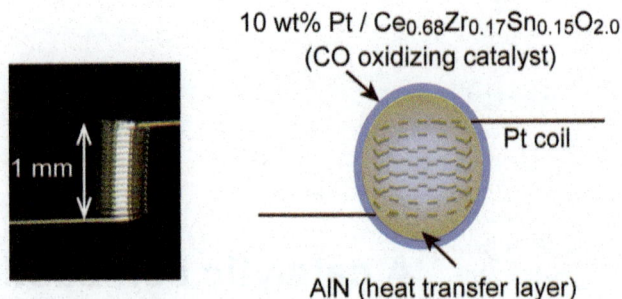

Figure 1. Photograph of the Pt coil employed in this work and an illustration of the catalytic combustion-type CO sensor incorporating an AlN interlayer and the $10\,\text{wt\%}$ Pt loaded $Ce_{0.68}Zr_{0.17}Sn_{0.15}O_{2.0}$ catalyst. $Ce_{0.68}Zr_{0.17}Sn_{0.15}O_{2.0}$ catalyst.

approximately $150\,^\circ$C, which was obtained by applying a direct-current voltage of 3 V to the Pt coil, for 30 s. Subsequently, the $10\,\text{wt\%}$ $Pt/Ce_{0.68}Zr_{0.17}Sn_{0.15}O_{2.0}$ solid catalyst was dispersed in ethylene glycol and was applied over the top of the AlN layer on the Pt coil. The coil was then heated at approximately $150\,^\circ$C for 30 s to drive off the ethylene glycol and sinter the catalyst. To allow an assessment of the efficiency of the AlN as an intermediate heat transfer layer, the total volume of AlN and catalyst which was applied was almost identical to the total volume of catalyst used in the previous version of the sensor without AlN. The CO sensing performance of the sensor was investigated using an electrometer (Advantest, R8240) to measure the DC voltage generated while passing a DC current of 90 mA through the sensor element to heat the cell up to $70\,^\circ$C. CO gas concentrations from 0 to 1000 ppm (parts per million) were produced by diluting 1000 ppm CO in air with dry air (O_2 gas concentration in the test gas: 20.95–21.00 %). Regardless of the CO gas concentration, the total gas flow rate passing over the sensor was kept constant at $40\,\text{mL}\,\text{min}^{-1}$. The sensor signal in response to exposure to CO gas was defined as $(R_{\text{gas}} - R_{\text{air}})/R_{\text{air}}$, where R_{gas} and R_{air} are the electrical resistances of the sensor when in contact with gas containing CO and with pure air, respectively. The sensor response time was defined as the time required for the electrical resistance of the device to reach 50 or 90 % of the equilibrium value eventually obtained at a given CO gas level.

3   Results and discussion

Figure 2 displays a representative response curve for the present sensor with the AlN intermediate heat transfer layer at $70\,^\circ$C, together with corresponding data for the sensor without the AlN layer (Hosoya et al., 2013). Although similar response curves are observed for both sensors, the sensor without AlN shows slight deviations in its signal even after the signal has plateaued at each CO concentration and, for this reason, over 200 s were required for this sensor to exhibit a 90 % response. In contrast, the sensor with AlN shows

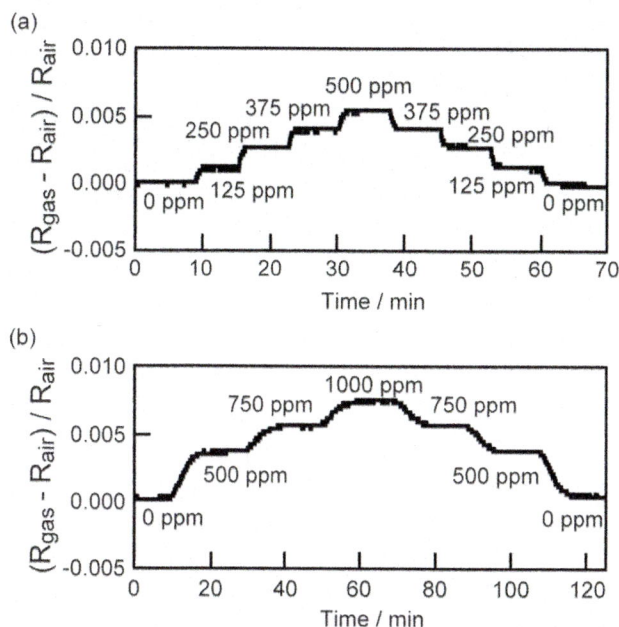

(a)

(b)

Figure 2. Representative response curves at 70 °C for sensors **(a)** with and **(b)** without the AlN intermediate heat transfer layer.

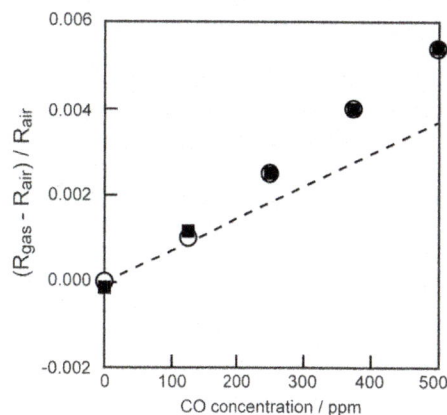

Figure 3. Signals at various CO concentrations ($\bigcirc$: increasing, $\blacksquare$: decreasing) from a catalytic combustion-type sensor incorporating AlN at 70 °C. Dashed line indicates the sensor signal obtained previously with a sensor without AlN (Hosoya et al., 2013).

more stable output such that the response time is drastically shortened to within 90 s. Furthermore, the time required for the AlN sensor to achieve a 50 % response (based on a meaningful sensor signal to noise or S / N ratio) is approximately 20 s, while the sensor without AlN required over 130 s for a 50 % response. In addition, while the previous sensor without the AlN layer also showed the response to CO at the concentration less than 500 ppm, its sensor signal was quite small compared to the present sensor, indicating that the sensitivity of the sensor was improved by applying the heat-conductive material of AlN having a low heat capacity.

Figure 3 depicts the equilibrium sensor signals obtained at 70 °C in response to various CO gas concentrations. As observed during our previous study of a sensor without AlN, the sensor with AlN produces a signal which varies in a strictly linear fashion with changing CO gas concentrations. Moreover, the sensitivity of the sensor with AlN is higher than that of the sensor without AlN; the response of the sensor with AlN to 500 ppm CO is 1.5 times that of the sensor without AlN. Since the sensor components, except for the presence of the AlN interlayer, are the same for both sensors, acceleration of the response time must be due to improved heat transfer from the CO oxidizing catalyst to the Pt coil. In the case of the sensor without the AlN interlayer, a portion of the heat generated by CO oxidation on the catalyst will be released to atmospheric air before it can be conducted to the Pt coil. As a result, a prolonged time period is necessary to reach a thermal equilibrium state in the Pt coil. In contrast, by applying AlN as an interlayer, due to its excellent thermal conductivity, a greater quantity of heat is effectively

and rapidly conducted to the Pt coil through the AlN. This results in both a steadier sensor signal and a more rapid response. The sensitivity was also improved by applying AlN as the heat transfer layer, because the heat capacity of AlN is less than that of the 10 wt% $Pt/Ce_{0.68}Zr_{0.17}Sn_{0.15}O_{2.0}$ catalyst. The temperature of a material with a lower heat capacity will increase to a greater extent in response to the same amount of heat energy and therefore the temperature of the Pt coil is further increased as the heat generated in the catalyst is conducted through the AlN layer. Because the sensor signal is dependent on the electrical resistance of the Pt coil, which in turn is directly related to the temperature of the Pt, the sensitivity of the sensor with AlN was improved due to the lower heat capacity of the AlN compared to that of the catalyst. The results in this paper indicate that the high thermal conductivity of AlN can improve CO sensing performance of the catalytic combustion-type sensor when AlN is applied as an intermediate heat transfer layer. This new sensor device should be further studied to prove its potential as a CO detector. The investigation so far lacks information on the influence of humidity and long-term stability, which will be addressed in a future study.

Acknowledgements. This study was partially supported by a Grant-in-Aid for Science Research (no. 24655193) from the Japan Society for the Promotion of Science and by The Iron and Steel Institute of Japan.

Edited by: A. L. Spetz
Reviewed by: two anonymous referees

References

Blurton, K. F. and Stetter, J. R.: Sensitive electrochemical detector for gas chromatography, J. Chromatogr., 155, 35–45, 1978.

Göpel, W. and Schierbaum, K. D.: SnO_2 sensors: current status and future prospects, Sensor Actuat. B-Chem., 26, 1–12, 1995.

Hosoya, A., Tamura, S., and Imanaka, N.: Low-temperature-operative Carbon Monoxide Gas Sensor with Novel CO Oxidizing Catalyst, Chem. Lett., 42, 441–443, 2013.

Korotcenkov, G.: Practical aspects in design on one-electrode semiconductor gas sensors: Status report, Sensor Actuat. B-Chem., 121, 664–678, 2007.

Ozawa, T., Ishiguro, Y., Toyoda, K., Nishimura, M., Sasahara, T., and Doi, T.: Detection of decomposed compounds from an early stage fire by an adsorption/combustion-type sensor, Sensor Actuat. B-Chem., 108, 473-477, 2005.

Sakaguchi, M., Ishikawa, A., and Hoshihara, I.: Development of New Catalytic Combustion Sensor for Breath Alcohol, J. Comb Soc. Jpn., 51, 129–133, 2009.

Tabata, S., Higaki, K., Ohnishi, H., Suzuki, T., Kunihara, K., and Kobayashi, M.: A micromachined gas sensor based on a catalytic thick film/SnO_2 thin film bilayer and a thin film heater Part 2: CO sensing, Sensor Actuat. B-Chem., 109, 190–193, 2005.

Xu, T., Hu, H., Luan, W., Qi, Y., and Tu, S.-T.: Thermoelectric carbon monoxide sensor using Co-Ce catalyst, Sensor Actuat. B-Chem., 133, 70–77, 2008.

Fabrication and characterization of a piezoresistive humidity sensor with a stress-free package

T. Waber[1], **M. Sax**[1], **W. Pahl**[2], **S. Stufler**[2], **A. Leidl**[2], **M. Günther**[3], **and G. Feiertag**[1]

[1]Munich University of Applied Sciences, Lothstraße 64, 80335 Munich, Germany
[2]EPCOS AG, a member of TDK-EPC Corporation, Anzingerstraße 13, 81617 Munich, Germany
[3] Technical University of Dresden, Helmholtzstraße 18, 01062 Dresden, Germany

Correspondence to: T. Waber (tobias.waber@hm.edu)

Abstract. A highly miniaturized piezoresistive humidity sensor has been developed. The starting point of the development was a $1 \times 1\,\mathrm{mm}^2$ piezoresistive pressure sensor chip. As sensing material, a polyimide was used that swells with increasing adsorption of water molecules. To convert the swelling into an electrical signal, a thin layer of the polyimide was deposited onto the bending plate of the pressure sensor. The humidity sensor was characterized in a climate chamber. The measurements show a sensitivity of 0.25 mV per percent relative humidity (%RH) and a non-linearity of 3.1 % full scale (FS) in the range of 30–80 % RH. A high cross-sensitivity to temperature of around $0.5\,\mathrm{mV}\,^\circ\mathrm{C}^{-1}$ was measured, so temperature compensation is necessary. For stress-free packaging of the sensor chip, a novel packaging technology was developed.

1 Introduction

More and more sensors are integrated into mobile phones (see e.g. Lane et al., 2010). In the last four years this trend has accelerated. In 2010 no humidity sensors and almost no pressure sensors were integrated into mobile phones. Now, several mobile phones with humidity and pressure sensors are on the market, e.g. the Galaxy S4 phone from Samsung with the SHTC1 humidity sensor from Sensirion (Mayer and Lechner, 2013), or the Galaxy S3 with the LPS331AP pressure sensor from STMicroelectronics. The Bosch Sensortec company went one step further and developed a combined humidity, pressure and temperature sensor (BME280) with a size of $2.5 \times 2.5 \times 0.93\,\mathrm{mm}^3$ (Bosch, 2014). The main reason for combining pressure and humidity sensors into one package is that both need a cavity package with a port to the surrounding air. Besides the move towards combined pressure and humidity sensors, the trend towards smaller footprint and height will continue.

To fulfill these requirements, a piezoresistive humidity sensor with a chip size of only $1 \times 1\,\mathrm{mm}^2$ was developed. This sensor could be assembled together with a pressure sensor chip and an application-specific integrated circuit (ASIC) in a package with a size of $3 \times 2 \times 0.67\,\mathrm{mm}^3$.

The humidity of the air has a significant influence on our well-being. Therefore, a humidity sensor can provide important information for the user of a smart phone. Very dry air is a burden on our respiration system; high humidity could lead to the growth of mold. Besides that, the data provided by the humidity sensor are interesting for the manufacturers of smart phones. They can be used to exclude warranty in case of water damage, so the interest in a small digital humidity sensor in mobile phones or navigation systems, best packaged with a temperature and pressure sensor, is very high.

In this paper, we first introduce the state of the art in humidity sensing. Commercial humidity sensors are shown here and the difference compared to our sensor is explained. We will then show the design of the new sensor chip and will explain the manufacturing process. In Sect. 4, the static responses of the sensor to different values of the humidity and temperature are presented. For different accuracy requirements, possible calibration algorithms for temperature compensation are shown. Finally, a packaging solution is proposed.

2   State of the art in humidity sensing

Several physical effects can be used to measure the humidity of the air. For capacity-type sensors, materials with a humidity-dependent dielectric constant are needed. Most polymers show this dependency. For impedance-type sensors, materials with a humidity-dependent resistance like salts or conductive polymers are used (Yamazoe and Shimizu, 1986; Sager et al., 1994). Also, resonant silicon cantilever sensors can detect the humidity. The resonance frequency of cantilevers decreases with increasing humidity due to the mass increase by adsorbed water molecules (Sone et al., 2004a, b; Wasisto et al., 2013). Furthermore, optical and gravimetric effects are used for humidity sensors (Lee and Lee, 2005).

Most commercial humidity sensors for mobile phones use the capacitive effect, e.g. the SHTC1 sensors from Sensirion, HTU21D from Measurement Specialties or Si7021 from Silicon Labs. The reasons for the choice of this working principle are cheap manufacturing, very low power consumption and a detection mechanism, which is very specific to humidity (Wagner et al., 2011; Lee and Lee, 2005).

An alternative to capacity-type sensors are piezoresistive-type sensors. Organic materials show a swelling when water from the surrounding air is absorbed. This effect is utilized in the hair tension hygrometer that was invented by de Saussure in 1783. The swelling can be converted into an electrical signal by making use of the piezoresistive effect. The first research work on piezoresistive humidity sensors was published by a group from the Technical University of Dresden (TUD) at the beginning of the 1990s (Sager et al., 1994; Gerlach and Sager, 1994; Buchhold et al., 1998a). A polymer layer on the top of a silicon-bending plate forms a bimorph that bends under the influence of humidity. Piezoresistors at the edge of the bending plate convert the mechanical stress caused by the humidity-induced bending into an electrical signal. Different polymer materials for piezoresistive humidity sensors were investigated at the Technical University of Dresden (Buchhold et al., 1998c; Buchhold et al., 1999; Guenther et al., 2001). However, this work did not result in a commercial sensor, and as far as we know, no further investigations for commercial piezoresistive humidity sensors were made. For use in mobile phones or navigation systems, the sensors have to be smaller, cheaper and suitable for production in bulk. Based on the prior work, we built a very small humidity sensor chip with a chip size of only $1 \times 1 \, mm^2$. Compared to prior work, the chip area was reduced by a factor of 10, while still maintaining excellent electrical performance. A challenge for the fabrication of piezoresistive humidity sensors is the stress-free packaging. Based on experiences with the packaging of piezoresistive pressure sensors, this problem was solved by developing a technology where the sensor chip is flip-chip bonded onto electroplated copper springs (Waber et al., 2013). The humidity sensor chip is compatible with this package.

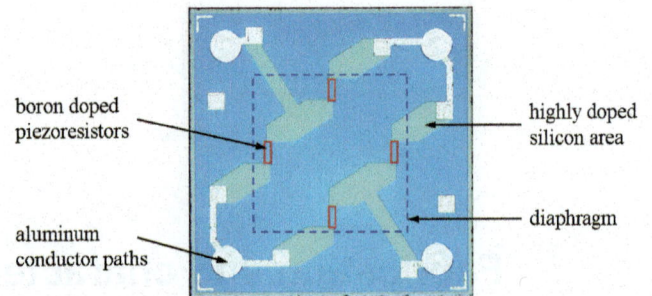

Figure 1. Pressure sensor chip with four piezoresistors interconnected as a Wheatstone bridge. The doping of the piezoresistors is too low to be visible in an optical microscope image. The location of the resistors is therefore indicated by red boxes. The footprint is $1 \times 1 \, mm^2$ and the height $150 \, \mu m$.

3   Design of the humidity sensor chip

A piezoresistive pressure sensor chip, which will be presented in Sect. 3.1, was the starting point of this work. In order to develop a humidity sensor, a thin layer of the polyimide was deposited onto the bending plate of the pressure sensor chip and was used as a sorption-mechanical transducer. Combining a polymer and a micro-fabricated pressure sensor chip made it possible to monitor the humidity-dependent swelling of a polymer continuously. The material properties, the layout and the processing of the polymer layer will be shown in Sect. 3.2. Calculations of the sensitivity and the noise of the humidity sensor are shown in Sect. 3.3.

3.1   Piezoresistive pressure sensor

Figure 1 shows the top view of the pressure sensor chip. This chip is applied as the sensing element of the T5400 pressure sensor module from EPCOS (Waber et al., 2013). The chip was fabricated using silicon on insulator (SOI) technology (see Fig. 2). First, cavities were etched into a silicon wafer by KOH etching (Fig. 2, step 2). A second silicon wafer was then bonded onto the cavity wafer (Fig. 2, step 3). The diaphragms (bending plates) were formed by grinding the second wafer down to a thickness of $10 \, \mu m$ (Fig. 2, step 4). On the bending plates, four piezoresistors were formed by doping. To achieve a high sensitivity, the piezoresistors were placed at the edge of the diaphragm, where the stress is at its maximum (Fig. 2, step 5). The resistors were connected as a Wheatstone bridge by highly doped areas (Fig. 2, step 6) and aluminum lines (Fig. 2, step 7). The sensor gives an output voltage that is proportional to the surrounding air pressure.

3.2   Polymer layer

Polymer layers absorb water molecules from the surrounding air. The absorption of water results in a swelling of the material. In layers with a thickness of a few μm, an equilibrium between the humidity of the surrounding air and the humidity

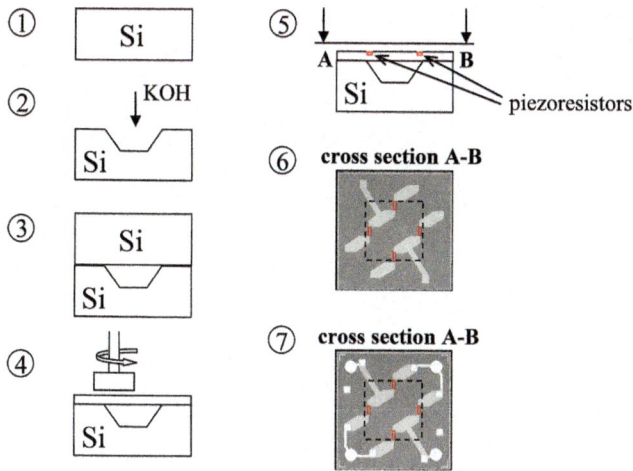

Figure 2. Pressure sensor chip fabricated in SOI technology. (1) A pure silicon wafer. (2) Cavities are etched by KOH etching. (3) A second silicon wafer is bonded onto the cavity wafer. (4) The bending plates are formed by grinding the second wafer. (5) Four piezoresistors are formed by doping on the bending plate. (6) The resistors are connected as a Wheatstone bridge by highly doped areas and (7) aluminum lines.

Figure 3. Sensor chip with a Durimide 7505 layer on top of the diaphragm. The patch has a size of $300 \times 300 \,\mu$m and a thickness of $3 \,\mu$m.

in the polymer is reached within minutes. It was shown earlier that the swelling is completely reversible (Buchhold et al., 1999), so thin polymer layers can be used to convert the relative humidity of the air into a swelling. The sensor incorporates the following chain of transducers:

- The polymer layer converts the humidity in the surrounding air into an in-plane expansion (sorption-mechanical transducer).

- Polymer layer and silicon-bending plate form a bimorph that transforms the expansion into a bending (mechanical transducer).

- The deformation of the bending plate results in a change in the stress in the piezoresistors.

- The piezoresistors transform the stress into a change in electrical resistance (mechano-electrical transducer).

As a humidity-responsive polymer material, the photosensitive Polyimide Durimide 7505, obtained from Fujifilm Electronic Materials Co., was used in the present work. In older publications the trade name Probimide is used for the same material. In the humidity sensor the Durimide 7505 layer has a thickness of $3 \,\mu$m. This thickness was chosen as a compromise between sensitivity and response time. A higher sensitivity could be reached by using a thicker polyimide layer. Unfortunately, the response time also increases with increasing thickness because more time is needed for the diffusion of the water molecules into the polymer layer (Buchhold et al., 1998d).

Patches with a size of $300 \times 300 \,\mu$m^2 were patterned in the center of the bending plate using a photolithographic process (see Fig. 3). To avoid cross-sensitivity to the air pressure, a port was etched in the bottom of the chip by deep reactive ion etching (DRIE) in order to open the reference volume (see Fig. 4).

For the characterization of the Durimide 7505, the absorption of water was first calculated and then measured.

The moisture loading capacity was characterized by a relation of the mass of the absorbed water to the mass of the dry polymer using the value of a saturation concentration c_s given by (Buchhold et al., 1998a)

$$c_s(\varphi, T) = \frac{m_{\text{water}}(\varphi, T)}{m_{\text{dry polymer}}} =$$

$$\frac{m_{\text{coated wafer}}(\varphi, T) - m_{\text{coated wafer}}(0\,\% \text{ rh})}{m_{\text{coated wafer}}(0\,\% \text{ rh}) - m_{\text{uncoated wafer}}}. \quad (1)$$

At 100 % RH, the saturation concentration reaches its maximum $c_{s,\text{max}}$.

3.2.1 Calculation

The datasheet value of the saturation concentration c_s of Durimide 7505 at 50 % RH and room temperature is 1.08 %. Assuming a linear relation between humidity and moisture absorption, the maximum saturation concentration $c_{s,\text{max}}$ is 2.16 %.

3.2.2 Measurement

For the measurement, first, an uncoated silicon wafer was weighed using a precision balance Sartorius BP210D. The wafer was then coated with a $3 \,\mu$m thick polyimide layer and was weighed in a chamber at defined values of temperature T and relative humidity φ of the surrounding air ($T = 19$–$25\,^\circ$C, $\varphi = 0$–85 % RH). Ambient conditions were controlled

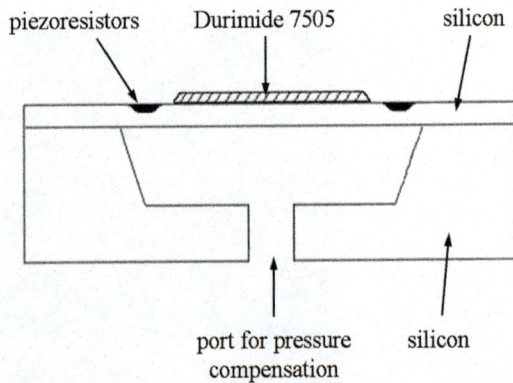

Figure 4. Schematic cross section of the sensor chip with a Durimide layer (not to scale). For pressure compensation, a port was etched onto the bottom of the chip.

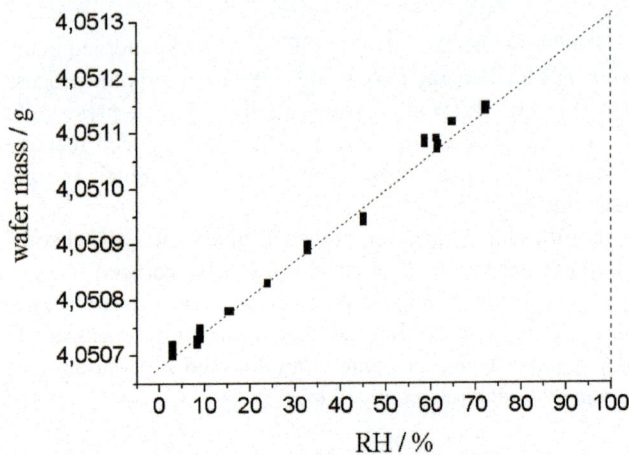

Figure 5. Relation between relative humidity and wafer mass. At 100 % RH, the wafer mass and the saturation concentration c_s (Eq. 1) reach the maximum. By averaging the results from four wafers, we found that the maximum loading capacity $c_{s,max}$ of Durimide 7505 was 3.2 %.

with a Vaisala HMP 230 humidity and temperature sensor with uncertainties of ± 0.5 K and ± 2 % RH, respectively. As a result, a linear relation between the relative humidity and the wafer mass was found, as shown in Fig. 5. By averaging the results from four wafers, we found that the maximum loading capacity of Durimide 7505 was 3.2 %.

The disagreement between the datasheet value and the measurements could be explained by different curing conditions of the polyimide layers. In Table 1, the measured result is compared with the loading capacity of other polymers that were investigated earlier by the TU Dresden in the 90s (Buchhold et al., 1998b). It can be seen that the maximum saturation concentration is comparable to other polymers used for humidity sensors. For humidity sensors, a high maximum loading capacity is desired, because it results in a high expansion and a high sensitivity.

Table 1. Maximum loading capacity of Durimide 7505 and the polymers measured in the reference (Buchhold et al., 1998b).

Polymer	$c_{s,max,lin}/\%$
PI2525	3.4
PI2560	3.3
PI2566	1.9
PI2611	2.0
Photo-BCB	< 0.3
Durimide 7505	3.2

4 Characterization of the piezoresistive humidity sensor

The response of the sensor to humidity and temperature was investigated. In the following sections, first the sensitivity and noise of the humidity sensor is calculated, and then the set-up and the results of the measurements are shown.

4.1 Calculations of sensitivity and noise

4.1.1 Calculation of the sensitivity

An equation for the calculation of the bridge voltage of a piezoresistive pressure sensor was given by Gerlach and Sager (Gerlach and Sager, 1994):

$$U_{out}(\varphi) = U_0 \cdot \frac{3}{2}(\pi_L - \pi_T) \cdot \frac{E_{PI}}{1 - \nu_{Si}} \cdot \frac{h_2}{h_1} \cdot \frac{b_1}{b_1 + b_2} \cdot \alpha_{l,\varphi} \cdot \varphi. \quad (2)$$

Here U_0 is the supply voltage; π_L, π_T the piezoresistive coefficients in the longitudinal and transversal directions; E_{PI} the Young modulus of polyimide; ν_{Si} the Poisson ratio of silicon; h_1 the thickness of the membrane and h_2 the thickness of polyimide; b_1 the width of the coated membrane and b_2 the width of the uncoated membrane; $\alpha_{l,\varphi}$ the humidity-induced length expansion coefficient; φ the relative humidity.

The derivative of the output voltage U_{out} of the humidity φ is the sensitivity Sens (see Eq. 3).

$$\text{Sens} = \frac{dU_{out}}{d\varphi} =$$
$$U_0 \cdot \frac{3}{2}(\pi_L - \pi_T) \cdot \frac{E_{PI}}{1 - \nu_{Si}} \cdot \frac{h_2}{h_1} \cdot \frac{b_1}{b_1 + b_2} \cdot \alpha_{l,\varphi}. \quad (3)$$

With the coefficient of humidity expansion $\alpha_{l,\varphi} = 32 \times 10^{-6}/\%$ RH, determined by Buchold (Buchhold et al., 1998b), the geometry parameters of the sensor chip and a supply voltage, $U_0 = 3$ V, we obtain for the sensitivity

$$\text{Sens} = 0.15 \frac{mV}{\% \, RH}. \quad (4)$$

4.1.2 Estimation of the noise

The EPCOS pressure sensor chip has a sensitivity of 0.018 mV/V/hPa. For a supply voltage of 3 V, the sensor has

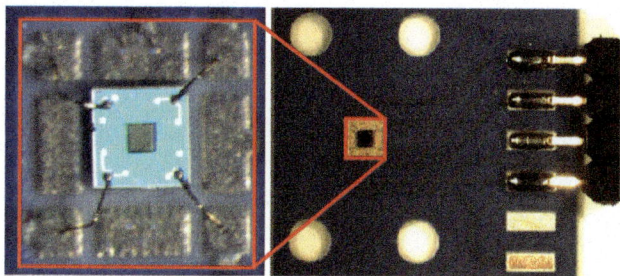

Figure 6. Humidity sensor chip is wire bonded onto a ceramic substrate. To avoid external mechanical stress, the sensor is not glued to the substrate. Mechanical stress would influence the measurement results.

a sensitivity of $0.054\,\mathrm{mV\,hPa^{-1}}$. The noise of the pressure sensor with ASIC is about $0.03\,\mathrm{hPa}$, so the noise of the humidity sensors can be estimated as follows:

$$\mathrm{RH_{noise}} = \frac{0.054\,\frac{\mathrm{mV}}{\mathrm{hPa}}}{0.15\,\frac{\mathrm{mV}}{\%\,\mathrm{RH}}} \cdot 0.03\,\mathrm{hPa} = 0.011\,\%\,\mathrm{RH}. \tag{5}$$

The uncertainty of the pressure sensor caused by external stress or thermally induced stress is less than $1\,\mathrm{hPa}$ (Waber et al., 2014). By converting this into humidity, an uncertainty of $< 0.4\,\%\,\mathrm{RH}$ can be estimated. This uncertainty and the noise are far below the typical accuracy requirements of around $3\,\%\,\mathrm{RH}$. In order to achieve short response times, the thickness of the polymer could be reduced, with only moderate impact on the accuracy.

4.2 Measurement set-up

The characterization of the sensor chip was carried out with a semi-automatic measurement set-up. The main components are a WEISS WKL 64 climate chamber and a Keithley 3706 system switch/multimeter with an extension multiplexer card. As reference sensors, a capacitive humidity sensor from the E+E company (accuracy at $20\,°\mathrm{C}$: $\pm 2\,\%\,\mathrm{RH}$ ($0\ldots90\,\%\,\mathrm{RH}$)) and a PT100 temperature sensor were used. As shown in Fig. 6, the sensor chips were connected to ceramic substrates by wire bonds. To avoid packaging stress, the sensors were not glued to the substrate. Only the bonding wires hold the chip in place. A die bonding material between the chip and the substrate would cause a mechanical stress, which would influence the measurement results.

4.3 Measurement results of the humidity sensor

In the next two sections, results of the measurements are presented. First, the response to humidity at constant temperature was investigated. The response to temperature at constant humidity was then measured.

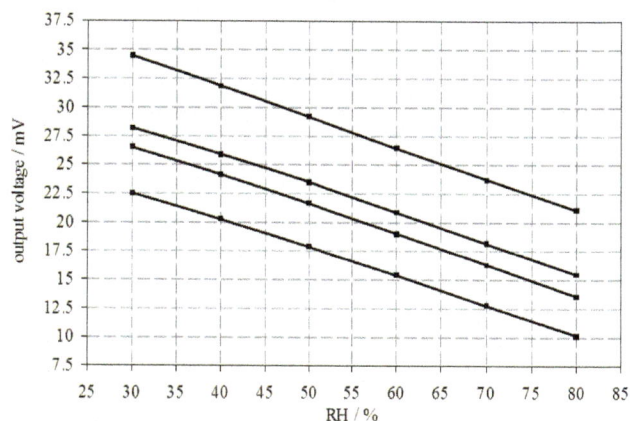

Figure 7. Relation between relative humidity and output voltage for four sensors. The sensitivity of the four sensors is between $0.24\,\mathrm{mV/\%\,RH}$ and $0.26\,\mathrm{mV/\%\,RH}$. The fact that all four curves are almost parallel is interesting for a one-point calibration (see Sect. 4.4).

4.3.1 Response to humidity

The measurement set-up described above was used. The supply voltage of the Wheatstone bridge was set to $3\,\mathrm{V}$. The temperature was kept constant at $25\,°\mathrm{C}$. The output bridge voltage of four different sensors was measured at different values of the relative humidity in the range from 30 to $80\,\%\,\mathrm{RH}$. The results are shown in Fig. 7. The output voltage is inversely proportional to the relative humidity. The output spans are around $12.5\,\mathrm{mV}$ for the humidity range from 30 to $80\,\%\,\mathrm{RH}$. The different offsets are mainly caused by manufacturing variations of the mechano-electrical transducer. Different resistances between aluminum and the highly doped silicon cause the variations.

For the four humidity sensor chips, the mean sensitivity was determined to be $0.25\,\mathrm{mV/\%\,RH}$. This result agrees quite well with the value of $0.15\,\mathrm{mV/\%\,RH}$ that was calculated in Sect. 4.1. All four sensors showed nearly similar sensitivities, so a one-point calibration could be sufficient for many applications. Possible calibration procedures are shown in Sect. 4.4.

The measured sensitivity is three times higher than the sensitivities obtained at the TU Dresden in the 1990s (Gerlach and Sager, 1994).

Linearity

The linearity between humidity and output voltage was investigated. Figure 8 shows the measurement results at different values of the relative humidity as well as a linear interpolation between the points at 30 and $80\,\%\,\mathrm{RH}$. The linearization error reaches a maximum at $55\,\%\,\mathrm{RH}$. For the four sensors, the linearization error was found to be in the range from 0.5 to $3.1\,\%$ full scale (FS). Prior results by the TU Dresden were also in the range from 0.5 to $5\,\%$ FS for

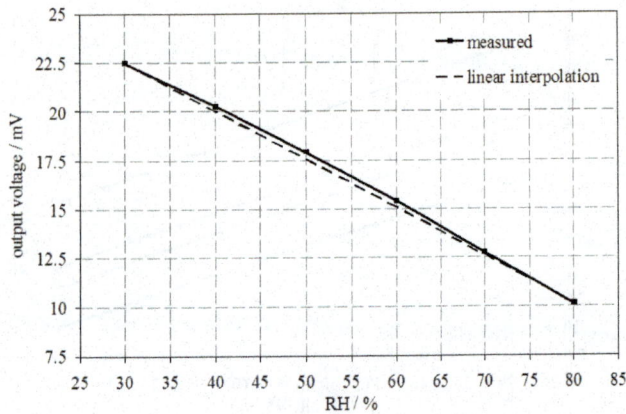

Figure 8. Comparison of the output voltage measured with the linear interpolation of the endpoints at 30 and 80 % RH. The maximum error for four sensors was 3.1 % FS, so linear interpolation is not sufficient if an accuracy of better than 3 % FS is required.

a humidity range from 10 to 95 % RH (Gerlach and Sager, 1994). The error caused by the non-linearity can be reduced significantly by a quadratic interpolation. If the accuracy requirement is high (< 3 % FS), a third measurement point at 55 % RH would be necessary for the calibration of the sensor.

Response time

To compare the dynamic behavior of our humidity sensor chip with the Sensirion SHT25 humidity sensor, a step response was measured at 25 °C for both. The time to reach 63 % of the saturation output voltage was 21 s for our sensor and 14 s for the SHT25. However, the polyimide layer is not optimized yet. The response time could be reduced by decreasing the thickness of the polyimide. Decreasing the thickness would also lead to an uncritical reduction of the sensitivity.

4.3.2   Cross-sensitivity to temperature

Piezoresistive sensors show a strong cross-sensitivity to temperature. At a constant humidity of 30 % RH, the output bridge voltage of the sensors was measured at temperatures from 25 to 40 °C. The results are shown in Fig. 9. A temperature change of 5 °C alters the output voltage by 2.5 mV. If this value is compared with the humidity sensitivity of 0.25 mV/ % RH, it becomes clear that temperature compensation is necessary for any practical use of the sensor. As this is also standard for piezoresistive pressure sensors, it is not an obstacle to an application of the sensor. Temperature sensors can easily be integrated into the ASIC of the sensor module. In Sect. 4.4, it is shown how cross-sensitivity to temperature can be compensated for.

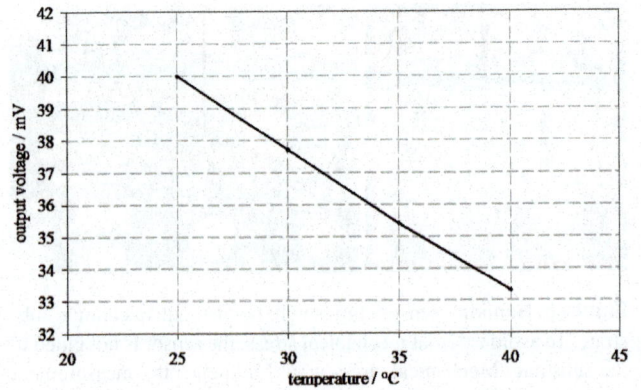

Figure 9. Bridge output voltage of the sensor at the relative humidity of 30 % RH and at different temperatures. As can be seen, the sensor has a strong cross-sensitivity to temperature. By changing the temperature by 5 °C, the output voltage alters by around 2.5 mV.

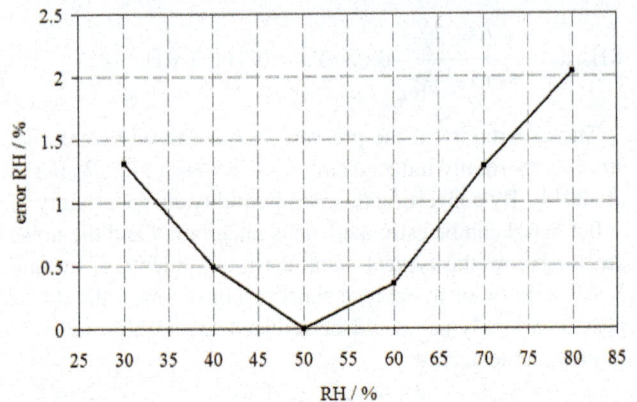

Figure 10. Error of the relative humidity using a one-point calibration at 50 % RH and 25 °C. The maximum error of four sensors is ± 2.2 % RH.

4.4   Calibration of the humidity sensor

In this section, different calibration procedures for the humidity sensor are proposed. In Sect. 4.3.1 it was shown that a quadratic function is necessary to describe the output voltage as a function of the humidity. The cross-sensitivity to temperature is also non-linear. A straightforward way to compensate for the non-linear sensor response is a calibration at three different temperatures and three different relative humidities. A total of nine calibration points is expensive, but will result in an accuracy much better than 3 % FS. If an accuracy of around 3 % FS is required, a two-point calibration is sufficient. The cheapest but less exact solution is a one-point calibration. Here the accuracy will be around 5 % FS.

For a better comparison to the values in datasheets of other sensors, the accuracy is specified in ($\pm$ % RH) absolute instead of (% FS) in the following (e.g. 1 % FS in the range of 30–80 % RH is absolute ± 0.25 % RH).

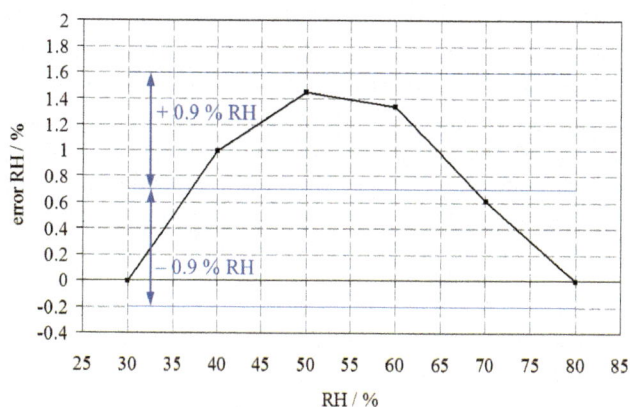

Figure 11. Error of the relative humidity using a two-point calibration at 30 % RH/25 °C and 80 % RH/25 °C. To get a high absolute accuracy over the whole range, a typical offset of 0.7 % RH needs to be subtracted from the original linear interpolation data. The maximum error of the four sensors is then ±0.9 % RH.

4.4.1 Calibration procedures

In Sect. 4.3.1 a sensitivity of (0.24–0.26) mV/ % RH was identified for the four sensors. The mean value of the sensitivity is 0.25 mV/ % RH. With this value as a constant and an offset obtained from a single calibration point at 50 % RH and 25 °C, the measured humidity can be calculated by

$$RH(U_{out}) = U_{out}/Sens + Off. \qquad (6)$$

Figure 10 shows the error of the relative humidity of one typical sensor at one-point calibration. For the four sensors investigated in this work, a maximum error of ±2.2 % RH in the range of 30–80 % RH is obtained. A typical tolerance of ±3 % RH between 20 and 80 % RH is given in the datasheet of the SHT20 humidity sensor. Also, the SHTC1 sensor from Sensirion, which is used in mobile phones, has the same tolerance. Therefore, a one-point calibration for humidity compensation of the new sensor is sufficient for many applications.

The best humidity sensor, SHT25 from Sensirion, has a typical tolerance of ±1.8 % RH in a larger humidity range between 10 and 90 % RH. All other values like hysteresis, repeatability and so on remained unchanged compared to the SHT20 sensor. This indicates that the only difference might be a calibration procedure with more calibration points.

To improve the accuracy of our humidity sensor, a two-point calibration at 30 % RH/25 °C and 80 % RH/25 °C was investigated. The offset and the sensitivity of each sensor can be determined with a two-point calibration. Figure 11 shows the error of the relative humidity of one sensor for a two-point calibration. To get a high absolute accuracy over the whole range, a typical offset of 0.7 % RH needs to be subtracted from the original linear interpolation data. The maximum error for all four sensors was then ±0.9 % RH, and

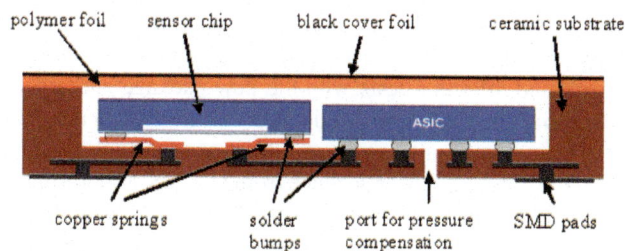

Figure 12. Novel flip-chip packaging technology for pressure and humidity sensor chips. High-temperature cofired ceramic (HTCC) is used for the package. The sensor chip is flip chip bonded on copper springs to obtain a stress-free package.

Figure 13. Copper springs in the cavity of the package. The springs are structured by photolithography. On the copper springs the sensor chip is flip-chip bonded.

therefore better than the SHT25 humidity sensor. However, the range between 30 and 80 % RH is smaller.

Temperature compensation in addition to humidity compensation is necessary. Two more additional points at different temperature therefore have to be measured. This means that, for temperature and humidity compensation, a four-point calibration at least should be used.

4.5 Discussion of the results

The measurements show that the new humidity sensor can be used to develop a commercial product. The accuracy will mainly depend on the effort spent on calibration. The measured sensitivity of 0.25 mV/ % RH is in quite acceptable agreement with the calculated value of 0.15 mV/ % RH. By reducing the thickness of the polymer layer, the response time of the sensor could be decreased. This would also lead to a lower but still sufficient sensitivity.

5 Outlook

Piezoresistive pressure and humidity sensors are very sensitive to package stress. The applications in mobile devices call for a small package size. For barometric pressure sensors, wire bonding type packages, e.g. BMP180 from Bosch or LPS331AP from ST Microelectronics, and low stress flip-chip packages, e.g., T5400 from EPCOS (Waber et al., 2013), have been developed that fulfill these requirements. Figure 12 shows the novel package with the copper springs and

Fig. 13 a scanning electron microscope picture of the springs. This package is also suitable for the new humidity sensor. Shock and vibration tests with the new package and a sensor chip have been done. Guided free fall tests from a height of two meters result in a mean offset of around 0.1 % RH. With the pressure sensor package from EPCOS, a footprint of 2.78×2.23 mm^2 and a height of 0.67 mm could be realized. ASICs, originally developed for barometric pressure sensors, could as well be used for the piezoresistive humidity sensor. Therefore, the development effort required to bring this new sensor onto the market is comparatively small. The sensor also offers the possibility of developing a combined pressure and humidity sensor. Most parts of the ASIC could be used for both measurements by adding a multiplexer between an analog-to-digital converter (ADC) and the sensor elements. With the integrated temperature sensor in the ASIC, we get an environmental sensor for mobile phones in one package. Furthermore, the development of an air density sensor can be interesting for hard disk manufacturers. Colin Johnson (Colin Johnson, 2009) wrote: "Without a measure of air density, hard drive heads can't run at minimum flight height without failing when barometric pressure changes dramatically, such as onboard an airborne plane."

The density of air ρ depends on the pressure p, the humidity φ and the temperature T. It applies Eq. (7):

$$\rho(p, \varphi, T) = p \cdot \frac{\left(1 - \varphi \cdot \frac{p_{svp}}{p} \cdot \left(1 - \frac{R_a}{R_{wv}}\right)\right)}{R_a \cdot T}, \quad (7)$$

with the gas constant of dry air $R_a = 287.058$ J kg^{-1} K^{-1}, the gas constant of water vapor $R_{wv} = 461.523$ J kg^{-1} K^{-1}, and the saturation vapor pressure p_{svp}.

There are two possibilities of obtaining the air density. First, the humidity sensor chip described above together with a pressure sensor chip and an ASIC with a temperature sensor could be mounted into one package. With the multiplexer in the ASIC, the three signals of humidity, pressure and temperature could be alternately measured, and then the air density is calculated. A pressure and humidity sensor must be integrated into the package for this solution. The second possibility is to use a pressure sensor chip with a polymer pad on the membrane. For a certain size and thickness of the polymer pad, the bridge voltage will be proportional to the air density. The advantage of this solution would be the low cost and small size of the sensor.

6 Conclusions

A very small humidity sensor chip with a footprint of only 1×1 mm^2 based on the piezoresistive effect has been developed. The chip size was reduced by a factor of 10 compared to previous developments. Our experiments showed that the sensor can be used to measure the relative humidity of the air with a sensitivity of 0.25 mV/ % RH.

Compared with humidity sensors applied in mobile phones today, the new sensor utilizing the current packaging technology of the EPCOS T5400 pressure sensor would be similar or smaller in size. Even in a first one-shot lab experiment, the performance of our piezoresistive sensor is at least comparable to current capacitive humidity sensors like the SHTC1 from Sensirion. With further optimization we see a good potential to meet whatever requirement is most important for the target application: either excellent accuracy or low costs by simple one-point calibration, or quick response times at still sufficient sensitivity by reducing the polymer thickness.

Edited by: U. Schmid
Reviewed by: two anonymous referees

References

Bosch: Sustainability in a Connected World is Bosch focus, International CES, Bosch press release, 2014.

Buchhold, R., Nakladal, A., Gerlach, G., Sahre, K., Müller, M., Eichhorn, K.-J., Herold, M., and Gauglitz, G.: A study on the microphysical mechanisms of adsorption in polyimide layers for microelectronic applications, J. Electrochem. Soc., 145, 4012–4018, 1998a.

Buchhold, R., Nakladal, A., Gerlach, G., Sahre, K., Eichhorn, K.-J., Herold, M., and Gauglitz, G.: Influence of moisture-uptake on mechanical properties of polymers used in microelectronics, MRS Proceedings, 511, 359–364, 1998b.

Buchhold, R., Nakladal, A., Gerlach, G., Sahre, K., Eichhorn, K.-J., and Müller, M.: Reduction of mechanical stress in micromachined components caused by humidity-induced volume expanison of polymer layers, Microsystem Technol., 5, 3–12, 1998c.

Buchhold, R., Nakladal, A., Gerlach, G., and Neumann, P.: Design studies on pieoresistive humidity sensors, Sensors Actuators B, 53, 1–7, 1998d.

Buchhold, R., Nakladal, A., Gerlach, G., Herold, M., Gauglitz, G., Sahre, K., and Eichhorn, K.-J.: Swelling behavior of thin anisotropic polymer layers, Thin Solid Films, 350, 178–185, 1999.

Colin Johnson, R.: MEMS barometers boost hard drives, GPS, Electronic Engineering Times, 2009.

Gerlach, G. and Sager, K.: A piezoresistive humidity sensor, Sensors Actuators A, 43, 181–184, 1994.

Guenther, M., Sahre, K., Suchaneck, G., Gerlach, G., and Eichhorn, K.-J.: Influence of ion-beam induced chemical and structural modification in polymers on moisture uptake, Surface Coating. Technol., 142–144, 482–488, 2001.

Lane, N. D., Miluzzo, E., Lu, H., Peebles, D., Choudhury, T., and Campbell, A. T.: A survey of mobile phone sensing, IEEE Communications Magazine, 48-9, 140–150, 2010.

Lee, C.-Y. and Lee, G. B.: Humidity sensors: a review, Sensor Letters, 3, 1–15, 2005.

Mayer, F. and Lechner, M.: Sensirion brings temperature and humidity sensors to the mobile phone, MEMS Trends, 15, 12–13, 2013.

Sager, K., Gerlach, G., and Schroth, A.: A humidity sensor of a new type, Sensors and Actuators B, 18–19, 85–88, 1994.

Sone, H., Fujinuma, Y., Hieida, T., Chiyoma, T., Okano, H., and Hosaka, S.: Picogram mass sensor using microcantilever, Proc of SICE Annual Conf., pp. 1508–1513, 2004a.

Sone, H., Fujinuma, Y., Hieida, T., Chiyoma, T., Okano, H., and Hosaka, S.: (Picogram mass sensor using microcantilever), Jap. J. Appl. Phys., 43, 4663–4666, 2004b.

Waber, T., Pahl, W., Schmidt, M., Feiertag, G., Stufler, S., Dudek, R., and Leidl, A.: Flip-chip packaging of piezoresistive barometric pressure sensors, Proc. SPIE, 8763, 876321, doi:10.1117/12.2016459, 2013.

Waber, T., Pahl, W., Schmidt, M., Feiertag, G., Stufler, S., Dudek, R., and Leidl, A.: Temperature characterization of flip-chip packaged piezoresistive barometric pressure sensors, Microsystem Technologies, 20, 861–867, 2014.

Wagner, T., Krotzky, S., Weiß, A., Sauerwald, T., Kohl, C.-D., Roggenbuck, J., and Tiemann, M.: A high temperature capacitive humidity sensor based on mesoporous silica, Sensors, 11-3, 3135–3144, 2011.

Wasisto, H. S., Merzsch, S., Huang, K., Stranz, A., Waag, A., and Peiner, E.: Simulation and characterization of silicon nanopillar-based nanoparticle sensors, Proc. SPIE, 8763, 87632D, doi:10.1117/12.2016970, 2013.

Yamazoe, N. and Shimizu, Y.: Humidity sensors: Principles and application, Sensors and Actuators, 10, 379–398, 1986.

4

Electrochemical analysis of water and suds by impedance spectroscopy and cyclic voltammetry

R. Gruden[1], A. Buchholz[1], and O. Kanoun[2]

[1]Seuffer GmbH & Co. KG, Bärental 26, 75365 Calw-Hirsau, Germany
[2]Technische Universität Chemnitz, Reichenhainer Strasse 70, 09126 Chemnitz, Germany

Correspondence to: R. Gruden (roman.gruden@seuffer.de)

Abstract. Optimum detergent dosage during a washing process depends on water quality, degree of pollution and quantity of laundry. Particularly, water quality is an important factor. Other parameters like carbonate- or non-carbonate hardness and calcium/magnesium (Ca/Mg) ratio in addition to total hardness of water have an impact on the amount of detergent. This work discusses the possibilities realizing a detergent sensor that measures important parameters for the washing process and assess the ideal necessary amount of detergent during the washing process. The approach is to combine impedance spectroscopy with cyclic voltammetry in order to determine both water quality and concentration of detergent in the suds which build up the basis for an optimum detergent dosage. The results of cyclic voltammetry show that it is possible to identify the Ca/Mg ratio and the carbonate hardness separately, which is necessary for the optimization of the washing process. Impedance measurements identify total hardness and detergent concentrations.

1   Introduction and motivation

This work is an extension of the conference article (Gruden and Kanoun, 2013a)[1].

In 2010 more than 580 000 tons of detergent and more than 1 billion liters of water were used in Germany for washing the laundry. On average, 30 % of the detergent is superfluous, resulting in water wasting during the rinsing process with an impact on environmental burden (Rüdenauer and Gensch, 2008).

Due to the higher concentration of detergent in the suds, more and longer rinsing processes are required to get rid of the detergent residues in the laundry. This requires not only higher water consumption but also more energy. If the rinsing process is adapted to the higher detergent concentration, the remaining residues can cause allergic reactions in sensitive individuals.

Furthermore, the washing result, customized to the individual properties of this load, is only optimal at one detergent quantity (Wagner, 2010; Smulders et al., 2002; Jakobi and Löhr, 1987).

At the present time automatic washing machines use only a control strategy that does not take the water quality or the actual properties of the laundry into consideration (Wagner, 2010; Smulders et al., 2002; Jakobi and Löhr, 1987). Moving to a regulation strategy by maintaining the washing result needs key parameters such as water quality, quantity of laundry (weight), concentration of dirt and concentration of detergent to be able to compute an optimal dosage.

The water parameter currently used for detergent dosage is total hardness. But it is also important to measure the carbonate hardness and the Ca/Mg ratio because these parameters also influence the washing performance, too. The hydrogen carbonate ion HCO_3^-, which is reasonable for the carbonate hardness, reacts to calcium carbonate which attaches to the heating elements in the washing machine and causes damages; calcium and magnesium form insoluble soaps which decrease the washing performance. The Ca/Mg ratio influences the washing performance because calcium soaps are more insoluble than magnesium soaps.

---

[1]Gruden, R. and Kanoun, O.: Water quality assessment by combining impedance spectroscopy measurement with cyclic voltammetry, AMA Sensor, Nürnberg, 2013.

These parameters should be measured online during the washing process with a robust low cost sensor (Gruden et al., 2012), in order to save resources, protect the environment and avoid allergic reactions (Tröltzsch, 2012).

2 State of the art and new approaches

Optical methods such as turbidity sensors are presently used for automatic detergent dosage (Tschulena and Lahrmann, 2006; Czyzewski et al., 1999), but they are sensitive to contamination and especially to the dirtiness water; furthermore, water parameters cannot be detected with turbidity sensors.

Base–acid titration is able to determine precisely the value of total hardness and the Ca / Mg ratio (Hütter, 1990) but this method is very expensive and works only under laboratory conditions.

Commercial online methods used for environmental investigation (Huang et al., 2005; Kräuter et al., 2006) and distribution systems (Verberk et al., 2006) focus on other water components and are too expensive for household low-cost applications. Voltammetric methods combined with ion-selective electrodes are expensive (Cammann and Galster, 1996) and not suitable for detergent determination because of complex and different detergent composition (Wagner, 2010). Low-cost applications need environmental harmful mercury electrodes to determine the water parameters (Pungor et al., 1977). Both methods are not feasible inside the washing machine during the washing process.

Electronic tongues are sensor arrays combined with voltammetric methods and multi-variable analysis (Winquist et al., 1997; Winquist, 2008) which are tested for large household appliances: applications (Winquist, 2011; Ivarsson, 2003; Eriksson et al., 2011). The existent problem is the complex data analysis. To have information about total hardness, conductive methods are more suitable, but they are generally not sufficiently accurate because the conductivity of the electrolyte χ_E depends on the total ion concentration including even ions which are not interesting for the washing process. Moreover χ_E is frequency dependent and the results of the established methods which use only one frequency is incorrect.

Suds investigations with surface acoustic wave (SAW) sensors have been tested successfully to detect single sufactant concentrations (Vivancos et al., 2012). The investigation of water parameters, surfactant mixtures and critical micelle concentration (CMC) are not shown. Moreover, frequencies of $f > 60$ MHz are needed for the SAW application and such electronic circuits are too expensive for the application in white goods.

Electrochemical impedance spectroscopy (EIS) and cyclic voltammetry (CV) are well-known analytical methods (Barsoukov and Macdonald, 2005; Brad and Faulkner, 2001) and deliver more specific information. For analysis of suds during the washing process they have not yet been applied.

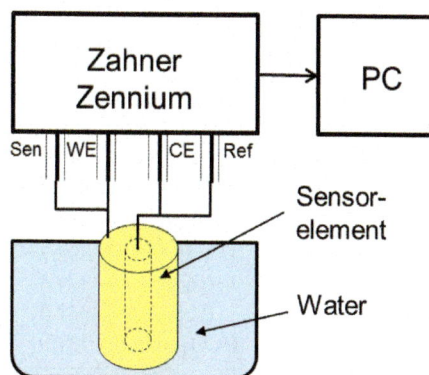

Figure 1. Measurement setup of Zahner Zennium for EIS and CV measurements.

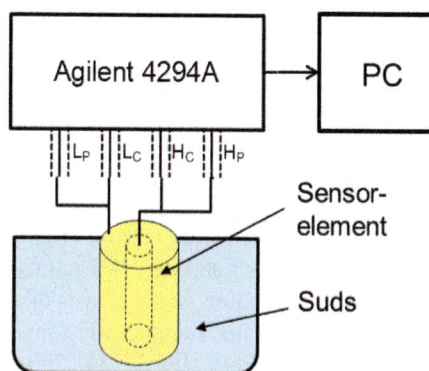

Figure 2. Measurement setup of Agilent 9294A for EIS measurements (Gruden and Kanoun, 2013b).

The combination of these measuring principles provides more data from different frequency ranges and allows more detailed investigations. The focus of this fluid analysis is the analysis of data and not to electrochemical effects of the sensor element. With this method, expensive sensors and environmentally harmful materials can be avoided. Due to the fact that frequencies of $f_{max} = 5$ MHz are needed, the cost of the electronic circuit is adequate for the planned application.

The focus of this paper is the investigation of predefined water samples and one standard detergent, and to ignore real water samples and commercial detergents because the feasibility of this new method should be tested. Tests with real water samples and commercial detergents can be investigated later.

3 Experimental

The experimental setup for electrochemical impedance spectroscopy (EIS) and cyclic voltammetry (CV) consisted of a Zahner Zennium impedance analyzer for frequency range 0.1 Hz to 4 MHz and an Agilent 9294A impedance analyzer (used for EIS only) with measurement adapter Agilent

Table 1. List of water samples with a total hardness of $1.48\,\mathrm{mmol\,L^{-1}}$ ($1.48\,\mathrm{mmol\,L^{-1}}$) and different carbonat hardness caused by different HCO_3^-/Cl^- ratio.

	Carbonate-hardness (mmol L^{-1})	HCO_3^-/Cl^- ratio	Conductivity χ_E ($\mu\mathrm{S\,cm^{-1}}$)
1	0	Cl^- only	411.5
2	0.25	1 : 1.35	392.7
3	0.5	1 : 0.63	374.3
4	0.69	1 : 0.25	347.4
5	0.93	HCO_3^- only	337.8

16048H for frequency range of 40 Hz to 110 MHz. All tests were carried out with a homemade cylindrical sensor element with gold electrodes and two-electrode configuration. Figure 1 shows the setup for the Zahner Zennium and Fig. 2 shows the setup for the Agilent 9294A.

For future application a low-cost electronic is planned and a prototype exists. The sensor element is made of ceramic with a 500 nm gold surface.

The temperature control was a Julabo LH46 Presto and a double-walled integral exchanger reactor. High purity water prepared with a Siemens LaboStar UV7 immediately before the experimental procedure was the basis of all applied solutions. The synthetic water samples were manufactured according to IEC 60734:2003 (IEC60734, 2003) and the custom-made Seuffer procedure.

The reference measurement of each water sample was carried out with the titrator TA20plus and the software TitriSoft 2.6 from SI-Analytics. Reference detergent IEC A/IEC 60456 (wfk, IEC60456[2]) of wfk-Testgewebe GmbH were used for the suds. The water samples for the experiments with different values of carbonate hardness (see Table 1) are equal except for the hydrogen carbonate / chloride ratio ($c\,(HCO_3^- + Cl^-) = 2.56\,\mathrm{mmol\,L^{-1}} = \mathrm{const.}$). All these samples have a total ion concentration of $c_{ion} = 4.2\,\mathrm{mmol\,L^{-1}}$ (Gruden and Kanoun, 2013a).

4   Results and discussion

The approach is to combine impedance spectroscopy and cyclic voltammetry in order to have a detailed water analysis as a basis for optimal detergent dosage. By means of impedance spectroscopy, the exact conductivity of the medium can be determined and changes in the composition of the medium can be detected. At the same time, the relative concentration α_{ion} (see Eqs. 1 and 2) of the species can be quantified by cyclic voltammetry. Many measurements have been carried out and show the feasibility of the combined measurement principle.

---

[2]wfk: IEC A* Referenzwaschmittel nach IEC 60456 Order Code: 88010-1, WFK IEC A*, wfk-Testgewebe GmbH, Brüggen-Bracht.

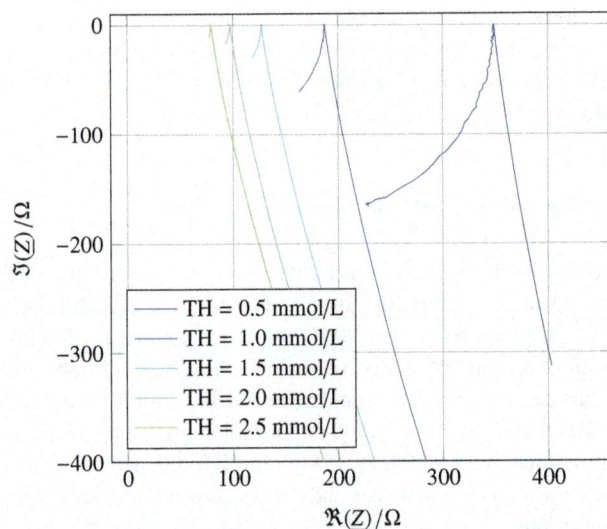

Figure 3. Dependence of the impedance spectra on total hardness of water.

Table 2. MTP frequencies of the impedance spectras of Fig. 3.

TH [mmol L^{-1}]	Frequency f [kHz]
0.5	3.5
1.0	7.0
1.5	17.8
2.0	20.0
2.5	35.5

$$c_{ion} = c_{Ca^{2+}} + c_{Mg^{2+}} + c_{Na^+} + c_{HCO_3^-} + c_{Cl^-} + c_{SO_4^{2-}} \quad (1)$$

$$1 = \alpha_{Ca^{2+}} + \alpha_{Mg^{2+}} + \alpha_{Na^+} + \alpha_{HCO_3^-} + \alpha_{Cl^-} + \alpha_{SO_4^{2-}} \quad (2)$$

Figure 3 shows the impedance spectra of water samples with different values of total hardness of water.

The impedance spectra are qualitatively equal because of the identical composition of the water samples. The maximum turning point (see Fig. 4) of the impedance spectra is equal to the resistance of the electrolyte R_E and is frequency dependent.

The frequency values of the MTP of the spectras of Fig. 3 are shown in Table 2.

The advantage of impedance spectroscopy compared to the usual single frequency conductance measurement is that the maximum turning point, and thus R_E, can be precisely determined by the multi-frequency method. The conductance G_E (reciprocal of R_E) correlates linearly with the total hardness of water (see Fig. 5) with a correlation coefficient of $r = 0.9997$.

Therefore, the resulting sensitivity of $S_{TH} = 4.85\,\mathrm{mS\,L\,mmol^{-1}}$ is very good for the planned application.

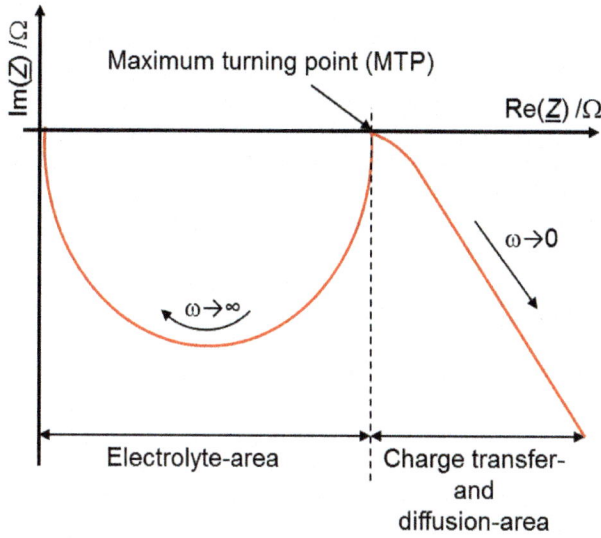

Figure 4. Qualitative plot of an impedance spectra.

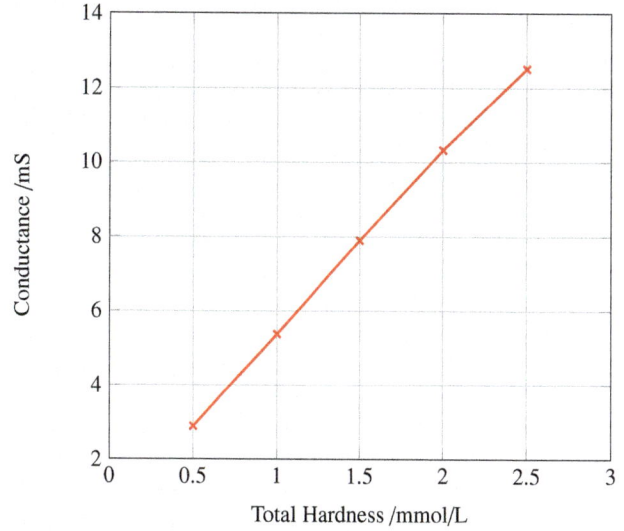

Figure 5. Dependence of the conductance on total hardness of water.

This high correlation is only achievable with synthetic water by use of the precise impedance spectroscopy method. A statistic of 124 real water samples of different German areas shows that the correlation coefficient of these water samples is $r = 0.8568$. The difference between synthetic and real water is caused by the different ion concentrations of the ions which do not contribute to total hardness (see Table 3, Eqs. 3 and 4).

$$\chi_E^0 = c_{Ca^{2+}} \cdot \Lambda_{Ca^{2+}}^0 + c_{Mg^{2+}} \cdot \Lambda_{Mg^{2+}}^0 + c_{Na^+} \cdot \Lambda_{Na^+}^0$$
$$+ c_{HCO_3^-} \cdot \Lambda_{HCO_3^-}^0 + c_{Cl^-} \cdot \Lambda_{Cl^-}^0 + c_{SO_4^{2-}} \cdot \Lambda_{SO_4^{2-}}^0 \quad (3)$$

$$\overline{\chi_{TH}} = \overline{\chi_E} - \chi_\varepsilon \rightarrow \overline{\chi_{TH}} \approx \overline{\chi_E} \quad (4)$$

The relationship between conductance G_E and conductivity χ_E is given by the cell constant K of the sensor element.

$$\chi_E = G_E \cdot K \quad (5)$$

The conductivity, χ_E, of the water sample is composed of the concentration of each species, c_{SP}, multiplied by its molar limit conductivity, Λ^0. The portion of conductivity which represents the total hardness of water, χ_{TH}, cannot be clearly defined because of the different combination possibilities of the hardness forming cations Ca^{2+} and Mg^{2+} with its possible corresponding anions HCO_3^-, Cl^- and SO_4^{2-}. All combinations between the cations and anions are possible and lead to different conductivities which do not correlate to the chemical effects. The arithmetic mean of χ_{TH} equals χ_E if the fault χ_ε is low or the requirements to the results are not strict. This results in the need for a differentiated analysis of ions in case of precise measurement requirements.

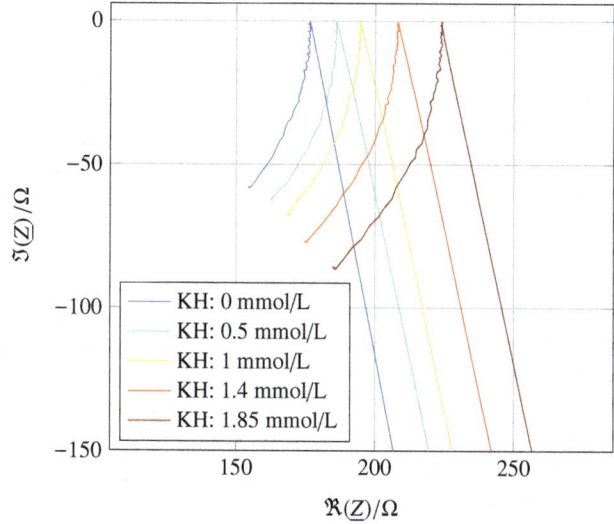

Figure 6. Impedance spectra of water samples with the same total hardness and different carbonate hardness.

Figure 6 shows the impedance spectra of five water samples, all of them has a total hardness of $1.48\,\mathrm{mmol\,L^{-1}}$ ($8.31°\,\mathrm{dH}$) but different carbonate hardness values (see Table 1).

All spectra are qualitatively equal. The frequency of the MTP is $f = 7\,\mathrm{kHz}$ at all spectras and the highest frequency point is $f = 4\,\mathrm{MHz}$. To distinguish them from the spectra of different total hardness values, further information by an additional measurement principle is necessary.

Figure 7 shows a schematic diagram of two cyclic voltammetry measurements.

At t_1 the different current responses (I_1 and I_2) are used to distinguish carbonate hardness. At I_t the different relaxation

Table 3. List of the ions which contribute mainly to the total ion concentration of water.

Ion	Contribution to	Remarks
Ca^{2+}	Total hardness (TH)	Together with Mg^{2+} main ion for TH
Mg^{2+}	Total hardness (TH)	Together with Ca^{2+} main ion for TH
Na^+	Only total ion concentration (TIC)	–
K^+	Only total ion concentration (TIC)	–
HCO_3^-	Carbonate hardness (KH)	
Cl^-	Non carbonate hardness (NKH)	All anions contribute to the conductivity of the total hardness, depending on the composition.
Cl^-	Non carbonate hardness (NKH)	All anions contribute to the conductivity of the total hardness, depending on the composition.
SO_4^{2-}	Non carbonate hardness (NKH)	

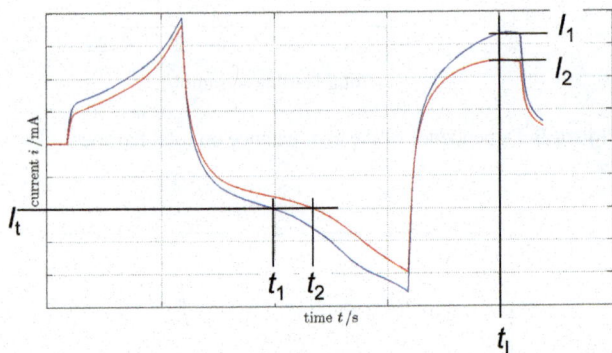

Figure 7. Two typical cyclic voltammetry measurements of different water samples.

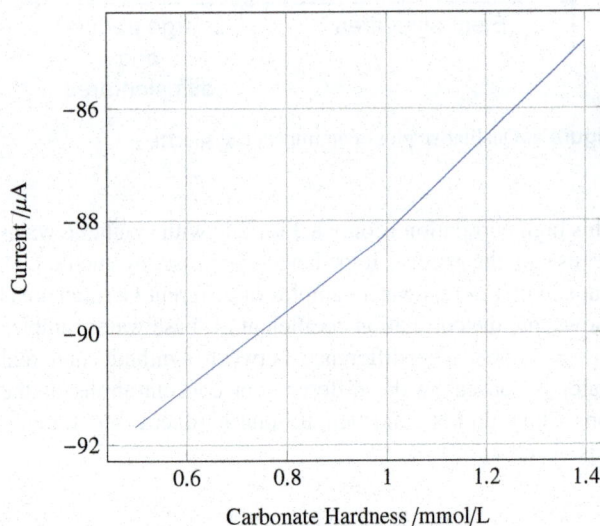

Figure 8. Current response as a function of carbonate hardness.

times (t_1 and t_2) are used to distinguish calcium-magnesium-ratios.

Figure 8 shows the results of the water investigation by cyclic voltammetry.

A sensitivity of $S_i = -6.7\,\mu A\,L\,mmol^{-1}$ was reached, which is sufficient for the target application. The absolute value of current I decreases linearly with the increasing carbonate hardness; the reason is that the ratio of the anions changes. The concentration of HCO_3^- increases and the concentration of Cl^- ions decrease. Cl^- ions have a higher contribution to the conductivity than HCO_3^-.

Figure 9 shows the impedance spectra of water samples with identical total and carbonate hardness but with a different calcium-magnesium-ratio.

The frequency of the MTP is $f = 4\,kHz$ at all spectras and the maximum frequency is $f = 4\,MHz$. The impedance spectra are qualitatively and quantitatively identical and thus it is not possible to distinguish different calcium-magnesium-ratios by impedance spectroscopy. The water samples are nearly identical from electrochemical point of view because the earth alkaline metals calcium and magnesium behave very similar.

Figure 10 shows the result of an investigation of the water samples by cyclic voltammetry.

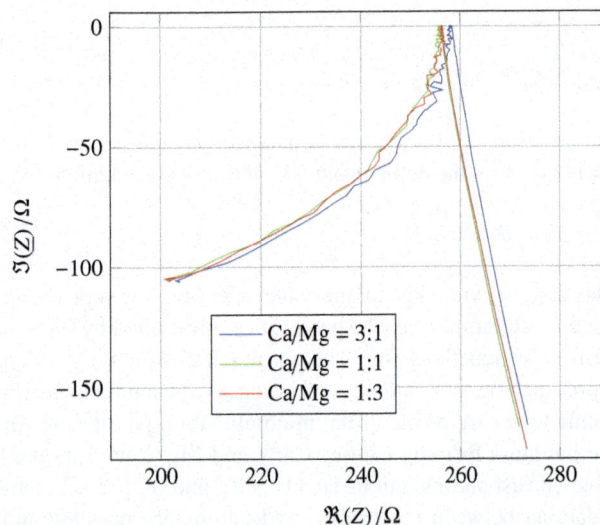

Figure 9. Impedance spectra of water samples with the same total and carbonate hardness but different Ca/Mg ratio ($mmol\,L^{-1}$).

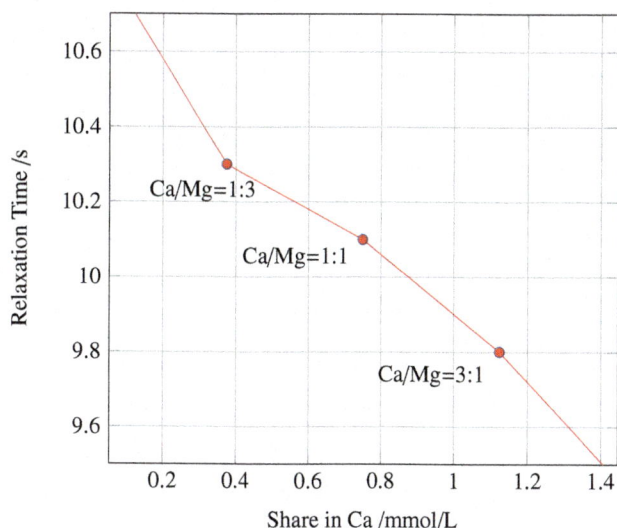

Figure 10. Relaxation time depending on calcium-magnesium-ratio.

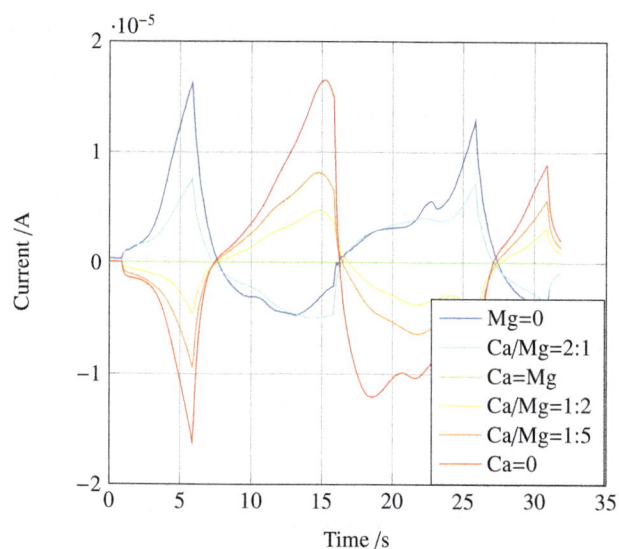

Figure 11. Relative current response of different Ca/Mg-ratios based on a balanced Ca/Mg-ratio.

A correlation of the relaxation time and the Ca/Mg ratio can be clearly seen. The relaxation time depends on the ionic radii of the species. Different radii mean different speeds of the ions. The resolution of the sensor signal to distinguish the different calcium-magnesium-ratios is an important factor for the planned application. For this experiment, solutions with different calcium-magnesium-ratios were used, the total concentration was constant.

$$c_{\text{tot}} = c(\text{Ca}) + c(\text{Mg}) = 1.5 \,\text{mmol L}^{-1} = \text{const.} \tag{6}$$

Figure 11 shows the results of the relative current responses of different calcium-magnesium-ratios. The curves are scaled that if the Ca/Mg ratio is 1 : 1 ($c(\text{Mg}) = c(\text{Cl})$) the current response is zero ($I = 0$). Differences from a balanced Ca/Mg ratio can be easily seen. Deviations lead to a current response $I \neq 0$ and the sensitivity of the sensor is $S \approx 0.9 \,\text{s L mmol}^{-1}$ of Ca which is adequate for the planned application.

The results of Ca/Mg ratio measurement could be confirmed only with synthetic water samples. Tests with real water samples could not confirm the laboratory results within the needed precision. The reasons for these different results could be the cross-sensitivity of the sensor system to other ions. A new approach to enlarge the measurement method by further tests to suspend the influence of other ions is in progress.

These three characteristics of water (total hardness of water, carbonate hardness and calcium-magnesium-ratio) are important for detergent dosage and have direct influence on the washing result (personal communication, 2012). The washing performance depends on the carbonate hardness and the Ca/Mg ratio because hydrogen carbonate ions react to insoluble calcium carbonate and attach to the laundry and on the heating elements of the washing machine, and calcium

and magnesium form salts with different solubility. All other ions, which also have impact on the washing process are not represented in this investigation. These ions will investigated later.

The new unit AQrate will be introduced to characterize water.

$$\text{AQrate} = f(\text{TH, KH, Ca/Mg}) \tag{7}$$

$$\text{TH} = c(\text{Ca}^{2+} + \text{Mg}^{2+}) \tag{8}$$

$$\text{KH} = c(\text{HCO}_3^-) \tag{9}$$

$$\text{Ca/Mg} = \frac{c(\text{Ca}^{2+})}{c(\text{Mg}^{2+})} \tag{10}$$

For detergent dosage recommendation AQrate will replace total hardness of water in a first step because it is much more precise and meaningful.

Figure 12 shows the impedance spectra of suds with different concentrations c_{det} and Table 4 shows the frequencies of the MTP.

The maximum turning points (MTP) correlate strongly with the concentration of the suds ($r \approx 1$) as long as they do not exceed a certain value (see Fig. 13).

Above this concentration c_{CMC}, the conductivity increases again linearly with the concentration, but with a lesser gradient. Thus there are two lines and their intersection point marks the critical micelle concentration (CMC) (Nakamura et al., 1998; Neto et al., 2006; Chang et al., 1998). There is a strong presumption that the optimum detergent concentration is close to the critical micelle concentration. On this issue intensive tests are currently being carried out (Gruden and Kanoun, 2013a).

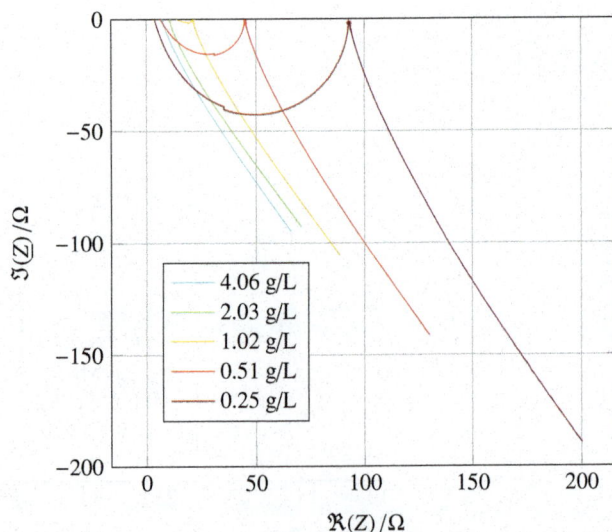

Figure 12. Dependence of the IS on detergent concentration.

Table 4. MTP frequencies of the impedance spectras of Fig.12.

c [g L^{-1}]	Frequency f [kHz]
0.25	48.4
0.51	71.4
1.02	153
2.03	172
4.06	191

5 Conclusion and outlook

The results indicate that it is possible to determine the detergent concentration of suds and the parameters of water which have impact on the washing process by a combination of impedance spectroscopy (EIS) and cyclic voltammetry (CV). The results of Ca / Mg ratio measurement could be confirmed only with synthetic water samples. Tests with real water samples could not confirm the laboratory results of Ca / Mg ratio within the needed precision. A big advantage of impedance spectroscopy as a multi-frequency method in comparison to conventional conductance measuring methods is the possibility to determine R_E precisely and thus form a solid basis for all further investigations. Moreover, further investigations are in progress to get more information from the impedance spectra to improve the planned application.

The detergent concentration of the suds can be measured precisely by impedance spectroscopy as well. The implementation of this method for the CMC detection and thus the optimization of the detergent amount and the washing process is in progress.

To determine the carbonate hardness and the calcium-magnesium-ratio, a combination of the results of impedance spectroscopy and cyclic voltammetry is necessary and the results show the feasibility.

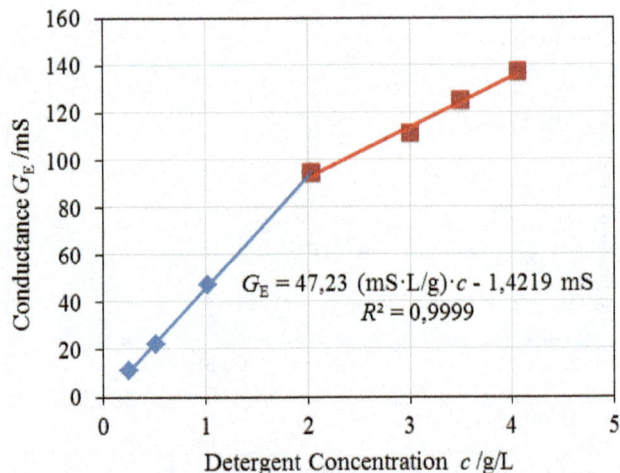

Figure 13. Dependence of the conductance on detergent concentration.

Both methods can be realized with the same sensor element and the same hardware. The planned application can be realized as a low cost solution for an online measurement during the washing process without environmentally harmful substances. The realized measurements indicate the feasibility of the planned application. For a final version some points like the online detection of the critical micelle concentration and the precise correlation of the measurement data to the water parameters still has to be clarified. Such tests are in progress.

The new unit AQrate has to be specified and will be extended by additional ion effects. With such adaptations, the new unit AQrate can consider region-specific characteristics of water.

Edited by: A. L. Spetz
Reviewed by: two anonymous referees

References

Barsoukov, E. and Macdonald, J. R.: Impedance Spectroscopy, Wiley-Interscience, 2005.

Brad, A. J. and Faulkner, L. R.: Electrochemical Methods, John Wiley & Sons, 2001.

Cammann, K. and Galster, H.: Das Arbeiten mit ionenselektiven Elektroden, Springer Verlag, 1996.

Chang, H.-C., Hwang, B.-J., Lin, Y.-Y., Chen, L.-J., and Lin, S.-Y.: Measurement of critical micelle concentration of nonionic surfactant solutions using impedance spectroscopy technique, Rev. Sci. Instrum., 69, 2514–2520, 1998.

Czyzewski, G., Schulze, I., and Engel, C.: Europäische Patentanmeldung EP 0 992 622 A2, 1999.

Eriksson, M., Lindgren, D., Bjorklund, R., Winquist, F., Sundgren, H., and Lundström, I.: Drinking Water monitoring with voltammetric sensors, Procedia Engineering, 25, 1165–1168, 2011.

Gruden, R., Kanoun, O., and Tröltzsch, U.: Influence of surface effects on the characterisic curves of detergent sensors, 9th International Multi-Conference on Signals, Sensors and Devices, Chemnitz: 20–23 März 2012.

Gruden, R. and Kanoun, O.: Water quality assessment by combining impedance spectroscopy measurement with cyclic voltammetry, AMA Sensor, Nürnberg, 2013a.

Gruden, R. and Kanoun, O.: Fast and Low-Cost Online Detection of Critical Micelle Concentration based on Impedance Specroscopy, The 7th International Conference on Sensing Technology, 3–5 December, Wellington, New Zealand, 2013b.

Huang, Z., Li, K., and Xu, H.: Research of an online measurement and control system of water-quality on FIA, Chinese Journal of Scientific Instrument 36, 343–346 + 385, 2005.

Hütter, L. A.: Wasser und Wasseruntersuchung, Otto Salle Verlag und Verlag Sauerländer, 1990.

IEC60734: Household electrical applicances Performance – Hard water for testing (IEC 60734:2003), CENELEC (European Committee for Electrotechnical Standardization), Brussels, 2003.

Ivarsson, P.: Electronic Tongues – New Sensor Technology in Household Appliances, Linköping, Sweden, 2003.

Jakobi, G. and Löhr, A.: Detergents and Textile Washing, VCH, 1987.

Kräuter, S., Lange, R., and Seifert, F.: Surface water quality measuring station adaptable to changing requirements, Water and Wastewater International, 21, p. 41, 2006.

Nakamura, H., Sano, A., and Matsuura, K.: Determination of Critical Micelle Concentration of Anionic Surfactants by Capillary Electrophoresis Using 2-Naphthalenemethanol as a Marker for Micelle Formation, Journal of Analytical Sciences, 14, 379–382, 1998.

Neto, J. M. G., da Cunha, H. N., Neto, J. M. M., and Ferreira, G. F. L.: Impedance spectroscopy analysis in a complex system: Sodium dodecyl sulfate solutions, J. Sol-Gel Sci. Techn., 38, 191–195, 2006.

Pungor, E., Nagy, G., and Feher, Z.: The flat surfaced membrane coated mercury electrode as analytical tool in the continuous voltammetric analysis, J. Electroanal. Chem., 75, 241–254, 1977.

Rüdenauer, I. and Gensch, C.-O.: Einsparpotentiale durch automatische Dosierung bei Waschmaschinen, Öko-Institut e.V., Studie im Auftrag der Miele & Cie. KG, 2008.

Smulders, E., Rähse, W., von Rybinski, W., Steber, J., Sung, E., and Wiebel, F.: Laundry Detergents, Wil, 2002.

Tröltzsch, U.: Anwendungspotential der Impedanzspektroskopie für die Waschlaugensensorik, 16. GMA/ITG-Fachtagung Sensoren und Messsysteme, 650–661, 2012.

Tschulena, G. and Lahrmann, A.: Sensors in Household Appliances, Wiley-VCH, 2006.

Verberk, J. Q. J. C., Hamilton, L. A., O'Halloran, K. J., Van Der Horst, W., and Vreeburg, J.: Analysis of particle numbers, size and composition in drinking water transportation pipelines: Results of online measurements, Wa. Sci. Technol., 6, 35–43, 2006.

Vivancos, J.-L., Rácz, Z., Cole, M., and Gardner, J. W.: Surface acoustic wave based analytical system for the detection of liquid detergents, Sensor. Actuat. B-Chem., 171, 469–477, 2012.

Wagner, G.: Waschmittel, Viley VCH, 2010.

Winquist, F.: Voltammetric electronic tongues – basic principles and applications, Microchim. Acta, 163, 3–10, 2008.

Winquist, F., Wide, P., and Lundström, I.: An electronic tongue based on voltammetry, Anal. Chim. Acta, 357, 21–31, 1997.

Winquist, F.: Multicomponent analysis of drinking water by a voltammetric electronic tongue, Anal. Chim. Acta, 683, 192–197, 2011.

5

A simple method to recover the graphene-based chemi-resistor signal

F. Fedi[1,4], F. Ricciardella[2,3], M. L. Miglietta[2], T. Polichetti[2], E. Massera[2], and G. Di Francia[2]

[1]CNR-Institute for Composite and Biomedical Materials, Portici (Naples), Italy
[2]ENEA UTTP-MDB Laboratory, R. C. Portici (Naples), Italy
[3]University of Naples "Federico II", Department of Physics, Naples, Italy
[4]Faculty of Physics, University of Vienna, Strudlhofgasse 4, 1090 Vienna, Austria

Correspondence to: F. Ricciardella (filiberto.ricciardella@enea.it)

Abstract. We present the development of a simple and fast method for restoring exhaust graphene-based chemi-resistors used for NO_2 detection. Repeatedly exposing the devices to gases or to air for more than 2 days, an overall worsening of the sensing signal is observed; we hypothesized that the poisoning effect in both cases is caused by the exposure to NO_2. Starting from this hypothesis and from the observation that NO_2 is soluble in water, we performed a recovery method consisting in the dipping of exhaust devices into ultrapure water at $100\,°C$ for $60\,s$. The device performances are compared with those obtained after the restoration is achieved using the typical annealing under vacuum method.

1 Introduction

A crucial point for solid-state gas sensor use at room temperature (RT) is the difficulty of recovery at the initial conditions after the sensing operation (Schedin et al., 2007; Yavari and Koratkar, 2012; Yuan and Shi, 2013). The drawback arises from the interaction energy between sensitive materials and gases, as reported by conventional transition state theory. At RT, in fact, the energies involved in the adsorption phenomena are in the range of eV, so that, once chemisorbed, reversibility is not thermodynamically favored. In order to allow the formation of a free interface on the sensitive layer, gas molecules generally have to be removed by recovery methods, supplying an external source of energy and UV irradiation, thermal treatment and electric field applications; all the above mentioned methods have been proposed for this purpose (Chen et al., 2001; Hyman and Medlin, 2005; Schedin et al., 2007; Charlier et al., 2009; Leghrib and Llobet, 2011). Chen et al. (2001) continuously apply cleaning in situ on the sensing material by pumping ultraviolet light, achieving the record gas detection of parts per trillion (ppt), although in an inert atmosphere. Even Schedin et al. (2007) found that ultraviolet irradiation or annealing at 150°C can restore the state of the devices after analyte exposure, although under vacuum. In Charlier et al. (2009), the carbon nanotube (CNT) based sensors were kept heated at 150°C to speed up gas desorption, while the test chamber was flushed with pure dry air for 1 h after each series of successive injections; when the airflow was interrupted, the sensors were left at ambient temperature for 12 h so that the full recovery of baseline resistance was reached. In this case, operating at RT, the NO_2 minimum concentration of 500 ppb was achieved. Leghrib and Llobet raised the temperature of the CNT-based sensors to 150°C, promoting the cleaning after the gas exposure, while dry air was injected in the chamber. Therefore, the standard approach is essentially based on the increase of the analyte molecule mobility by providing the energy to improve the adsorbate desorption from the sensitive layer. For the specific sensor device investigated in this work, we propose a new refreshing method that relies on the capability of removing the NO_2 through water, the most suitable solvent for this target analyte (Tan and Piri, 2013). The chemi-resistor performances towards 350 parts per billion (ppb) of NO_2 in wet N_2 environment (relative humidity = 50 %, temperature = 22 °C) were compared before and after the restoration process. The proposed method

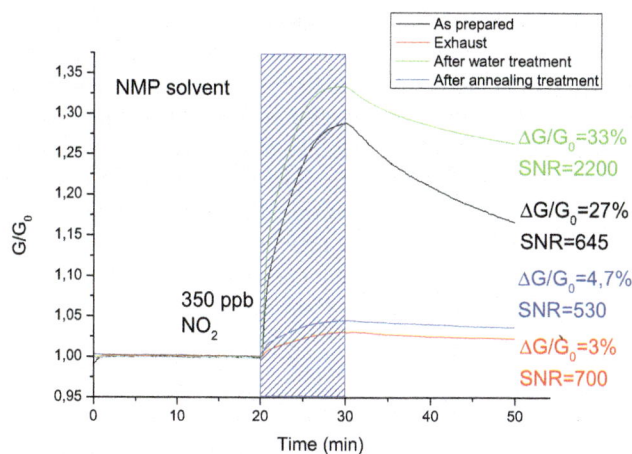

Figure 1. Normalized electrical conductance behaviors for chemi-resistors towards 350 ppb of NO_2 in a wet N_2 environment as soon as prepared (black line), after one month (red line), after the restoration by the dipping method (green line), and after the restoration by the annealing method (blue line). The standard protocol adopted in these tests consists of the baseline of 20 min in wet N_2, the window exposure of 10 min to NO_2 (blue dashed color area in the figures), and the recovery phase of 10 min. The graphene was exfoliated using NMP as a solvent.

Figure 2. Normalized electrical conductance behaviors for chemi-resistors towards 350 ppb of NO_2 in a wet N_2 environment as soon as prepared (black line), after the restoration by the dipping method (red line), and after the restoration by the annealing method (green line). The standard protocol adopted in these tests consists of the baseline of 20 min in wet N_2 and the window exposure of 10 min to NO_2 (blue dashed color area in the figures). In this case, the graphene solution was prepared by exfoliating graphite flakes in a mixture of IPA/n-BuOH solvent.

was tested on devices prepared starting from two different graphene solutions, and the results were also compared with those obtained by using the thermal treatment at 130 °C in vacuum.

2 Materials and methods

Colloidal graphene suspensions were prepared by the liquid phase exfoliation (LPE) method. Graphite flakes (Sigma-Aldrich, product no. 332461) at $2.5\,\mathrm{g\,L^{-1}}$ were dispersed in NMP (N-methyl-pyrrolidone, Sigma-Aldrich, product no. 328634) or in a mixture of isopropanol and n-butanol (IPA/n-BuOH). Then, a mild sonication treatment was required for 168 h at a low power aiming to promote the graphite exfoliation (Fedi et al., 2014; Khan et al., 2010). Films prepared from the colloidal suspension were characterized as reported in Khan et al. (2010), confirming the presence of a few layers of graphene (FLG).

In order to fabricate the chemi-resistor devices, a few microliters of the colloidal dispersion were deposited by drop-casting directly onto alumina substrates with interdigitated Au electrodes (Fedi et al., 2014).

Chemi-resistor testing

The chemi-resistors were tested in a Gas Sensor Characterization System (GSCS, Kenosistec equipment) under N_2 gas flow at atmospheric pressure, temperature and relative humidity (RH) set at 22 °C and 50 %, respectively. The standard exposure protocol adopted for each measurement consists

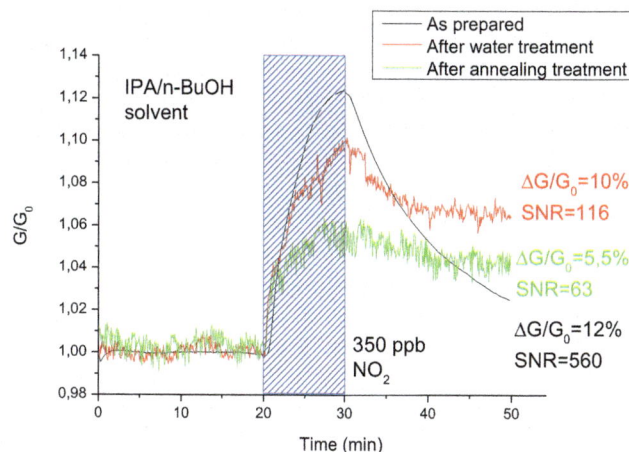

basically of three steps: (a) the baseline, during which only the carrier gas is fluxed with the aim of stabilizing the current; (b) the exposure window to the target gas; and (c) the recovery phase, during which the analyte flow is stopped and again only the carrier gas is fluxed in order to return the device to the initial conditions (Figs. 1–2) (Massera et al., 2014; Ricciardella et al., 2014).

The devices were tested at three different steps: as soon as prepared, after about one month, during which they were exposed in air, and finally after applying the two described restoring methods, namely the device annealing at 130 °C in vacuum for 120 min and the newly developed method that encompasses the dipping of exhaust devices into ultrapure water at 100 °C for 60 s, followed by a drying step on the hot plate at 150 °C for 5 min.

3 Results and discussion

To remove the adsorbed molecules from the sensitive sites, the proposed method takes advantage of the strong solubility of NO_2 in H_2O. This dissolution mechanism is well known in the literature, as explained by Tan and Piri (Tan and Piri, 2013). NO_x is a mixture of all nitrogen oxides (N_2O, NO, NO_2, N_2O_3, N_2O_4, and N_2O_5), most of which immediately react with water upon dissolution, resulting in HNO_3 and HNO_2 formation. Only N_2O (nitrous oxide) and NO (nitric oxide) do not react and hardly dissolve in water according to the following equilibrium reaction: $3NO_2 + H_2O \leftrightarrow 2HNO_3 + NO$.

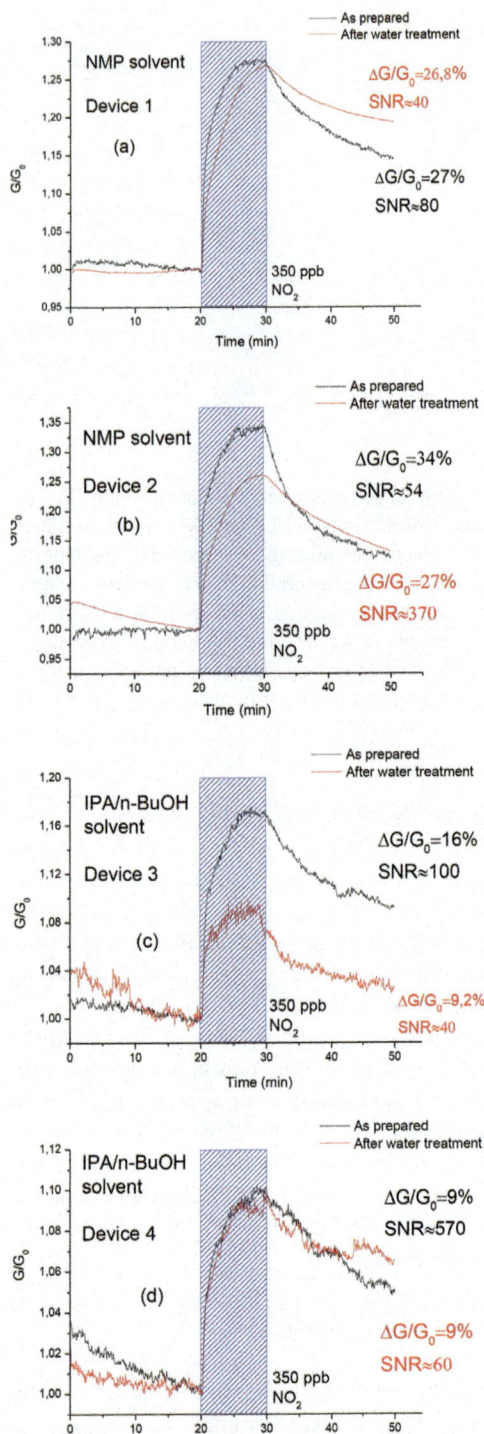

Figure 3. Normalized electrical conductance behaviors for another four chemi-resistors that have been realized as described in Fig. 1 (devices 1 and 2) and in Fig. 2 (devices 3 and 4). The black and red lines in each panel report the detected device signal as soon as prepared and after the restoration by the dipping method, respectively. The protocol employed in the tests is the standard one adopted for the previous measurements.

In Fig. 1, the normalized conductance behaviors of chemi-resistors exposed to NO_2 vs. the acquisition time are reported, G_0 being the initial value during the gas inlet (Fedi et al., 2014). Tests in Fig. 1 referred to devices prepared with graphene exfoliated in NMP. The stacked curves are related to tests performed at different times on the same device.

It is straightforward to observe the effects of the restoring approaches on the graphene-based devices. After the first test, which was carried out as soon as the solution was spread onto the transducers (black line), the chemi-resistor was stored in a covered petri dish and left in air. When it was tested again after one month, the variation of the device conductance towards the same analyte concentration and test conditions dramatically decreased (3 % vs. 27 %) (red line), maybe due to the strong poisoning effect that occurred in that period by the exposure to contaminants present in the atmosphere, NO_2 included. The device was refreshed by using the water-based restoration approach proposed here and then exposed again using the same measurement protocol. A definite increase in $\Delta G/G_0$ (green line) was then observed with respect to the behavior exhibited by the exhaust device (33 % vs. 3 %). Moreover, the conductance variation resulted in being even higher than the value provided by the freshly prepared device. The response (Fig. 1c) clearly shows how the refresh method has beneficial effects on the device, suggesting water-removing effects on the NO_2 molecules.

Our approach was compared with the method usually adopted in the literature (Schedin et al., 2007). The restoration in vacuum at 130 °C for 120 min was carried out after leaving the device another month in air. Figure 1 (blue line) reports the behavior exposing the chemi-resistor at 350 ppb of NO_2. Since the conductance variations are similar to those measured with the exhaust device (red line in Fig. 1), it is straightforward to see the validity of the first recovery approach with respect to annealing under vacuum.

An identical procedure was accomplished on a chemi-resistor based on graphene colloidal dispersion prepared by using IPA/n-BuOH solvent instead of NMP, and in Fig. 2 their normalized conductance is reported.

Figure 2 (black line) shows the conductance of the chemi-resistors exposed to NO_2 as soon as prepared. Differently from the previous case, when the device was left in air and then tested again, an overall absence of signal (not reported) was observed, indicating the occupancy of all sensing sites by the adsorbates. The red and green lines in Fig. 2, respectively, report on the tests after dipping and vacuum-heat restoring methods, respectively. Once more, the result (red line in Fig. 2) confirms that the developed method is able even to restore sensing layers fully insensitive to NO_2. At the same time, the green line in Fig. 2 shows that the proposed method restores the sensing layer performances more efficiently than the thermal one. In some cases, the water approach can not only restore the exhaust devices, but it is even able to enhance the sensing capability towards analyte.

The comparison between Fig. 1 (green line) and Fig. 2 (red line) clearly shows a remarkable difference related to both the conductance and the signal-to-noise ratio (SNR) values. As a matter of fact, since NMP has been demonstrated to be the most efficient solvent for exfoliating graphene (Khan et al., 2011), a larger yield of graphene flakes is actually induced by this solvent with respect to any other. As a result, SLG and FLG interconnections also increase, resulting in the enhancement of signals and an improvement in the SNR.

In Fig. 3, the tests performed on other two series of devices are reported. Figure 3a–b and c–d refer to devices based on graphene suspension dissolved in NMP and IPA/n-BuOH solvent, respectively. The reproducibility of the restoration process for the developed approach is confirmed in all four experiments. In some cases, the full restoration can be obtained and the initial conditions recovered, as can be observed for Device 4 (Fig. 3d).

4 Conclusions

In summary, an easy, original method to refresh the exhaust graphene-based chemi-resistor after NO_2 exposure or after storage in air has been introduced. This novel approach basically derives from the ability of water to remove the adsorbed NO_2 molecules. The method effectiveness is demonstrated by the fact that devices fully unreactive to the analyte, after the restoration, show performances not only comparable to but in some cases even better than those obtained when the device is freshly prepared.

Edited by: M. Meyyappan
Reviewed by: two anonymous referees

References

Charlier, J. C., Arnaud, L., Avilov, I. V., Delgado, M., Demoisson, F, Espinosa, E. H., Ewels, C. P., Felten, A., Guillot, J., Ionescu, R., Leghrib, R., Llobet, E., Mansour, A., Migeon, H. N., Pireaux, J. J., Reniers, F., Suarez-Martinez, I., Watson, G. E., and Zanolli, Z.: Carbon nanotubes randomly decorated with gold clusters: from nano^{2}hybrid atomic structures to gas sensing prototypes, Nanotechnology, 20, 375501, doi:10.1088/0957-4484/20/37/375501, 2009.

Chen, R. J., Franklin, N. R., Kong, J., Cao, J., Tombler, T. W., Zhang, Y., and Dai, H.: Molecular photodesorption from single-walled carbon nanotubes, Appl. Phys. Lett., 79, 2258–2260, 2001.

Fedi, F., Ricciardella, F., Polichetti, T., Miglietta, M. L., Massera, E., and Di Francia, G.: Exfoliation of Graphite and Dispersion of Graphene in Solutions of Low-Boiling-Point Solvents for Use in Gas Sensors, in: Sensors and Microsystems, Springer, 2014.

Hyman, M. P. and Medlin, J. W.: Theoretical study of the adsorption and dissociation of oxygen on Pt (111) in the presence of homogeneous electric fields, J. Phys. Chem. B, 109, 6304–6310, 2005.

Khan, U., O'Neill, A., Lotya, M., De, S., and Coleman, J. N.: High-Concentration Solvent Exfoliation of Graphene, Small, 6, 864–871, 2010.

Khan, U., Porwal, H., O'Neill, A., Nawaz, K., May, P., and Coleman, J. N.: Solvent-exfoliated graphene at extremely high concentration, Langmuir, 27, 9077–9082, 2011.

Leghrib, R. and Llobet, E.: Quantitative trace analysis of benzene using an array of plasma-treated metal-decorated carbon nanotubes and fuzzy adaptive resonant theory techniques, Anal. Chim. Acta, 708, 19–27, 2011.

Massera, E., Miglietta, M. L., Polichetti, T., Ricciardella, F., and Di Francia, G.: Reproducibility of the Performances of Graphene-Based Gas-Sensitive Chemiresistors, in: Sensors and Microsystems, Springer, 2014.

Ricciardella, F., Massera, E., Polichetti, T., Miglietta, M. L., and Di Francia, G.: A calibrated graphene-based chemi-sensor for sub parts-per-million NO_2 detection operating at room temperature, Appl. Phys. Lett., 104, 183502, doi:10.1063/1.4875557, 2014.

Schedin, F., Geim, A., Morozov, S., Hill, E., Blake, P., Katsnelson, M., and Novoselov, K.: Detection of individual gas molecules adsorbed on graphene, Nat. Mater., 6, 652–655, 2007.

Tan, S. P. and Piri, M.: Modeling the Solubility of Nitrogen Dioxide in Water Using Perturbed-Chain Statistical Associating Fluid Theory, Ind. Eng. Chem. Res., 52, 16032–16043, 2013.

Yavari, F. and Koratkar, N.: Graphene-Based Chemical Sensors, J. Phys. Chem. Lett., 3, 1746–1753, 2012.

Yuan, W. and Shi, G.: Graphene-based gas sensors, J. Materials Chem. A, 1, 10078–10091, 2013.

6

Aerosol-deposited $BaFe_{0.7}Ta_{0.3}O_{3-\delta}$ for nitrogen monoxide and temperature-independent oxygen sensing

M. Bektas, D. Hanft, D. Schönauer-Kamin, T. Stöcker, G. Hagen, and R. Moos

Department of Functional Materials, University of Bayreuth, 95447 Bayreuth, Germany

Correspondence to: R. Moos (functional.materials@uni-bayreuth.de)

Abstract. The gas sensing properties of resistive gas sensors of $BaFe_{0.7}Ta_{0.3}O_{3-\delta}$ (BFT30) prepared by the so-called aerosol deposition method, a method to manufacture dense ceramic films at room temperature, were investigated. The electrical response of the films was investigated first under various oxygen concentrations and in a wide temperature range between 350 and 900 °C. Between 700 and 900 °C, the conductivity of $BaFe_{0.7}Ta_{0.3}O_{3-\delta}$ (BFT30) depends on the oxygen concentration with a slope of almost 1/4 in the double-logarithmic plot vs. oxygen partial pressure. In addition, the sensor response is temperature independent. BFT30 responds fast and reproducibly to changing oxygen partial pressures even at 350 °C. The cross-sensitivity has been investigated in environments with various gases (C_3H_8, NO, NO_2, H_2, CO, CO_2, and H_2O) in synthetic air between 350 and 800 °C. BFT30 exhibits good sensing properties to NO between 350 and 400 °C in the range from 1.5 to 2000 ppm with a high selectivity to the other investigated gas species. Thus this semiconducting ceramic material is a good candidate for a temperature-independent oxygen sensor at high temperatures with the application in exhausts and for a selective nitrogen monoxide (NO) sensor at low temperatures for air quality monitoring.

1 Introduction

Selective oxygen sensors should respond only to changes in the oxygen partial pressure (pO_2) but neither to temperature variations nor to other noise factors like interfering gas components (Moseley and Williams, 1989). At the same time, oxygen sensors have to withstand harsh ambient conditions (Alkemade and Schumann, 2006; Moos et al., 2011). Besides established zirconia-based sensors (Riegel et al., 2002; Ivers-Tiffée et al., 2001), semiconducting oxides are often discussed as suitable sensor materials for that purpose (Gerblinger et al., 1995; Fleischer and Meixner, 1991). Since their pO_2-dependent defect disorder leads to changes in their conductivity, simple resistance measurements can indicate oxygen concentrations in exhausts or in flue gases. Due to their high melting and decomposition temperatures, perovskite oxides were extensively studied. Besides that, the perovskite structure has doping flexibility to optimize sensor performance for particular applications (Fergus, 2007). Up to now, only a few materials have been found that show a low or even negligible temperature dependency of conductivity like $Co_{1-x}Mg_xO$ (Park and Logothetis, 1977),

$SrMg_{0.4}Ti_{0.6}O_3$ (Yu et al., 1986), La_2CuO_4 (Blase et al., 1997), and $SrTi_{0.65}Fe_{0.35}O_{3-\delta}$ (STF35) (Menesklou et al., 1999; Moos et al., 2003). Originally, Williams et al. (1982) suggested STF as a temperature-independent oxygen sensor material. Over the past decade, it has attracted a lot of attention from the research point of view. Unfortunately, STF has some problems in real exhaust applications like sulfur oxide poisoning (Rettig et al., 2004) and instability under fuel-rich conditions. Moseley and Williams (1989) suggested $BaFe_{0.8}Ta_{0.2}O_{3-\delta}$ as a temperature-independent oxygen sensor material. In our previous study, the oxygen sensing properties of bulk ceramic samples of the entire series of $BaFe_{1-x}Ta_xO_{3-\delta}$ (BFT) with $0.2 \leq x \leq 0.7$ were investigated as a function of pO_2 and temperature between 400 and 900 °C (Bektas et al., 2014). The conductivity of $BaFe_{0.7}Ta_{0.3}O_{3-\delta}$ (BFT30) responds rapidly and reproducibly to oxygen partial pressure changes, and it shows a negligible temperature dependency between 800 and 900 °C.

So far, many resistive thick film oxygen sensors have been produced by screen printing method, e.g., Mg-doped $SrTiO_3$ (Zhou et al., 1997), $SrTi_{0.65}Fe_{0.35}O_{3-\delta}$ (Menesklou et al., 1999), $LaCu_{0.3}Fe_{0.7}O_{3-\delta}$, (Sahner et al., 2006a) and $La_{0.05}Sr_{0.95}Ti_{0.65}Fe_{0.35}O_{3-\delta}$ (LSTF) (Sahner et al., 2006b). Moos et al. (2003) investigated LSTF thick films that were produced by screen printing method as temperature-independent resistive oxygen sensor. The authors found that the highest possible sintering temperature is $1100\,°C$, since otherwise the thick films can deteriorate during sintering due to reactions between the sensitive film and the alumina substrate. To avoid such reactions and to produce stable LSTF thick films, a screen-printed $SrAl_2O_4$ diffusion barrier layer has to be inserted between gas sensitive layer and alumina substrate (Moos et al., 2003; Sahner et al., 2005). In order to overcome this drawback, Sahner et al. (2009) suggested applying the so-called aerosol deposition method (ADM) for the preparation of the sensor films. ADM is a novel and powerful method to manufacture dense ceramic layers at room temperature directly from ceramic powders. ADM is a completely cold method, at which neither carrier gases, nor powders or substrates must be heated (Akedo, 2006). Akedo (2006) suggested that, during the deposition process, submicron ceramic particles in a carrier gas are driven by a pressure difference between an aerosol and a coating chamber. These particles are accelerated by a nozzle to several hundred meters per second and ejected onto a substrate where they form a dense coating with nanosized particles (Akedo, 2008; Lebedev et al., 2005). So, the great advantage of ADM is that no diffusion barrier layer is required since interactions between film and substrate cannot occur, as no high temperature step is involved. Initial tests for STF clearly showed that the unique temperature-independent conductivity behavior can be fully retained using ADM (Sahner et al., 2009).

This study investigates the oxygen sensing properties of the thick film resistive sensor material $BaFe_{0.7}Ta_{0.3}O_{3-\delta}$, processed by ADM. In contrast to earlier studies of STF and BFT, also the temperature range down to $350\,°C$ is investigated as well as influences of interfering exhaust gas components like NO, NO_2, H_2, CO, CO_2, and H_2O.

2 Experimental

$BaFe_{0.7}Ta_{0.3}O_{3-\delta}$ powder was prepared from commercially available precursor powders $BaCO_3$ (Merck), Fe_2O_3 (Alfa Aesar, 99 %) and Ta_2O_5 (Alfa Aesar, 99 %). Conventional mixed-oxide technique was used to produce BFT30 powder for ADM. To obtain homogeneous mixtures, the precursor powders were ball-milled in cyclohexane for 4 h and dried in air for 1 day. The powders were then calcined at $1350\,°C$ for 15 h. In order to reduce the particle size, the powders were milled again and dried in air atmosphere. Finally, to remove large agglomerates the dried powders were sieved with $90\,\mu m$ meshes. Before aerosol deposition, the powder was dried again for 1 day at $200\,°C$.

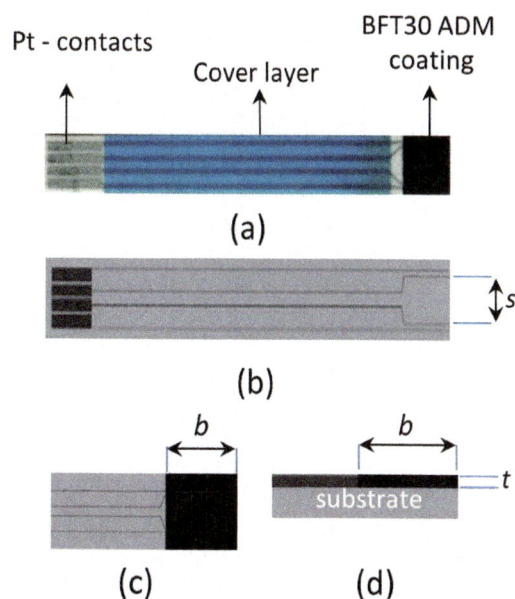

Figure 1. Sketch of the sensor setup, **(a)** top view of sensor showing Pt contacts, cover layer and BFT30 ADM coating, **(b)**, **(c)** and **(d)** indicating the dimension for the calculation of the conductivity.

Before starting the aerosol deposition, alumina substrates with four screen-printed platinum electrodes (electrode space s) were prepared. Four platinum wires were contacted to the electrodes with platinum paste (LPA 88–11S, Heraeus). The contacted substrates were fired in air at $1000\,°C$ for 10 min. On this transducer, the ADM layer was applied. Figure 1 sketches the setup and explains the meaning of the variables that appear in Eq. (1).

The aerosol deposition experiment setup has already been depicted in detail by Sahner et al. (2009). Nitrogen served as the carrier gas and the gas flow rate was $4\,L\,min^{-1}$. The shaking frequency of the vibrating table was set to $400\,min^{-1}$. The distance between nozzle and substrate was adjusted to 4 mm. The substrate holder moved with a velocity $1\,mm\,s^{-1}$. By adjusting the coating time, samples with a thickness, t, of around $12\,\mu m$ were produced. In order to increase electrical and mechanical properties of coated samples, they were heat-treated at $950\,°C$ for 2 h. The temperature of this "degreening process" is far below the typical sintering temperatures of about $1350\,°C$ but a little higher as the highest sensor operation temperature of $900\,°C$. This degreening process was selected to obtain reproducible results.

Phase purity of BFT30 powder was investigated by an X-Ray diffraction (PANalytical Xpert Pro) at room temperature using CuK_α radiation ($1.541874\,Å$). Micrographs of sensors were determined by scanning electron microscope (SEM; Zeiss Ultra plus FE-SEM). For the gas sensing measurements, a four-wire technique with a Keithley 2700 digital multimeter in the offset-compensated resistance measuring mode was used. The ADM-coated samples were mounted on

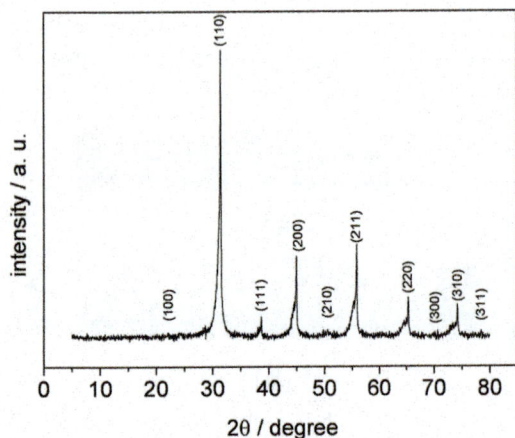

Figure 2. XRD pattern of the BFT30 powder as it was used for the ADM process.

a sample holder and inserted into a tubular furnace. The oxygen concentration was varied in the range from 1 to 100 % ($pO_2 \approx 0.01$–1 bar) in the temperature range from 350 to 900 °C. The total gas flow was adjusted to 200 mL min^{-1}. The cross-sensitivity of coated samples was investigated in environments with various gases (C_3H_8, NO, NO_2, H_2, CO, CO_2, and H_2O) in flowing synthetic air between 350 and 800 °C. The conductivity was calculated from the resistance, R, according to Eq. (1):

$$\sigma = \frac{1}{R} \cdot \frac{s}{b \cdot t}. \tag{1}$$

3 Results and discussion

3.1 Characterization

Figure 2 shows the X-ray diffraction (XRD) graph of the $BaFe_{0.7}Ta_{0.3}O_{3-\delta}$ (BFT30) precursor powder that was calcined at 1350 °C for 15 h. The XRD patterns indicate that the phase structure of BFT30 has a tetragonal to cubic perovskite phase transition when calcined at 1350 °C. In other words, the sensor material has a non-cubic perovskite phase (as indexed). Second phases cannot be observed.

The BFT30 powders were ADM-processed onto substrates as described above. One specimen was cut and prepared for SEM micrographs. A top view image and a cross section (polished) are shown in Fig. 3a and b, respectively. The resulting film is found to be dense and homogeneous with a film thickness of about 12 µm. A closer look into the microstructure revealed a crystallite size in the range of some tens of nanometers, as it is typical for ADM-processed ceramic materials since the grains fracture during impaction on the substrate (Lebedev et al., 2005; Akedo, 2006).

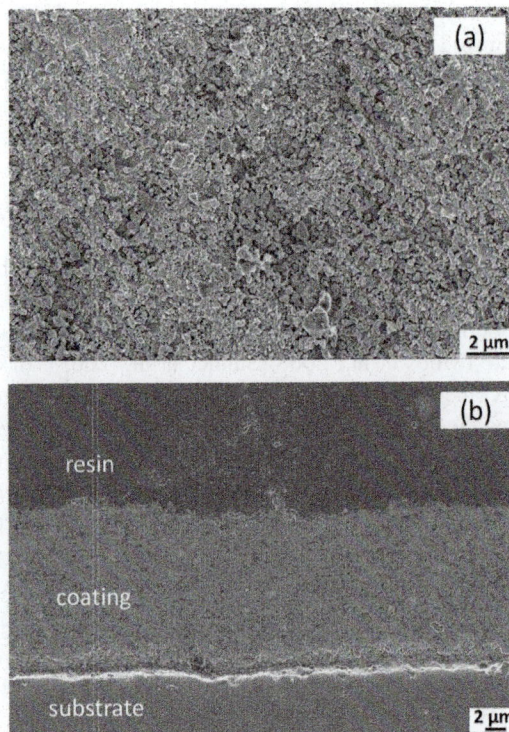

Figure 3. SEM images of ADM-coated BFT30 sensor **(a)** top view, **(b)** cross-section micrograph.

3.2 Oxygen sensing properties

The oxygen sensing properties of BFT30 thick films were investigated by varying gradually the oxygen partial pressure from 10^{-2} to 1 bar at a fixed temperature. The oxygen concentration was held for 20 min at each step. The conductivity reached a stable final value within these 20 min as shown for 900 °C in Fig. 4a, and an almost stable final value for 350 °C in Fig. 4b. Similar results were observed for BFT30 bulk samples between 500 and 900 °C in our previous study (Bektas et al., 2014). It can be seen from Fig. 4b that an ADM-processed BFT30 sensor shows a fast and reversible response to oxygen partial pressure changes even at low temperatures of 350 °C. At such a low temperature, one usually cannot expect a full equilibration of the oxygen equilibrium as given by defect chemical constants, due to the usually slow oxygen kinetics at low temperatures. However, if oxygen diffusion within the crystallite is the time-limiting step of the sensor response, the oxygen equilibration time should be proportional to the square of the grain size (Schönauer, 1991). Therefore, the nanosized crystallite structure may explain this unexpected fast response.

In Fig. 5 the conductivity of the BFT30 sensors is shown in a double-logarithmic representation vs. oxygen partial pressure. Since BFT30 is a p type conductor, its conductivity increases with increasing oxygen partial pressure. Very often, and in agreement with defect chemistry (Merkle and Maier,

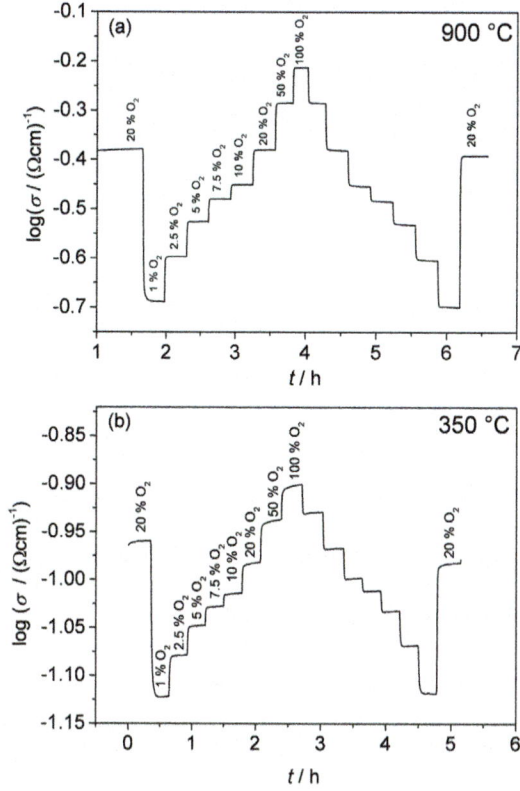

Figure 4. Conductivity of the BFT30 sensor as a function of pO_2 **(a)** at 900 °C and **(b)** at 350 °C.

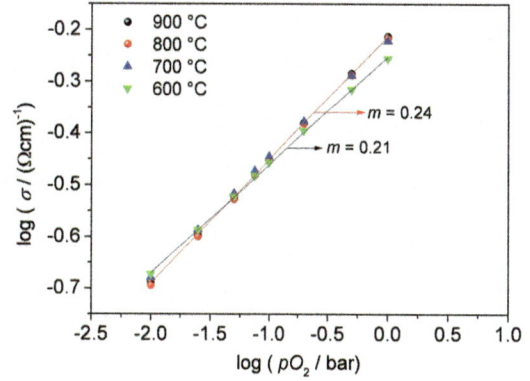

Figure 5. Double-logarithmic plot of conductivity vs. oxygen partial pressure for a BFT30 ADM-coated sensor. Please note that the curves of 700, 800, and 900 °C can hardly be distinguished.

2008), one finds a behavior for the conductivity according to Eq. (2), with the exponent m being in the range between $1/6$ and $1/4$:

$$\sigma = \text{const.} \times pO_2{}^m. \tag{2}$$

In a double-logarithmic representation, the sensitivity, S_{O_2}, is the slope of the curve (Eq. 3):

$$S_{O_2} = \frac{d \lg(\sigma)}{d \lg(pO_2)} = m. \tag{3}$$

The sensitivity of the sensor increases with temperature from 0.21 at 600 °C to 0.24 between 700 and 900 °C. It is remarkable that between 700 and 900 °C the curves are almost identical, indicating an almost perfect temperature independency. Only below 700 °C, the sensor becomes temperature dependent. According to Bektas et al. (2014), the increasing tantalum content reduces the amount of oxygen vacancies. The oxygen deficiency, δ, moves from 0.5 towards zero. At the same time, the conductivity decreases with tantalum content, x, since the holes are compensated. The tantalum stabilizes the phase by reducing oxygen deficiency. This may lead to the temperature independency.

3.3 Sensitivities to the other gases

In order to determine the behavior of BFT30 ADM-coated thick film sensors to other exhaust components, the resistance of the sensor was measured under the presence of defined concentrations of the following components: C_3H_8 (500 ppm), NO (1000 ppm), NO_2 (500 ppm), H_2 (5000 ppm), CO (1000 ppm), CO_2 (5 %), and H_2O (5 %) between 350 and 800 °C. Base gas was synthetic air. The resistances, R, were measured, and together with the values in base gas, R_0, the response to each component, $S_{component}$, was calculated according to Eq. (4). The results are summarized in Table 1 for all investigated temperatures.

$$S_{component} = \frac{R - R_0}{R_0}. \tag{4}$$

One can see from Table 1 that the sensor responds only in a negligible way to other exhaust components above 600 and 800 °C. This has been expected, since we assume at such high temperatures a full equilibration between oxide ceramic material and oxygen atmospheres, an effect that becomes clear if one considers that both the oxygen incorporation kinetics and the oxygen (vacancy) diffusion are fast at such temperatures. Therefore, at high temperatures between 600 and 800 °C the resistance of BFT30 ADM-coated samples depends mainly on the oxygen partial pressure. The dynamic response curve (here expressed as the normalized resistance R/R_0) from the cross-sensitivity test at 800 °C confirms this (Fig. 6). One can see that most gases have a negligible influence on the sensor response. Even at lower temperature, only minor cross-effects were found, with sensitivities mainly below 10 %. However, a prominent influence of NO at 400 and 350 °C is observed, with only a marginal effect of NO_2. This astonishing behavior may offer the chance to use BFT as a resistive NO sensor for air quality control or other applications where the oxygen concentration remains constant. This will be investigated in the next paragraph more in detail.

Table 1. Response, $S_{component}$, as defined in Eq. (4) of the BFT30 ADM thick film sensor to other exhaust components.

$T/°C$	$S_{C_3H_8}/\%$	$S_{NO}/\%$	$S_{NO_2}/\%$	$S_{H_2}/\%$	$S_{CO}/\%$	$S_{CO_2}/\%$	$S_{H_2O}/\%$
350	51.1	234.1	15.3	5.58	1.06	0.21	9.82
400	12.5	119.4	15.0	4.48	2.89	1.52	8.82
600	2.77	2.81	1.13	1.28	2.84	8.66	2.97
800	2.63	0	0.11	0.33	1.11	1.55	1.20

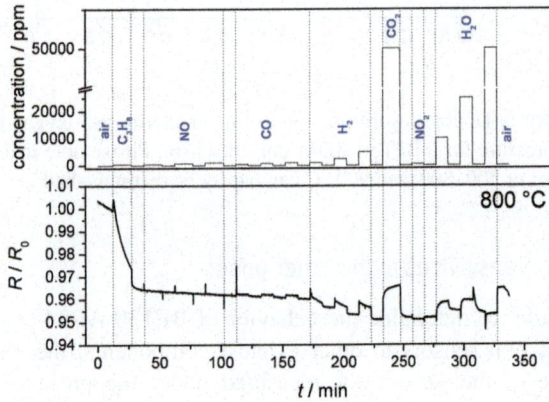

Figure 6. Cross-sensitivity result of a BFT30 ADM-coated sensor at 800 °C (base gas is synthetic air).

3.4 NO sensing properties

According to today's increasing environmental standards, toxic and harmful gases such as NO_x (NO and NO_2), H_2S, CO, and NH_3 that are dangerous for environment and/or human health have to be detected by highly selective gas sensors (Huusko et al., 1993; Satake et al., 1994; Marr et al., 2014). Besides that, for automotive in-cabin applications NO_x sensors may be used for air quality control, by determining the amount of NO_x in the passenger compartment (Kim et al., 2008). In these cases, the oxygen concentration of the ambience remains almost constant. With respect to the high NO response, it may therefore make sense to investigate whether the ADM-processed BFT sensor can be used as an NO sensor. Therefore, response to NO was investigated more in detail. According to Table 1, 400 °C seems to be the first choice, since at that temperature the effect of NO exceeds all others largely. Figure 7a shows the normalized resistance, R/R_0, of a BFT30 ADM-coated sensor under the presence of NO in concentrations between 250 and 2000 ppm in flowing synthetic air. As can be seen, the sensor has a fast, stable and reversible response to NO. Especially low NO concentrations seem to affect the sensor response strongly. At higher NO concentrations, a saturation behavior occurs. Therefore, lower NO concentrations were observed. Figure 7b shows the NO dependence of the ADM BFT30 sensor at lower NO concentrations from 1.5 to 100 ppm. The absolute lower dosing limit in the used test bench is 1.5 ppm. Again, one finds a

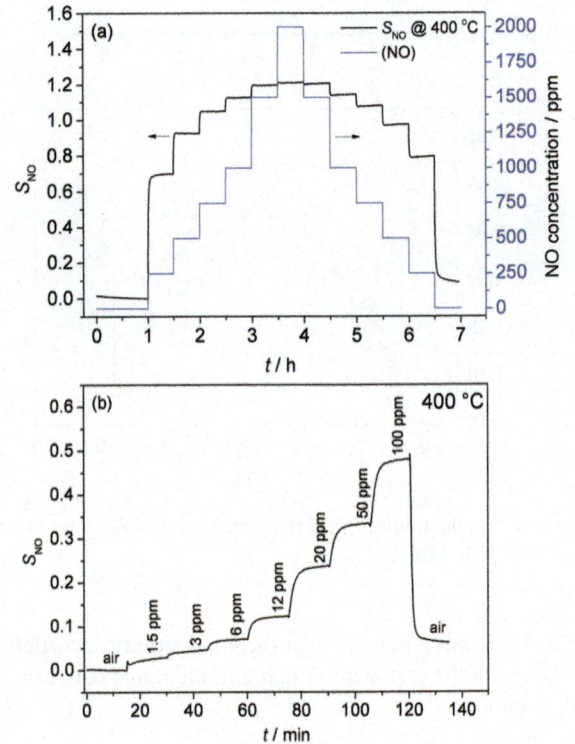

Figure 7. Response of an ADM BFT30 sensor at 400 °C. **(a)** NO concentrations from 250 to 2000 ppm, **(b)** NO concentrations between 1.5 and 100 ppm.

fast response but with some irregularities at 1.5 ppm NO that may originate from mixing problems in the test bench. During recovery, it takes some time until the steady-state value is reached. Nevertheless, we consider this a promising approach, especially due to the high selectivity.

These promising results shall be compared with literature data for other semiconducting NO sensors. Sayago et al. (1995) studied the sensitivity of undoped, Pt-, In- and Al-doped SnO_2 as NO_x and CO sensors. They found response times of around 20 min for Al- or In-doped SnO_2 sensors when exposed to NO_x. Akiyama et al. (1993) investigated WO_3 as NO and NO_2 sensor material in the temperature range between 200 and 500 °C. They found that WO_3 responds quite well to NO and NO_2 in air. According to their suggestions, the NO sensitivity can be further promoted by electrode metals and addition of noble metals like Ru and

Au. Besides that, Penza et al. (1998) investigated WO_3-based thin film sensor devices with activator layers of Pd, Pt and Au. The response time to NO for WO_3 : Pt sensors decreased to 2.8 min. Despite that, one has to state that the response of most other materials to NO is higher, and the ADM BFT30 sensor has a faster response to NO (less than 1 min) and a good selectivity to many gases that are relevant for air quality monitoring. Especially NO, CO and H_2O are relevant in this respect (Wiegleb and Heitbaum, 1994; Kim et al., 2008).

4 Conclusions

BFT30 dense and thick film sensors were manufactured by the aerosol deposition method at room temperature. The resistance change of the sensors was investigated as a function of oxygen partial pressure and temperature. BFT30 sensors show a fast, reproducible, and temperature-independent response to pO_2 between 700 and 900 °C. Cross-sensitivity tests have been conducted under interfering gases in synthetic air. The results demonstrate that the sensor has no crucial sensitivity to the other test gases in the temperature range from 600 to 800 °C. At 400 °C, an interesting selective response to NO in the concentration range from 1.5 to 2000 ppm is observed.

Consequently, the BFT30 ADM-coated thick film sensors may be used as temperature-independent oxygen sensors in exhausts between 700 and 900 °C. On the other hand, they can be used as NO sensors for air quality control applications at 400 °C, when the oxygen concentration remains constant.

Acknowledgements. The authors would like to thank German Research Foundation (DFG) for funding under the project number MO1060/22-1. The authors also thank B. Putz for XRD measurements and A. Mergner for SEM/EDX characterizations, both from the University of Bayreuth.

Edited by: M. Penza
Reviewed by: two anonymous referees

References

Akedo, J.: Aerosol deposition of ceramic thick films at room temperature: densification mechanism of ceramic layers, J. Am. Ceram. Soc., 89, 1834–1839, 2006.

Akedo, J.: Room Temperature Impact Consolidation (RTIC) of Fine Ceramic Powder by Aerosol Deposition Method and Applications to Microdevices, J. Thermal Spray Technol., 17, 181–198, 2008.

Akiyama, M., Zhang, Z., Tamaki, J., Miura, N., and Yamazoe, N.: Tungsten oxide-based semiconductor sensor for detection of nitrogen oxides in combustion exhaust, Sensors and Actuators B: Chemical, 13–14, 619–620, 1993.

Alkemade, U. G. and Schumann B.: Engines and exhaust after treatment systems for future automotive applications, Solid State Ionics, 177, 2291–2296, 2006.

Bektas, M., Schönauer-Kamin, D., Hagen, G., Mergner, A., Bojer, C., Lippert, S., Milius, W., Breu, J., and Moos, R.: $BaFe_{1−x}Ta_xO_{3−δ}$- a material for temperature independent resistive oxygen sensors, Sensors and Actuators B: Chemical, 190, 208–213, 2014.

Blase, R., Härdtl, K., and Schönauer, U.: Oxygen Sensor Based on Non-Doped Cuprate, United States Patent Specification, US 5, 792, 666, 1997.

Fergus, J. W.: Perovskite oxides for semiconductor-based gas sensors review, Sensors and Actuators B: Chemical, 123, 1169–1179, 2007.

Fleischer, M. and Meixner, H.: Oxygen sensing with long-term stable Ga_2O_3 thin films, Sensors and Actuators B: Chemical, 5, 115–119, 1991.

Gerblinger, J., Hauser, M., and Meixner, H.: Electric and Kinetic Properties of Screen-Printed Strontium Titanate Films at High Temperatures, J. Am. Ceram. Soc., 78, 1451–1456, 1995.

Huusko, J., Lantto, V., and Torvela, H.: TiO_2 thick-film gas sensors and their suitability for NOx monitoring, Sensors and Actuators B: Chemical, 15–16, 245–248, 1993.

Ivers-Tiffée, E., Härdtl, K. H., Menesklou, W., and Riegel, J.: Principles of solid state oxygen sensors for lean combustion gas control, Electrochimica Acta, 47, 807–814, 2001.

Kim, J. S., Hwang, I. S., Kim, S. J., Lee, C. Y., and Lee, J. H.: CuO nanowire gas sensors for air quality control in automotive cabin, Sensors and Actuators B: Chemical , 135, 298–303, 2008.

Lebedev, M., Akedo, J., and Ito, T.: Substrate heating effects on hardness of an $α$-Al_2O_3 thick film formed by aerosol deposition method, J. Crystal Growth, 275, e1301–e1306, 2005.

Marr, I., Groß, A., and Moos, R.: Overview on conductometric solid-state gas dosimeters, J. Sens. Sens. Syst., 3, 29–46, doi:10.5194/jsss-3-29-2014, 2014.

Menesklou, W., Schreiner, H. J., Härdtl, K. H., and Tiffée, E. I.: High temperature oxygen sensors based on doped $SrTiO_3$, Sensors and Actuators B: Chemical, 59, 184–189, 1999.

Merkle, R. and Maier, J.: How is oxygen incorporated into oxides? A comprehensive kinetic study of a simple solid-state reaction with $SrTiO_3$ as a model material, Angewandte Chemie-International Edition, 47, 3874–3894, 2008.

Moos, R., Rettig, F., Hürland, A., and Plog, C.: Temperature-independent resistive oxygen exhaust gas sensor for lean-burn engines in thick film technology, Sensors and Actuators B: Chemical, 93, 43–50, 2003.

Moos, R., Izu, N., Rettig, F., Reiß, S., Shin, W., and Matsubara, I.: Resistive Oxygen Gas Sensors for Harsh Environments, Sensors, 11, 3439–3465, 2011.

Moseley, P. T. and Williams, D. E.: Gas sensors based on oxides of early transition metals, Polyhedron, 8, 1615–1618, 1989.

Park, K. and Logothetis, E. M.: Oxygen sensing with $Co_{1−x}Mg_xO$ ceramics, ECS J. Solid State Sci. Technol., 124, 1143–1446, 1977.

Penza, M., Martucci, C., and Cassano, G.: NOx gas sensing characteristics of WO_3 thin films activated by noble metals (Pd, Pt, Au) layers, Sensors and Actuators B: Chemical , 50, 52–59, 1998.

Riegel, J., Neumann, H., and Wiedenmann, H.-M.: Exhaust gas sensors for automotive emission control, Solid State Ionics, 152–153, 783–800, 2002.

Rettig, F., Moos, R., and Plog, C.: Poisoning of temperature independent resistive oxygen sensors by sulfur dioxide, J. Electroceram., 13, 733–738, 2004.

Sahner, K., Moos, R., Matam, M., Tunney, J., and Post, M.: Hydrocarbon sensing with thick and thin film p-type conducting perovskite materials, Sensors and Actuators B: Chemical, 108, 102–112, 2005.

Sahner, K., Straub, J., and Moos, R.: Cuprate-ferrate compositions for temperature independent resistive oxygen sensors, J. Electroceramics, 16, 179–186, 2006a.

Sahner, K., Moos, R., Izu, N., Shin, W., and Murayama, N.: Response kinetics of temperature-independent resistive oxygen sensor formulations: a comparative study, Sensors and Actuators B: Chemical, 113, 112–119, 2006b.

Sahner, K., Kaspar, M., and Moos, R.: Assessment of the novel aerosol deposition method for room temperature preparation of metal oxide gas sensor films, Sensors and Actuators B: Chemical, 139, 394–399, 2009.

Satake, K., Katayama, A., Ohkoshi, H., Nakahara, T., and Takeuchi, T.: Titania NOx sensors for exhaust monitoring, Sensors and Actuators B: Chemical, 20, 111–117, 1994.

Sayago, I., Gutiérrez, J., Arés, L., Robla, J. I., Horrillo, M. C., Getino, J., Rino, J., and Agapito, J. A.: The effect of additives in tin oxide on the sensitivity and selectivity to NOx and CO, Sensors and Actuators B: Chemical, 26–27, 19–23, 1995.

Schönauer, U.: Response times of resistive thick-film oxygen sensors. Sensors and Actuators B: Chemical, 4, 431–436, 1991.

Wiegleb, G. and Heitbaum, J.: Semiconductor gas sensor for detecting NO and CO traces in ambient air of road traffic, Sensors and Actuators B: Chemical, 17, 93–99, 1994.

Williams, D. E., McGeehin, P., and Tofield, B. C.: Oxygen Sensors, US Patent 4, 454–494, 1982.

Yu, C., Shimizu, Y., and Arai, H.: Investigation on a lean-burn oxygen sensor using perovskite-type oxides, Chem. Lett., 4, 563–566, 1986.

Zhou, X., Sørensen, O. T., and Xu, Y.: Defect structure and oxygen sensing properties of Mg-doped $SrTiO_3$ thick film sensors, Sensors and Actuators B: Chemical, 41, 177–182, 1997.

7

Metal oxide semiconductor gas sensor self-test using Fourier-based impedance spectroscopy

M. Schüler, T. Sauerwald, and A. Schütze

Laboratory for Measurement Technology, Department of Mechatronics, Saarland University,
Saarbrücken, Germany

Correspondence to: M. Schüler (m.schueler@lmt.uni-saarland.de)

Abstract. For the self-test of semiconductor gas sensors, we combine two multi-signal processes: temperature-cycled operation (TCO) and electrical impedance spectroscopy (EIS). This combination allows one to discriminate between irreversible changes of the sensor, i.e., changes caused by poisoning, as well as changes in the gas atmosphere. To integrate EIS and TCO, impedance spectra should be acquired in a very short time period, in which the sensor can be considered time invariant, i.e., milliseconds or less. For this purpose we developed a Fourier-based high-speed, low-cost impedance spectroscope. It provides a binary excitation signal through an FPGA (field programable gate array), which also acquires the data. To determine impedance spectra, it uses the ETFE (empirical transfer function estimate) method, which calculates the impedance by evaluating the Fourier transformations of current and voltage. With this approach an impedance spectrum over the range from 61 kHz to 100 MHz is acquired in ca. 16 µs.

We carried out TCO–EIS measurements with this spectroscope and a commercial impedance analyzer (Agilent 4294A), with a temperature cycle consisting of six equidistant temperature steps between 200 and 450 °C, with lengths of 30 s (200 °C) and 18 s (all others). Discrimination of carbon monoxide (CO) and methane (CH$_4$) is possible by LDA (linear discriminant analysis) using either TCO or EIS data, thus enabling a validation of results by comparison of both methods.

1 Introduction

Metal oxide semiconductor (MOS) gas sensors are highly sensitive to a broad range of reducing and oxidizing gases, and they are available at relatively low cost. Their operation is based on resistance measurements of a sensitive layer, which in most cases consists of a granular metal oxide. The resistance of this layer is strongly determined by the adsorption of oxygen, which creates a depletion region at the metal oxide surface leading to an energy barrier between grains (Tricoli et al., 2010; Morrison, 1982; Kohl, 1989).

The interaction of adsorbed oxygen and reducing or oxidizing gases on the grain surfaces depends strongly on temperature and shows different behavior, depending on the gases and temperatures (Clifford and Tuma, 1982, 1983; Morrison, 1987). Temperature is usually controlled by an integrated heater, e.g., a microstructured platinum resistor on the substrate.

By variation of temperature, it is possible to obtain a virtual multi-sensor or virtual sensor array, i.e., to evaluate the sensor resistance at different temperatures and thereby gain selectivity, in a manner similar to the use of multi-sensory arrays (Stetter and Penrose, 2002; Schütze et al., 2004). This method, which we denote as temperature-cycled operation (TCO), can be used to increase the selectivity and sensitivity of metal oxide gas sensors considerably (Heilig et al., 1997; Lee and Reedy, 1999).

Another way to increase the selectivity is the measurement of the complex sensor impedance by electrical impedance spectroscopy (EIS). One of the underlying effects is the change in capacity at the grain boundaries, which is caused by gas, primarily oxygen. The dielectric properties of the chemical species present in the sensing layer also influence the capacitance properties of the sensor layer – their measurement by EIS can thus increase selectivity (Weimar and Göpel, 1998; Bârsan and Weimar, 2003).

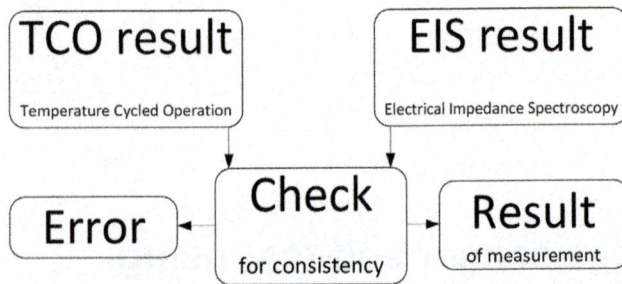

Figure 1. Scheme of the sensor self-test strategy (cf. Schüler et al., 2014).

Besides selectivity, long-term sensor stability is a challenge for the use of SC gas sensors, especially in safety-critical applications. A variety of factors influence this stability, including structural and phase transformations, poisoning, degradation of contacts and heaters, bulk diffusion, sensor design, humidity, temperature changes and interference effects (Korotcenkov and Cho, 2011). By increasing selectivity, TCO – combined with an optimized signal processing (Fricke et al., 2008) – addresses especially the influence of interference gases and changes in humidity, as well as temperature changes (provided the actual sensor temperature is controlled, not just heater voltage). The combination of TCO and EIS can increase the reliability of MOS gas sensors further, acquiring data which may give additional information on the sensor properties. These data not only reflect poisoning, but also may contain information about structural and phase properties, contacts and heater state (although the heater is usually isolated from the heater electrodes, heater properties may affect the measured data by capacitive coupling), bulk diffusion and humidity (Bârsan and Weimar, 2003). These different properties affect the acquired data in different ways, and the generalization of the method described here requires further research. This paper is confined to the study of poisoning induced by HMDSO (hexamethyldisiloxane), a compound which is present as a solvent in many polymers and causes significant changes in the properties of MOS gas sensors. Previous works have shown that both TCO and EIS can be used to discriminate different gases by multivariate analysis (Conrad et al., 2007). By checking the consistency of results acquired with both methods, the reliability of the measurement can be increased (Reimann et al., 2008). This approach is illustrated in Fig. 1.

To perform EIS and TCO measurements simultaneously, impedance spectra should be acquired in a very short time period (milliseconds or less) during which the sensor temperature (and other properties) can be considered constant, i.e., the sensor as a time-invariant system. For this purpose we developed a Fourier-based high-speed, low-cost impedance spectroscope (Schüler et al., 2014).

2 Impedance measurement

The complex impedance $Z(\omega)$ can be defined as follows (Barsoukov and Macdonald, 2005):

$$Z(j\omega) = |Z| \exp(j\theta) = \frac{F\{v(t)\}}{F\{i(t)\}}. \tag{1}$$

Here, $F\{v(t)\}$ denotes the Fourier transform of the voltage, and $F\{i(t)\}$ denotes the Fourier transform of the current.

The measurement of impedance can be carried out either in the time domain, which requires a transformation of the measured values to the frequency domain, or it can be carried out in the frequency domain, e.g., by setting amplitude and phase of a driving current and measuring the phase shift and amplitude of the resulting voltage drop across the device under test (DuT). Most commercial high-frequency impedance analyzers work in the frequency domain – this approach promises good accuracy and does not require sophisticated signal processing (Barsoukov and Macdonald, 2005; Agilent 4294A). On the other hand, the measurements take inherently longer, since only one frequency is measured at a given time. For example, the instrument used for the reference measurements presented here requires at least 3 ms to measure the impedance at one frequency (Agilent 4294A); a high-resolution impedance spectrum with, e.g., 200 tested frequencies requires at least 0.6 s. This duration is even higher if frequencies below 100 kHz are considered or high precision is required. On the other hand, when measurements are carried out in the time domain, a broad range of frequencies can be analyzed simultaneously; thus, measurement times can be greatly reduced. However, determination of the impedance spectrum requires signal processing, e.g., Fourier transformation, to transfer data recorded in the time domain to the frequency domain (Barsoukov and Macdonald, 2005).

To achieve a compact measurement system with acquisition times (well) below 1 ms, we built an impedance spectroscope which carries out the measurement in the time domain and transforms it into the frequency domain using FFT (fast Fourier transform), in order to calculate the impedance. This method is known as ETFE (empirical transfer function estimate) and is widely used in system identification to estimate transfer functions of LTI (linear time invariant) systems (Ljung and Glad, 1994). In a general form, the ETFE approach can be written as follows (Ljung and Glad, 1994):

$$G_S(\omega) = \frac{Y_S(\omega)}{U_S(\omega)}. \tag{2}$$

$G_S(\omega)$ is the transfer function of a system, $Y_S(\omega)$ is the output in the frequency domain, and $U_S(\omega)$ is the corresponding input. We can consider the impedance of a sensor as a transfer function describing the relationship between voltage and current across the sensor:

$$Z(\omega) = \frac{U(\omega)}{I(\omega)} = \frac{U_{\text{Sensor}}(\omega)}{U_{\text{ref}}(\omega)/R_{\text{ref}}}. \tag{3}$$

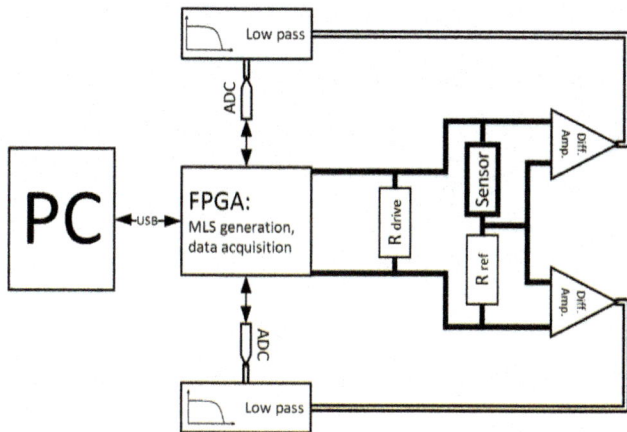

Figure 2. Hardware setup of the FoBIS measurement system (cf. Schüler et al., 2014).

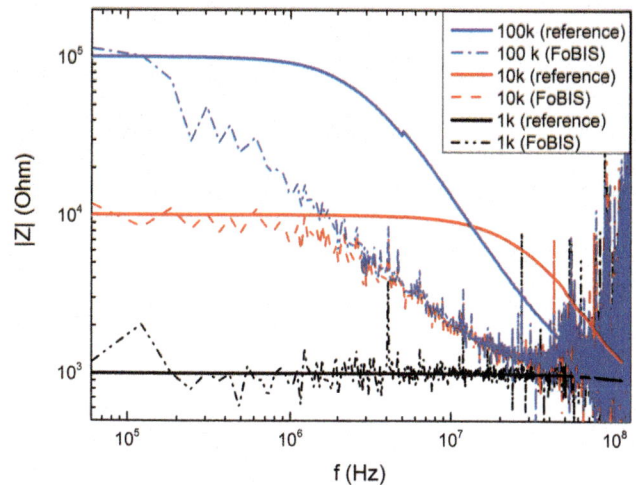

Figure 3. Measurements of test impedances ($1/10/100\,k\Omega$ high-precision resistors with small parasitic capacitances ($< 0.4\,\text{pF}$)), carried out with a commercial high-precision impedance analyzer (Agilent 4294A) and the FoBIS measurement system.

In Eq. (3), $Z(\omega)$ is the impedance, $U(\omega)$ resp. $U_{\text{Sensor}}(\omega)$ the voltage, and $I(\omega)$ the current. $I(\omega)$ is not set or directly measured, but determined from the voltage $U_{\text{ref}}(\omega)$ across a resistor of known impedance (R_{ref}).

Figure 2 shows the corresponding hardware setup: a binary signal of broad frequency range is generated by a field-programmable gate array (FPGA) and transmitted using the BLVDS_25 differential signaling standard, which generates a current of defined value. This current causes a voltage drop of $\pm 1.2\,\text{V}$ across the resistor R_{drive} ($75\,\Omega$), and thereby also across the voltage divider consisting of the sensor and R_{ref} which is connected in parallel (Reimann et al., 2008). Thus, due to the much higher impedance of the voltage divider compared to R_{drive}, a voltage of $\pm 1.2\,\text{V}$ is applied across the voltage divider to drive the sensor. The branches of this voltage divider are connected to high-speed ADCs (analog-to-digital converters; model ADS62P49EVM, Texas Instruments) via differential amplifiers (AD8130, Analog Devices), which enable a robust signal transmission, as well as low-pass filters, which eliminate frequencies above ca. 100 MHz. We use an MLBS (maximum length binary sequence) excitation signal with a length of $16.384\,\mu\text{s}$, which is sampled at a frequency of 250 MHz. The bandwidth of such a signal ranges from 61 kHz to 125 MHz (Schüler et al., 2014). However, since frequencies above 100 MHz are removed in order to prevent aliasing effects, the bandwidth of the measurement system ranges from 61 kHz to 100 MHz. (Schüler et al., 2014).

The introduced measurement method is denoted as FoBIS (Fourier-based impedance spectroscopy). Figure 3 shows measurements of test impedances carried out with the FoBIS system on the one hand, and with the commercial high-precision impedance analyzer Agilent 4294A on the other hand. Obviously, the FoBIS system cannot compete in terms of absolute accuracy. Due to parasitic capacitances, the signal intensity decreases at higher frequency and thereby the SNR

(signal-to-noise ratio) decreases inherently. On the other hand, the quickness of the measurement principle (acquisition time $\approx 16.4\,\mu\text{s}$) enables measurements in temperature-cycled operation even for microstructured sensors with thermal time constants in the millisecond range (Elmi et al., 2008).

3 Experimental

To evaluate the self-test strategy sketched in Fig. 1, measurements were carried out with the UST Umweltsensortechnik GGS 1330, an SnO_2-based sensor sensitive towards a broad range of reducing gases. At first, we characterized the unimpaired sensor. The measurements were carried out in temperature-cycled operation with the Agilent 4294A, and the FoBIS system. To enable temperature-cycled operation with the Agilent 4294A, we implemented the temperature cycle shown in Fig. 4, consisting of six equidistant temperature steps between 200 and 450 °C, with lengths of 30 s (200 °C) and 18 s (all others). The impedance analyzer Agilent 4294A requires 10.8 s to record one impedance spectrum with a range from 200 Hz to 110 MHz with 201 supporting points with logarithmic distribution at the second-highest precision available ("measurement bandwidth" = 4). Acquisition of the impedance spectra was triggered 5 s after each temperature change, when the sensor has almost reached a steady state (the impedance relaxation induced by the temperature change is mostly complete, cf. Fig. 7). After these initial characterization measurements, the sensor was exposed to 70 ppm HMDSO for 10 min in temperature-cycled operation. After this poisoning, the characterization measurements were repeated. Figures 5 and 6 show a section of the characterization measurements carried out before and

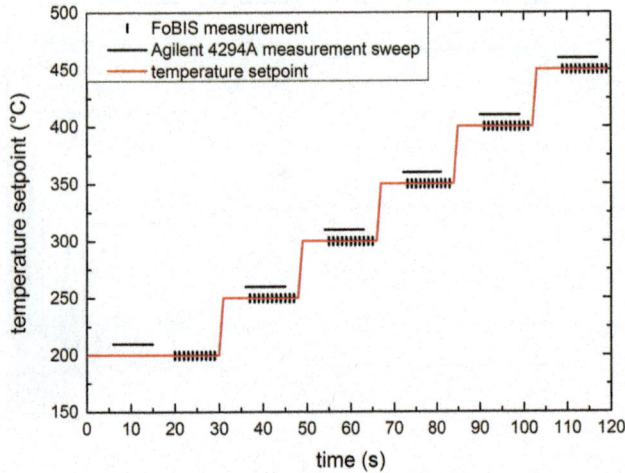

Figure 4. Temperature set points and timing of impedance measurements during temperature cycle.

after poisoning. The presented values are quasi-static sensor responses, i.e., sensor response values which were acquired at defined time/temperature points within the temperature profile. The sensor response is defined as follows:

$$S = \frac{|Z_{\mathrm{air}}|}{|Z_{\mathrm{gas}}|} - 1, \tag{4}$$

where $|Z_{\mathrm{air}}|$ is the magnitude of the impedance in air, acquired for each temperature in the first temperature cycle, and $|Z_{\mathrm{gas}}|$ is the impedance value at a given time during the measurement.

Figure 5 shows quasi-static sensor responses at 200 Hz and 62.8 kHz, acquired with the Agilent 4294A before and after poisoning. The gas concentrations are shown in the bottom row of Figs. 5 and 6. In this paper, we consider two concentrations each of methane (CH_4: 550 and 1100 ppm) and carbon monoxide (CO: 50 and 100 ppm); whereas the gas profile also contained hydrogen (H_2: 5 and 10 ppm) and ethanol (C_2H_5OH: 5 and 10 ppm) (Schüler et al., 2014). All gases were applied in zero air with 50 %rh (relative humidity).

The sensor reactions differ for the different gas–temperature combinations in a typical manner (at 350 °C, the sensor reactions of CH_4 and CO are comparably high, while at 200 °C, carbon monoxide causes a significantly higher sensor reaction), which enables selective measurements. The quasi-static sensor responses of the unimpaired sensor are very high during the first cycles in presence of methane, especially at high temperatures. This phenomenon might be caused by a reaction of a surface species taking place at higher temperatures in presence of methane, in which the species causing the reaction is being consumed or desorbed. The measurement after poisoning does not exhibit this behavior.

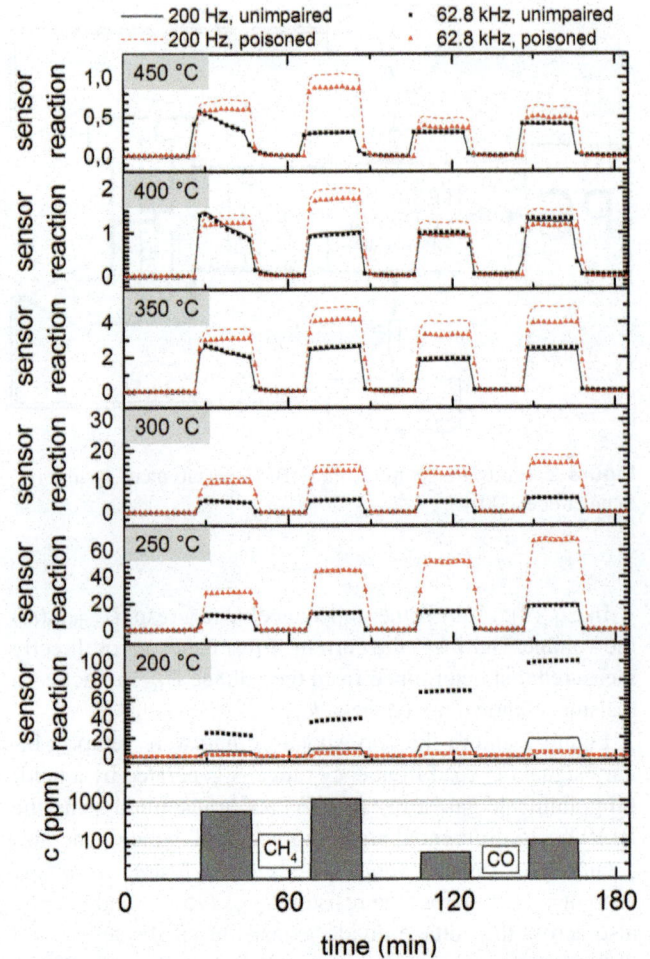

Figure 5. Laboratory measurement, carried out with commercial impedance analyzer Agilent 4294A. The graphs show sensor reactions for each step of the temperature cycle at 200 Hz and 62.8 kHz with the unimpaired as well as with the poisoned sensor.

For temperatures of 250 °C and above, the sensor reactions match well for both frequencies, confirming earlier research which has shown that the response at frequencies up to 100 kHz corresponds to the DC resistance which is usually measured (Reimann, 2011). However, at 200 °C, the sensor reaction measured at 62.8 kHz differs significantly from the one measured at 200 Hz. The reason for this difference is a typical resonance effect in the frequency spectrum of the UST GGS 1330, which has been described in Reimann (2011). The frequency, at which this resonance peak appears, depends on the measurement setup and the resistance of the sensor, it shifts towards lower frequencies for high resistances. At 200 °C the resonance peak is at ca. 100 kHz in air, and the impedance measured is strongly influenced by the resonance. When resistivity drops due to the presence of a reducing gas, the shift of the resonance peak towards higher frequencies is added to the actual drop in resistance, leading to an extraordinarily high sensor response at

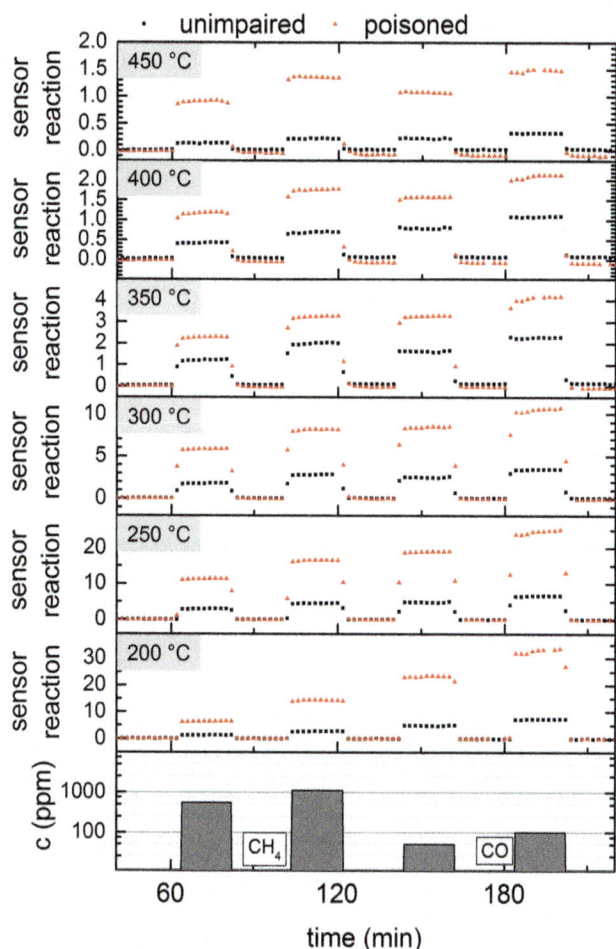

Figure 6. Laboratory measurement, carried out with FoBIS system. The graphs show sensor reactions for each step of the temperature cycle at 61 kHz for the unimpaired as well as for the poisoned sensor.

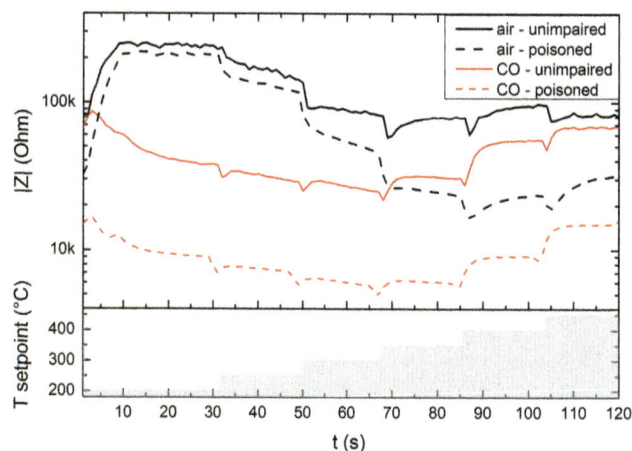

Figure 7. Sensor impedance of UST 1330 at 61 kHz, in air (50 %rh, black lines) and in air with 50 ppm CO (red lines) during temperature cycle. The solid lines show the impedance of the unimpaired sensor; the dashed lines show the impedance values of the same sensor after its exposure to 70 ppm HMDSO (hexamethyldisiloxane) during 10 min.

Table 1. Features used for EIS-based LDA.

Quantity/frequency range (all at 250 °C)	Mean value	Slope
Abs. value of impedance $|Z|$ (61–183 kHz)		X
Abs. value of impedance $|Z|$ (456 kHz–4.6 MHz)	X	X
Abs. value of impedance $|Z|$ (4.9–6.4 MHz)	X	
Abs. value of impedance $|Z|$ (6.4–8.9 MHz)	X	X
Abs. value of impedance $|Z|$ (8.9–27.4 MHz)	X	
Abs. value of impedance $|Z|$ (40.8–103 MHz)	X	X
Impedance angle $\Theta(Z)$ (594 kHz–5.3 MHz)	X	

62.8 kHz. The poisoned sensor has a significantly lowered resistance, at which the resonance peak is shifted towards much higher frequencies, which explains the lower sensor response measured at 200 °C after poisoning. Overall, the sensor reactions are higher for the poisoned sensor. A decrease in sensitivity is not what one would implicitly expect from a sensor poisoning mechanism which deactivates surface states on the sensor. However, some reflections on the functioning of the sensor give plausibility to the phenomenon: the poisoning reaction of HMDSO with sensor surfaces takes place primarily at the locations which are exhibited most directly to the sensor poison, deactivating these in the first place. The metal oxide layer is contacted at its very bottom by electrodes; thus, the measured resistance is influenced most strongly by the sensor reactions taking place near those electrodes. The poisoning of the upper sensor layers inhibits sensor reactions on the more "electrically insensitive" outer surface and enables the target gas to penetrate to the more "electrically sensitive"

region in direct vicinity of the electrodes, where the sensor reaction takes place and is measured with high sensitivity.

Figure 6 shows the quasi-static sensor responses of the measurement carried out with the FoBIS system before poisoning. Here, the shown values were calculated by averaging 10 measurements at the end of each temperature step. The reason that 10 data points per temperature step were averaged was to achieve better comparability of the FoBIS measurements with those carried out with the Agilent 4294A, which can only acquire one spectrum per temperature step. Figure 6 shows the sensor responses at 61 kHz. At this frequency, the SNR of the FoBIS spectroscope is relatively high and the sensor responses towards all gases are clearly recognizable at all temperatures.

Figure 7 shows sensor impedance values at 61 kHz over the temperature cycle, in pure air with 50 %rh, as well as in presence of 50 ppm CO. The solid lines represent the unimpaired sensor; the dashed lines represent the measurement carried out after poisoning. In air, the poisoning has little influence on the impedance values at lower temperatures, whereas at high temperatures, the impedance of the poisoned

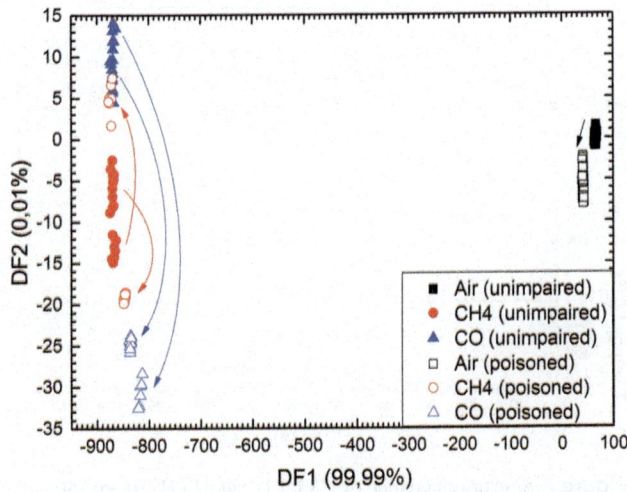

Figure 8. LDA of 10 EIS features (mean values/slopes) for unimpaired (solid symbols) and poisoned (open symbols) sensor, acquired with Agilent 4294A impedance analyzer.

Table 2. Results of leave-one-out cross validation using a kNN classifier ($k = 3$) with Euclidian distance.

Measurement method	Sensor state	Percentage of correct classifications		
		Air	CH$_4$	CO
EIS-IA	unimpaired	100 %	100 %	100 %
	poisoned	100 %	62.5 %	0 %
EIS-FoBIS	unimpaired	100 %	81.25 %	87.5 %
	poisoned	100 %	0 %	100 %
TCO-IA	unimpaired	100 %	100 %	100 %
	poisoned	100 %	100 %	0 %
TCO-FoBIS	unimpaired	100 %	100 %	100 %
	poisoned	100 %	50 %	100 %

sensor is strongly reduced compared to the unimpaired sensor. This results in a clearly recognizable change of the shape of the impedance over the temperature profile. In presence of 50 ppm of carbon monoxide, the shape of the impedance over the temperature cycle changes less strongly, whereas there is a strong decrease in the magnitude of the impedance after poisoning, which actually leads to an increase in sensitivity vs. CO at low sensor temperatures.

4 Feature extraction, signal processing and results

To evaluate the measurements and enable a classification of the results, we use LDA (linear discriminant analysis), a supervised algorithm for dimensionality reduction. LDA calculates a linear transformation, which maps a multidimensional feature vector to a vector with $N - 1$ dimensions, N being the number of classes to be discriminated (Backhaus et al., 2000). In our case, the classes correspond

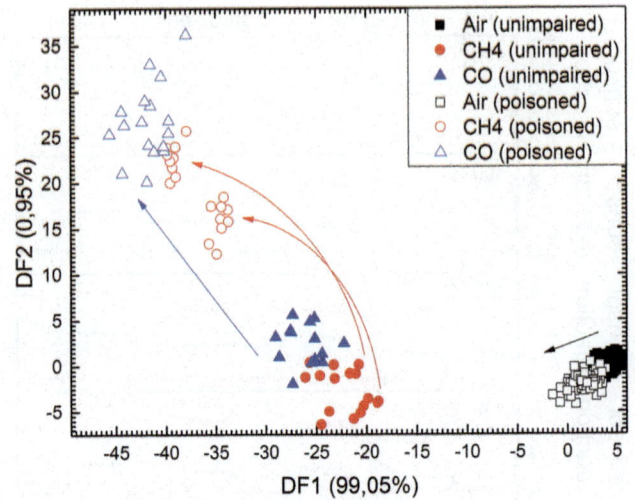

Figure 9. LDA of 10 EIS features (mean values/slopes) for the unimpaired (solid symbols) and the poisoned (open symbols) sensor, acquired with FoBIS impedance spectroscope.

to the different gases to allow discrimination of the different gases independent of the gas concentration. The transformation calculated by the LDA maximizes the distance between data sets from different classes and minimizes the variance (spread) within the single classes. Although LDA reduces the dimensionality, the number of dimensions in the feature vector should be limited to a number much smaller than the number of measurements in order to achieve a stable discrimination, i.e., to prevent overfitting (Luo et al., 2011).

We carried out LDAs with EIS and TCO data, using the commercial impedance analyzer Agilent 4294A and the FoBIS impedance spectroscope. Hereafter, the LDA evaluations are denoted as follows: EIS-IA (EIS data, commercial impedance analyzer), EIS-FoBIS (EIS data, FoBIS impedance spectroscope), TCO-IA (TCO data, commercial impedance analyzer), and TCO-FoBIS (TCO data, FoBIS impedance spectroscope).

The feature vectors for the EIS-IA and EIS-FoBIS LDA were generated by extracting mean values and slopes in selected frequency intervals. Table 1 outlines the features used for these LDAs, which were obtained from the 250 °C temperature step only. At this temperature, the signal in air is only slightly influenced by the poisoning. Figure 8 shows the EIS-IA LDA. The solid symbols represent measurements carried out with the unimpaired sensor; the open symbols represent the measurements after poisoning. Air is clearly discriminated from both gases by the first discriminant function (DF1), which contains 99.99 % of the discriminatory information. For the unimpaired sensor, CO and CH$_4$ are discriminated in the second discriminant function (DF2), which contains only 0.01 % of the discriminatory information, and is therefore very sensitive to small changes in the sensor signal. In DF1, there is only a small shift for all groups after

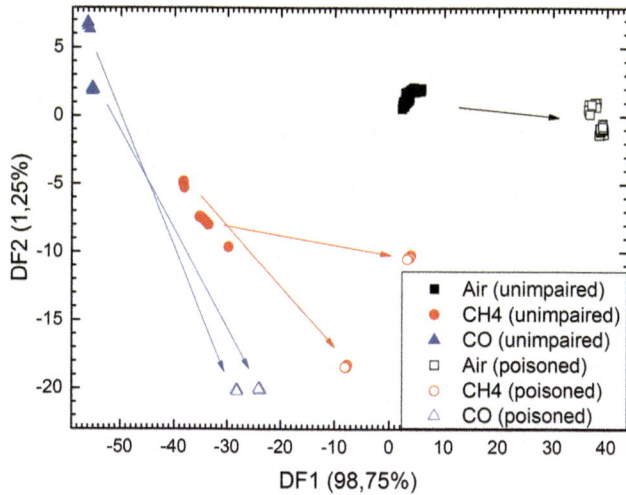

Figure 10. LDA of five TCO features (absolute impedance value differences) for unimpaired (solid symbols) and poisoned (open symbols) sensor, acquired with Agilent 4294A impedance analyzer.

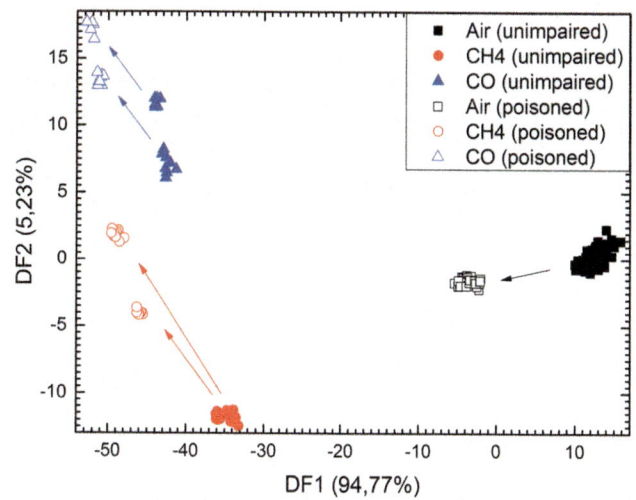

Figure 11. LDA of five TCO features (absolute impedance mean value differences) for unimpaired (solid symbols) and poisoned (open symbols) sensor, acquired with the FoBIS impedance spectroscope.

poisoning. Along DF2, however, the shifts are relatively bigger for the gas measurements. These shifts are different for CO and CH4: the CO measurements shift relatively far, those measured at 100 ppm further than the ones taken at 50 ppm. The different methane concentrations (550 ppm/1100 ppm) shift in crosswise manner. These shifts might result from the long measurement time of the Agilent 4294A, during which the sensor characteristics might change, combined with a higher sensitivity of DF2 to small changes in the sensor signal. We carried out a leave-one-out cross-validation, using a k nearest neighbor (kNN, $k = 3$) Euclidian distance classifier for the different LDAs (Francois et al., 2011). The results are shown in Table 2. It shows that after poisoning discrimination between CH4 and CO is no longer possible. However, air is classified correctly even after poisoning, and can thus be discriminated from both gases.

Figure 9 shows the EIS-FoBIS LDA. Similar to the EIS-IA LDA, this evaluation shows only a small shift in the measurements acquired in pure air caused by poisoning. In the presence of CO or CH4 the measurements shift along both discriminant functions after poisoning. For the EIS-FoBIS LDA, the shifting directions are rather parallel, unlike in the EIS-IA LDA, where a crosswise shift was observed. In this LDA, the discrimination between CH4 and CO is imperfect even before poisoning, with 81.25 % and 87.5 %, respectively, of measurements classified correctly. After poisoning, all CH4 measurements are classified incorrectly. Although the evaluation results are not completely equivalent for both measurement systems, the differences between the gas measurements of unimpaired and poisoned sensors can be recognized in the LDAs carried out with impedance data from either system, and pure air can be discriminated from gas even after poisoning.

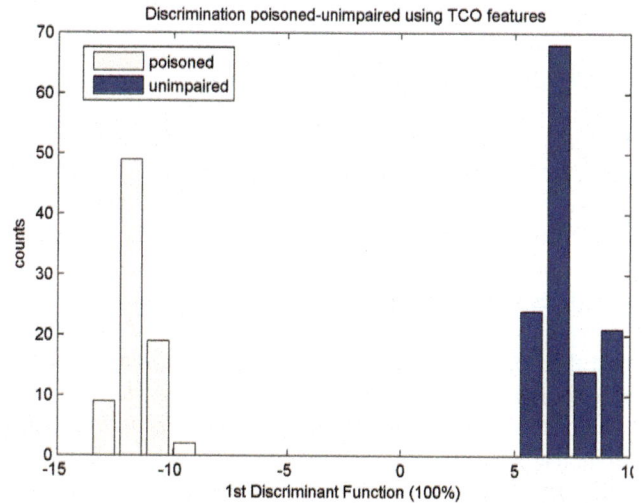

Figure 12. LDA discriminating measurements carried out with the poisoned (bright bars) and unimpaired (dark bars) sensor using five TCO features (absolute impedance mean value differences). The data were acquired using the FoBIS impedance spectroscope; measured conditions are pure air plus two concentrations each of CH4 (550, 1100 ppm) and CO (50, 100 ppm), all at 50 %rh.

To extract the features for the TCO-based LDAs shown in Figs. 10 and 11, we calculated the differences of the normalized absolute impedance values at 61 kHz at adjacent temperature steps, i.e.,

$$|Z(61\,\text{kHz}, T_i)| - |Z(61\,\text{kHz}, T_{i-1})| \qquad (5)$$
$$i = 1\ldots5, \ T_0 = 200\,°C, \ T_1 = 250\,°C, \ T_2 = 300\,°C,$$
$$T_3 = 350\,°C, \ T_4 = 400\,°C, \ T_5 = 450\,°C.$$

Again, 10 measurements were averaged for each FoBIS-based data point.

Figure 10 shows the TCO-IA LDA: for the unimpaired sensor, the different gases are well separated, allowing a perfect classification. For the poisoned sensor, the LDA yields results which have strongly shifted. Unlike the EIS-based LDA, these shifts are of similar direction and value for the different gas exposures. This result would not enable a correct classification of either gas. In the TCO-FoBIS LDA (Fig. 11), the poisoning causes significant shifts as well, although the cross-validation shows a partially correct classification even after poisoning. Thus, a check of this result, which takes into account the EIS-based result (as outlined in Fig. 1), can increase the reliability in the detection of sensor changes.

Figure 12 shows the result of a TCO-based LDA carried out with data from all measurements shown in Fig. 11, but separated into two different classes: poisoned sensor and unimpaired sensor. The result shows that a detection of sensor impairments is also possible from TCO data acquired using the FoBIS impedance spectroscope for the limited test gas selection that was examined.

5 Conclusion and outlook

We developed a compact high-speed impedance measurement system, which enables the combination of EIS and TCO even for sensors with small thermal time constants, i.e., MEMS-based sensors. Sensor changes affect the results of both measurement methods differently, which enables a reliable detection of sensor impairments for improved reliability of gas sensor systems. The shifts in the EIS-based LDAs suggest that the impedance data contain information useful for recognition of sensor impairments. The long measurement time of the commercial impedance analyzer is an important challenge in temperature-cycled operation, which shows the necessity of faster impedance measurement systems for combined TCO–EIS measurements. The evaluation of TCO features indicates that a detection of sensor impairments can be possible from TCO data only. Therefore, further research will consider the influence of sensor changes to the dynamic behavior of the sensors and also examine the influence of various impairments (e.g., sulfur compounds), as well as broader ranges of test gases and gas concentrations.

Acknowledgements. We gratefully acknowledge funding by the German ministry of Economic Affairs and Energy (BMWi, grant no. 16962N). The authors would like to thank 3S GmbH, Saarbrücken, Germany, for hardware support.

Edited by: M. Meyyappan
Reviewed by: one anonymous referee

References

Agilent 4294A: Precision Impedance Analyzer Data Sheet, available at: http://cp.literature.agilent.com/litweb/pdf/5968-3809E.pdf, last access: 25 November 2013.

Backhaus, K., Erichson, B., Plinke, W., and Weiber, R.: Multivariate Analysemethoden, Springer-Verlag, ISBN 3540-67146-3, 2000.

Barsoukov, E. and Macdonald, J. R.: Impedance Spectroscopy – Theory, Experiment, and Application, John Wiley & Sons, ISBN 0-471-64749-7, 2005.

Bârsan, N. and Weimar, U.: Understanding the fundamental principles of metal oxide based gas sensors; the example of CO sensing in the presence of humidity, J. Phys. Condens. Matter, 15, R813–R839, 2003.

Clifford, P. K. and Tuma, D. T.: Characteristics of semiconductor gas sensors II. Transient response to temperature change, Sensors Actuators, 3, 255–281, 1982, 1983.

Conrad, T., Trümper, F., Hettrich, H., and Schütze, A.: Improving the Performance of Gas Sensor Systems by Impedance Spectroscopy: Application in Under-Ground Early Fire Detection, Proc. SENSOR Conference 2007, Volume I, AMA Service GmbH, ISBN 978-3-9810993-1-7, 169–174, 2007.

Elmi, I., Zamponelli, S., Cozzani, E., Mancarella, F., and Cardinali, G. C.: Development of ultra-low-power consumption MOX sensors with ppb-level VOC detection capabilities for emerging applications, Sensors and Actuators B, 135, 342–351, 2008.

Francois, D., Wertz, V., and Verleysen, M.: Choosing the Metric: A Simple Model Approach, Meta-Learning in Computational Intelligence Studies in Computational Intelligence, 358, 97–115, 2011.

Fricke, T., Reimann, P., Horras, S., Leonhardt, E., Sahm, P., Schütze, A.: A systematic approach for the automatic signal processing, evaluation and optimization of T-cycled gas sensors, IEEE International Instrumentation and Measurement Technology Conference I^2 MTC, 2008.

Heilig, A., Bârsan, N., Weimar, U., Schweizer-Berberich, M., Gardner, J. W., and Göpel, W.: Gas identification by modulating temperatures of SnO_2-based thick film sensors, Sensors and Actuators B, 43, 45–51, 1997.

Kohl, D.: Surface processes in the detection of reducing gases with SnO2-based devices, Sensors Actuators, 18, 71–113, 1989.

Korotchenkov, G. and Cho, B. K.: Instability of metal oxide-based conductometric gas sensors and approaches to stability improvement (short survey), Sensors and Actuators B, 156, 527–538, 2011.

Lee, A. P. and Reedy, B. J.: Temperature modulation in semiconductor gas sensing, Sensors and Actuators B, 60, 35–42, 1999.

Ljung, L. and Glad, T.: Modeling of dynamic systems, PTR Prentice-Hall, Upper Saddle River, NJ, USA, 1994.

Luo, D., Ding, C., and Huang, H.: Linear Discriminant Analysis: New Formulations and Overfit Analysis, Proceedings of the 25th AAAI Conference on Artificial Intelligence, 2011.

Morrison, S. R.: Semiconductor gas sensors, Sensors Actuators, 2, 329–341, 1982.

Morrison, S. R.: Selectivity in semiconductor gas sensors, Sensors Actuators, 12, 425–440, 1987.

Reimann, P.: Gasmesssysteme basierend auf Halbleitergassensoren für sicherheitskritische Anwendungen mit dem Ansatz der Sensorselbstüberwachung, Dissertation, Shaker Verlag, ISBN 978-3-8440-0232-4, 2011.

Reimann, P., Dausend, A., and Schütze, A.: A self-monitoring and self-diagnosis strategy for semiconductor gas sensor systems, Proc. IEEE Sensors Conference 2008, Lecce, Italy, 27–29 October, 2008.

Schüler, M., Sauerwald, T., Walter, J., and Schütze, A.: High speed impedance spectroscope for metal oxide gas sensors, Lecture Notes on Impedance Spectroscopy, 5, accepted, 2014.

Schütze, A., Gramm, A., and Rühl, T.: Identification of Organic Solvents by a Virtual Multisensor System with Hierarchical Classification, IEEE Sensors, 4, 857–863, 2004.

Stetter, J. R. and Penrose, W. R.: Understanding Chemical Sensors and Chemical Sensor Arrays (Electronic Noses): Past, Present, and Future, Sensors update 10.1, 10, 189–229, 2002.

Tricoli, A., Righettoni, M., and Teleki, A.: Semiconductor gas sensors: dry synthesis and application, Angew. Chem. Int. Ed., 49, 7632–7659, doi:10.1002/anie.200903801, 2010.

Weimar, U. and Göpel, W.: A.C. measurements on tin oxide sensors to improve selectivities and sensitivities, Sensors and Actuators B, 26, 13–18, 1995.

8

Humidity measurement with capacitive humidity sensors between −70 °C and 25 °C in low vacuum

A. Lorek

German Aerospace Center (DLR), Berlin, Germany

Correspondence to: A. Lorek (andreas.lorek@dlr.de)

Abstract. At the German Aerospace Center (DLR), capacitive humidity sensors are used to measure relative humidity in experiments under extreme atmospheric conditions such as on Mars or in the coldest regions on Earth. This raises the question whether such experiments can be performed using low-cost humidity sensors with a tolerable measurement uncertainty. As part of the standardizing project SMADLUSEA (project no. SF11021A), nine capacitive humidity sensors (Sensirion SHT75) were investigated for pressure ranging from 10 to 1000 hPa (low vacuum) and temperatures from −70 to 25 °C. It has been shown that these sensors worked reliably and with reproducibly measured values over the entire investigated pressure and temperature range. There was no aging of the sensors observable. In addition to the known strong temperature dependency, the SHT75 also shows a pressure dependency below −10 °C. A characteristic curve for the SHT75 was calculated with an expanded uncertainty of 7 % of the measured values.

In conclusion, low-cost capacitive humidity sensors offer the option to obtain reliably measured values even under extreme conditions with comparatively little effort.

1 Introduction

What are the most useful sensor principles and their potential measurement ranges under Martian conditions? This was a fundamental question for the development of the in situ trace humidity measuring system called MiniHUM, designed for the ExoMars lander to measure the humidity of the near-surface Martian atmosphere (Koncz, 2012). The coulometric and capacitive sensor principles were chosen to develop a lightweight and low-energy device. The coulometric sensors (Lorek et al., 2010), which we will not further described in this paper, measure the absolute humidity and have the potential to detect trace humidity below frost points of −100 °C. However, our own measurements have shown low chemical activity at temperatures less than −50 °C which leads to a limited functionality for this type of sensor.

At standard environmental conditions (25 °C and 1013 hPa), the capacitive sensors measure reliably in the range from 10 to 90 % relative humidity ($U_{w,i}$). However, there are only insufficient data, even for coulometric sensors, about the behavior in low vacuum (1 to 1000 hPa) and temperatures between −70 and 25 °C. Evidence for the ability of the capacitive sensor principle is the Phoenix Mars mission (Zent et al., 2010). Capacitive humidity sensors are also used successfully in meteorology, for example, on radiosondes in weather balloons. While some of them have been investigated for temperatures down to −70 °C (Miloshevich et al., 2001; Hudson et al., 2004), there are not sufficient data on their behavior in a low vacuum especially down to 10 Pa.

Our measurements of nine capacitive humidity sensors type SHT75 from Sensirion AG (2011) provide a database which demonstrates function and measurement uncertainty of off-the-shelf sensors in the already mentioned pressure and temperature range. The experiments were performed at the Martian Simulation Facility (MSF) at DLR (Lorek and Koncz, 2013).

2 Theoretical background

The SHT75-sensors measure the relative humidity $U_{w,i}$ (Eq. 1), defined as the ratio in per cent of the water vapor partial pressure e [Pa] to saturation vapor pressure under saturation conditions above a planar water e_w [Pa] or ice surface e_i [Pa] at the same total pressure p [Pa] and temperature T [K] (derived from WMO, 2012):

$$U_{w,i} = \left(\frac{e}{e_{w,i}}\right)_{p,T} 100 \%. \tag{1}$$

In this paper conditions of pure phase e were assumed and used for calculations and not the water vapor partial pressure in a real gas e', because the difference between both pressures is <0.5 % and negligible for this investigation (VDI/VDE 3514 Part 1, 2007; Bögel, 1977; WMO, 2012).

The following Eqs. (2) for e_w [hPa] and (3) for e_i [hPa] are recommended by WMO (2012). Equation (2) is related within the temperature range from −50 to 100 °C

$$\lg e_w = 10.79574 \left(1 - \frac{T_1}{T}\right) 5.028 \lg\left(\frac{T_1}{T}\right)$$
$$+ 1.50475 \times 10^{-4} \left[1 - 10^{-8.2969\left\{\frac{T}{T_1}\right\}}\right]$$
$$+ 0.42873 \times 10^{-3} \left[10^{+4.76955\left\{1-\frac{T}{T_1}\right\}} - 1\right]$$
$$+ 0.78614 \tag{2}$$

and Eq. (3) from −100 to 0 °C with $T_1 = 273.16$ K (triple point temperature of water):

$$\lg e_i = -9.09685 \left(\frac{T_1}{T} - 1\right) - 3.56654 \lg\left(\frac{T_1}{T}\right)$$
$$+ 0.87682 \left(1 - \frac{T_1}{T}\right) + 0.78614. \tag{3}$$

The WMO (2012) recommends calculating the relative humidity relative to U_w. One reason is that for temperatures below 0 °C, the relative humidity in clouds is often supersaturated with respect to ice ($U_i > 100 \%$). A further reason is a better comparability of the most meteorological measurements witch are often displayed in U_w.

However, the measurements discussed in this paper were partially performed down to −70 °C. This is out of the range of Eq. (2). A super-saturation of the sample gas seems to be unlikely because there are enough inner surfaces in pipes and measurement cells for the condensation of surplus water vapor. Therefore, in this paper, the relative humidity below 0 °C is obtained in relation to ice (U_i) according to the technical definition for the relative humidity (VDI/VDE 3514 Part 1, 2007).

3 Experimental procedure

3.1 Experimental setup

3.1.1 Sensors and measured value acquisition

The investigated SHT75 is a polymer based capacitive humidity sensor manufactured in CMOS technology. In addition it has a band-gap temperature sensor and electronic units for signal processing (Sensirion AG, 2011). The humidity sensor measures the permittivity of the hydrophilic polymer depending of the adsorbed water content which is affected by the partial water vapor pressure of the surrounding atmosphere. An integrated A/D converter generates a serial output from the analog measurement signal at the digital interface. The dimensions without electrical connectors are $(6.4 \times 3.7 \times 3.1)$ mm (length × width × height).

A nine-channel measuring device for SHT sensors from dr. wernecke Feuchtemesstechnik GmbH was used to process the serial data and output an integer rough humidity value (SO_{RH}) and temperature value. From these, the device calculates the relative humidity value $U_{w(SHT75)}$, based on Eq. (4).

The measuring range for U_w specified by the manufacturer is from 0 to 100 % and for the temperature from −40 to 123.8 °C.

Equation (4) is the manufacturer formula (Sensirion AG, 2011) for the calculation of the temperature compensated relative humidity:

$$U_{w(SHT75)} = (t - 25)(0.01 + 0.00008 SO_{RH})$$
$$- 2.0468 + 0.0367 SO_{RH}$$
$$- 1.5955 \times 10^{-6} SO_{RH}^2 \tag{4}$$

using t [°C] as sensor temperature.

The expanded uncertainties (U_{99}) for $U_{w(SHT75)}$ is $U_{99} = 1.8 \%$ within U_w 10 to 90 % at 25 °C and increase to the measuring range limits of $U_{99} = 4 \%$. The expanded uncertainty for the temperature increases from $U_{99} = 0.3$ K at 25 °C to a maximum of 1.5 K at −40 °C (Sensirion AG, 2011).

The expanded uncertainties U_{99}, given in this paper, result from a multiplication of the standard uncertainty u (type A or B) with factor of 3 ($k = 3$) and include 99 % of the measured values (GUM, 2008).

Using SHT75 has an historical background. To measure in situ the near-surface atmospheric humidity on Mars in 2003 our laboratory was looking for a humidity sensor. Information from the Open University about this sensor working under extreme conditions and its special features (e.g., digital output, low hysteresis, inexpensive, low power consumption, off-the-shelf product and certified) led to the choice of the SHT75. Since then, the laboratory has been using this sensor.

The investigated temperatures range down to −70 °C which is lower than the threshold of −40 °C for the SHT75 measuring range. Therefore and as a reference, 15 Pt100 temperature sensors were used. Twelve of them were Pt100 thin-film resistors 6W 538 from IST AG (2012) with the dimensions $5 \times 3.8 \times 0.65$ mm (length × width × height) and three were Pt100-wire-wound resistors from Service für Messtechnik Geraberg GmbH, all with classification A. The wire-wound resistors were cased in a copper block to prevent damage by the mounting on the exterior measuring cell housing. The block dimensions are $10 \times 3 \times 3$ mm (length × width × height). The Pt100 sensors were calibrated for temperatures between -75 and $50\,^\circ\text{C}$ (Deep-Well-Bad 7831, Fluke Europe B.V) with a resulting expanded uncertainty for all Pt100 of $U_{99} = 0.1$ K.

The pressure measurement was done using an Active Capacitive Transmitter CMR 361 (Pfeiffer Vacuum GmbH). It has an effective range from 0.1 to 1100 hPa and an $U_{99} = 0.01\,\%$ (full scale) in the range from 1 to 40 000 Pa and an $U_{99} = 0.1\,\%$ for $p > 40\,000$ Pa.

A Keithley-digital multimeter DMM 3706 with integrated Dual 1×20 Multiplexer-card (Model 3721 with 3721-ST screw terminal) was used for data acquisition from CMR 361 and all Pt100.

A dew point mirror hygrometer S8000RS (Michell Instruments GmbH) was the reference for the humidity of the sample gas. It has a measuring range from ca. $-100\,^\circ\text{C}$ frost point temperature (t_f) to $20\,^\circ\text{C}$ dew point temperature (t_p) with an expanded uncertainty of $U_{99} = 0.3$ K at $20\,^\circ\text{C}$ to 0.51 K at $-60\,^\circ\text{C}$.

3.1.2 Experimental setup

The experimental setup, basically described by Lorek and Koncz (2013), consists of a gas mixing system, a temperature test chamber which contains the measuring cells, a membrane vacuum pump (Fig. 1) and the measurement equipment mentioned in Sect. 3.1.1. The pipes, measuring cells and connection components between gas mixing system and vacuum pump consisted of stainless steel to minimize adsorption at and permeation through the pipe walls.

The sample gas can be nearly continuously humidified in an absolute humidity range from $-73\,^\circ\text{C}$ (t_f) to $5\,^\circ\text{C}$ (t_p).

It was found that for the pipes which transport the sample gas inside the chamber to the measuring cells, a diameter of DN 16 is sufficient to ensure constant pressure and prevent water condensation in pipes and cells inside the chamber (Fig. 2). The dimensions of the measuring cells are $180 \times 70 \times 70$ mm (length $\times$ width $\times$ height). Each cell has a centered 38 mm diameter hole over the entire length and two blind holes of the same dimensions down to the vertical center with a distance of 80 mm. These blind holes form the measuring chambers holding the sensors and exposing them to the sample gas (Fig. 2a).

3.1.3 Sensor setup

Each of the CF-DN40 flanges sealing the first measuring chamber of each measuring cell (Fig. 2) is equipped with a 2 mm stainless steel pipe for the pressure measurement and a 32-pin male connector. Both were glued into the flange using ECCOBOND 286 (2013). The connectors provide the electrical feedthrough for 4 Pt100 (four-wire configuration) and three SHT75 on each measuring cell.

The SHT75 measurement range for temperature is limited at $-40\,^\circ\text{C}$. Therefore, a calibrated Pt100 was glued onto each SHT75 (Sect. 3.1.1). The fourth Pt100 was placed in

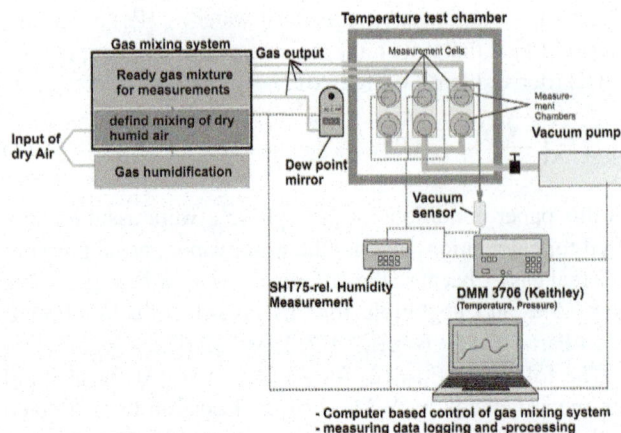

Figure 1. Experimental setup.

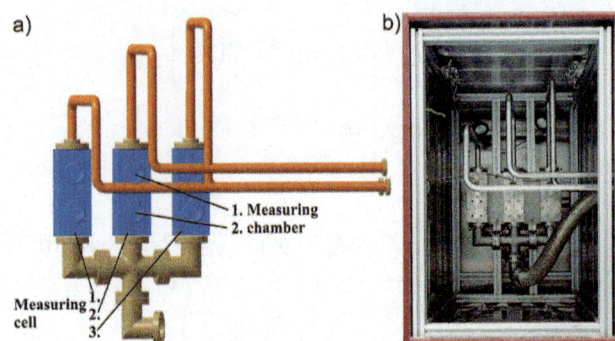

Figure 2. (a) Illustration of measuring cells and pipe routing inside the temperature test chamber; **(b)** temperature test chamber with measuring cells.

the center of the three mounted SHT75/Pt100 combinations (Fig. 3). This Pt100 measured the undisturbed gas temperature to reveal temperature differences between the Pt100–SHT75 combinations and the housing temperature. In order to place sensors in the center of the gas flow, they were soldered to the connector with 15 mm enameled copper wires (diameter 150 µm).

A 1.8 m four-wire cable with PFA-coating (usable down to $-200\,^\circ\text{C}$) was used to connect each sensor with the measurement device. To minimize heat influences from the outside, cables were routed mainly inside the temperature test chamber. Finally, for every measuring cell, one Pt100 (Sect. 3.1.1) was mounted outside every measurement cell to measure the housing temperature.

The quantity of nine SHT75 is given by the Pt100 measuring equipment and the selected 32-pin male connector.

In the following each SHT75 and Pt100 will be designated using the nomenclature in Table 1.

Table 1. Nomenclature for the Pt100 and SHT75 used.

Nomenclature	SHT75	Pt100 at SHT75	Pt100 in the middle	Measuring cell Pt100 outside at
Measuring cell 1	SHT75_1	Z1PtS1	Z1Pt1	Z1B1
	SHT75_2	Z1PtS2		
	SHT75_3	Z1PtS3		
Measuring cell 2	SHT75_4	Z2PtS1	Z2Pt1	Z2B1
	SHT75_5	Z2PtS2		
	SHT75_6	Z2PtS3		
Measuring cell 3	SHT75_7	Z3PtS1	Z3Pt1	Z3B1
	SHT75_8	Z3PtS2		
	SHT75_9	Z3PtS3		

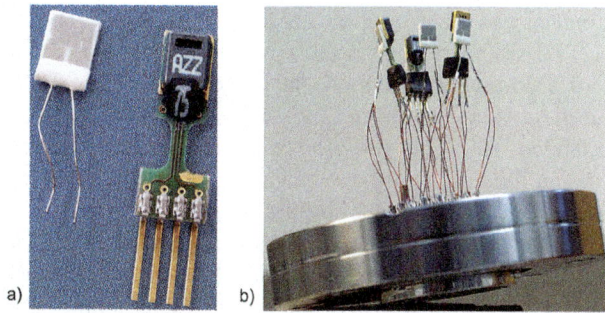

Figure 3. (a) Left: Pt100 (6W 538), and right: SHT75; **(b)** three SHT75/Pt100 combinations with the fourth Pt100 placed in the center.

3.2 Experimental procedure

After a leakage test, a continuous volume flow of $\dot{V}_N = 30\,\mathrm{L\,h^{-1}}$ (at 20 °C and 1013 hPa) sample gas was passed through each cell (Figs. 1 and 2a). At 10 hPa, the flow has to be reduced to $\dot{V}_N = 15\,\mathrm{L\,h^{-1}}$ to ensure not to exceed the performance of the vacuum pump. When at the required temperature thermal equilibrium was reached inside the measuring chamber, the measurements were started at the highest pressure for this temperature (Table 2) and, if possible, the relative humidity steps between $U_{\mathrm{w,i(ref)}}$ ca. 5 and 95 % were generated. The achievable relative humidity was limited due to the chamber temperature and the generated humidity of the gas mixing system. The pressure was kept constant over 0.5 to 15 h depending on the expected duration to reach humidity equilibrium at each humidity step of the measuring chambers and sensors.

The investigated temperatures, pressures and relative humidities are listed in Table 2. The calculation of the relative reference humidity $U_{\mathrm{w,i(ref)}}$ inside the measuring chambers using Eq. (1) is based on the assumption that

$$\frac{e_{\mathrm{ref}}}{e_{(\mathrm{ref}N)}} = \frac{p}{p_N} \quad \text{and thereby}$$

Table 2. Range of the relative humidity under investigation for the different temperature and pressure conditions. Relative humidity is calculated with respect to water U_{w} or ice U_{i} (marked by brackets).

	P in hPa					
	1000	800	500	200	100	10
25 °C	28 2	18 3	11 4	4		
10 °C	71 1	57 4	36 4	14 5	7	
0 °C	91 6	89 6	72 6	29 5	14 5	
−10 °C	90 (99) 5 (6)	88 (97) 5 (6)	84 (93) 5 (6)	62 (68) 5 (6)	29 (32) 5 (6)	4 (5)
−20 °C	79 (96) 4 (5)	79 (96) 5 (6)	82 (100) 4 (5)	77 (94) 4 (5)	66 (79) 5 (6)	7 (9) 4 (5)
−30 °C	74 (99) 4 (5)	74 (99) 4 (5)	74 (99) 4 (5)	74 (99) 4 (5)	66 (89) 3 (4)	18 (24) 3 (4)
−40 °C	65 (96) 7 (10)	55 (81) 6 (9)	67 (99) 3 (4)	67 (98) 3 (4)	60 (89) 3 (4)	53 (78) 3 (4)
−50 °C	(85) (5)		(93) (7)	(93) (14)	(89) (20)	(99) (8)
−60 °C	(75) (10)		(71) (5)	(90) (17)		(99) (17)
−70 °C			(86) (18)	(79) (14)		(99) (15)

$$U_{\mathrm{w,i(ref)}} = \left(\frac{e_{\mathrm{ref}}}{e_{\mathrm{w,i}}}\right)_{p,T} 100\,\%, \tag{5}$$

where $e = e_{\mathrm{ref}}$ is the partial water vapor pressure inside the measuring chamber and $e_{(\mathrm{ref}N)}$ is the partial water vapor pressure at normal pressure p_N [101 325 Pa]. Depending on the $t_{\mathrm{p,f}}$ values given by the dew point hygrometer and the measured values p from the vacuum sensor CMR 361 (Sect. 3.1.1, Fig. 1), the value $e_{(\mathrm{ref}N)}$ was calculated from Eqs. (2) or (3).

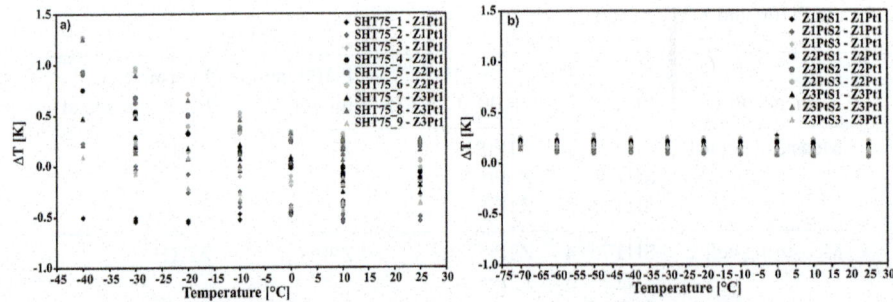

Figure 4. ΔT of the measured values relative to the Pt100 in the middle of each measuring chamber. In **(a)** temperature is measured by the Pt100 glued onto SHT75, while in **(b)** temperature is measured by the SHT75 itself.

3.3 Experimental results

3.3.1 Temperature

Figure 4 shows the average values of the temperature differences (ΔT), calculated from the measurements of the Pt100 glued onto the SHT75 (see Table 1, column 3, and Fig. 3b) and the SHT75 temperature values relative to the Pt100 placed in the center (see Table 1, column 4, and Fig. 3b) at each temperature (see Table 2, column 1). The expanded uncertainty of ΔT is in Fig. 4a $U_{99} = 0.32$ K at 25 °C to $U_{99} = 1.5$ K at −40 °C and in Fig. 4b $U_{99} = 0.15$ K.

Figure 4a shows the aberration development for the SHT75 band-gap temperature sensors from 0.5 K at 25 °C to 1.3 K at −40 °C. This increase of aberration to the Pt100 reference is due to the reduction of accuracy at low temperatures (Sensirion AG, 2011).

Figure 4b shows clearly that the temperature measurement with Pt100 compared to the measurement with the SHT75 band-gap temperature sensor (Fig. 4a) leads to a significantly smaller and temperature-independent variability of the measured values. This has a direct influence on the precision and accuracy of the relative humidity measurements. The observable offset of ca. 0.2 K in Fig. 4b probably results from the emitted heat of the active SHT75 (see Sensirion AG, 2011) and is, due to the strong scattering of the ΔT values, in Fig. 4a not clearly observable.

A comparison of the temperature measured by the Pt100 mounted outside the cells (see Table 1, column 5) with the Pt100 placed in the center (see Table 1, column 4) shows nearly identical temperature values with deviations within the uncertainty of the measurement (see Sect. 3.1.1). Therefore, it can be assumed that the sample gas temperature was identical to the cell housing temperature and that the active SHT75 sensors indeed generate the temperature increase of 0.2 K.

3.4 Relative humidity

In Fig. 5a–d, we show the signal of the SHT75_5 (see Table 2) representative for the other eight SHT75. $U_{w(SHT75)}$ was calculated from Eq. (4) based on the temperature and

humidity measurements of the SHT75 and $U_{w(ref)}$ from Eqs. (2) and (5) based on the $t_{p,f}$ values of the dew point hygrometer and the measurements of the Pt100 glued onto the SHT75 (see Table 1, column 3). Note that the U_w values for −70 °C in Fig. 5d lie outside the range of validity of Eq. (2) but these values are irrelevant for the demonstration of the reliability. The peaks in Fig. 5d result from the automatic calibration function of the dew point hygrometer. Temperature and pressure influence the sorption rate of water molecules at the sensors and at the walls of pipes and measuring cells. This effect changes the equilibration time of the entire experimental system. A comparison of the sensor response time ($T90$) of Fig. 5a with b at the same pressure of 1000 hPa shows a significant delay of some minutes at 10 °C and nearly an hour at −40 °C and therefore a strong dependency on temperature. For this reason, we did not perform measurements at −70 °C and 1000 hPa because the response time of the sensors for each humidity step would have required several days with a resulting increase of the scatter of the measured values and a decrease of the sensitivity (see the 500 hPa-fit in Sect. 3.4.3). The situation was different at lower pressure. A comparison at 1000 hPa (Fig. 5b) with 10 hPa (Fig. 5c) at −40 °C shows a reduction of the $T90$ time to < 10 min.

The nine SHT75 sensors were tested in several test series over 136 days. On 74 days of this time, the sensors were subjected to temperatures < −40 °C, which is outside their specifications (see Sect. 3.1.1). Nevertheless, the sensors were still functioning (Fig. 5d) and showed reproducible results (see Sects. 3.4.3 and 3.4.4) over the entire temperature and pressure range (Table 2).

3.4.1 Comparison with manufacturer's data of the SHT75

Figure 6 shows the $U_{w(SHT75)}$ values from Eq. (4) at temperatures between −40 and 10 °C and a pressure of 1000 hPa based only on SHT75 measurements. The calculation of $U_{w(ref)}$ has been described in Sect. 3.4. The area marked in grey illustrates the expanded uncertainty U_{99} calculated using the standard uncertainty (type B GUM, 2008) with the

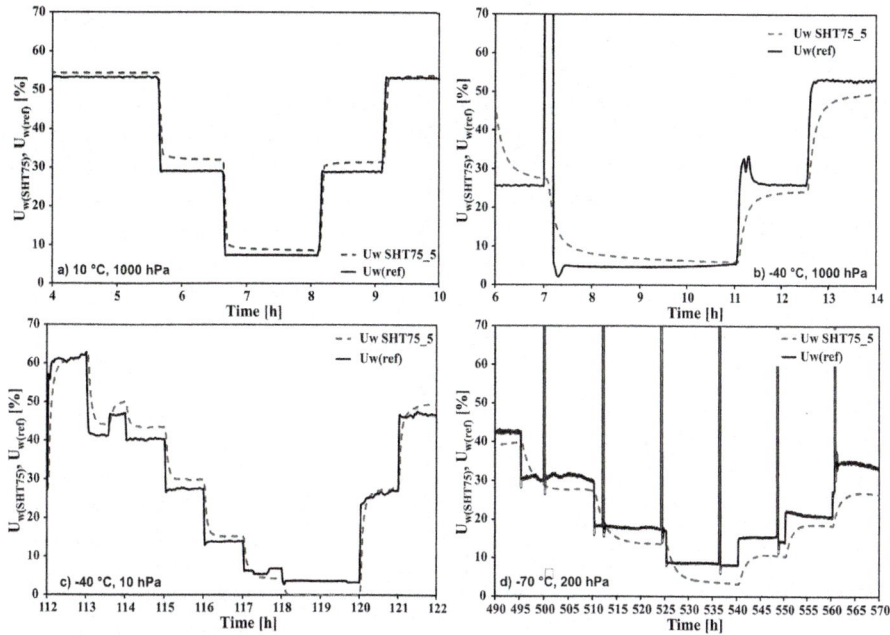

Figure 5. Signal of SHT75_5 (representative for all SHT75 sensors) at different temperatures and pressures.

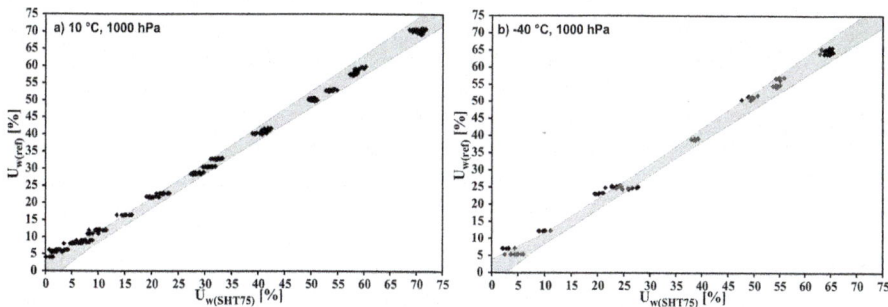

Figure 6. $U_{w(SHT75)}$ values of all SHT75 sensors at 1000 hPa for **(a)** 10 °C and for **(b)** −40 °C.

assumption that the measured values are normally distributed around the diagram diagonal.

For both temperatures, the expanded uncertainty of $U_{w(ref)}$ is $U_{99} = U_{w(ref)} \cdot 0.05$ and $U_{99} = 1.8\%$ for $U_{w(SHT75)}$ with an increase to 4 % at most for U_w at 0 or 100 % (see Sect. 3.1.1) were calculated based on the uncertainties for sensors and measuring instruments in Sect. 3.1.1.

The $U_{w(SHT75)}$ values in Fig. 6a are at $U_{w(ref)}$ values > 20 % within the calculated uncertainty area. Below this value, the measured $U_{w(SHT75)}$ values differ more and more to one side of the diagonal and partially are not inside the calculated uncertainty range. For −40 °C (Fig. 6b), a similar behavior with stronger variation can be observed. It appears that not all error sources of the measuring system are sufficiently known at lower humidity and temperatures to explain the observed systematic error. A leakage is unlikely because the $U_{w(SHT75)}$ values are lower than the $U_{w(ref)}$ values. Further possibilities are the inaccuracy of the manufacturer's fit or, more likely, deviations from this fit through gluing the

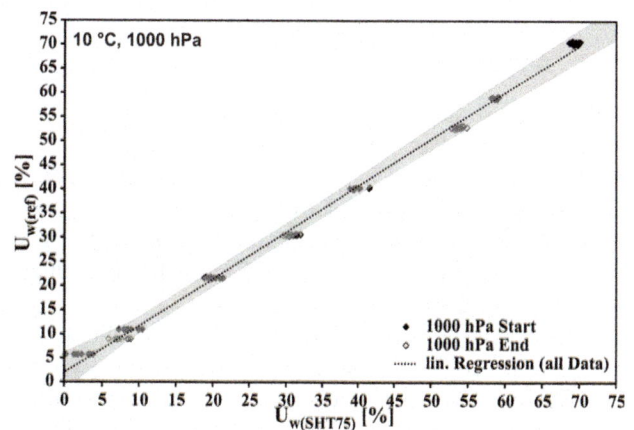

Figure 7. $U_{w(SHT75)}$ values at 10 °C and 1000 hPa of all SHT75 at the start and end of the 136 days test time; expanded uncertainty area marked in grey.

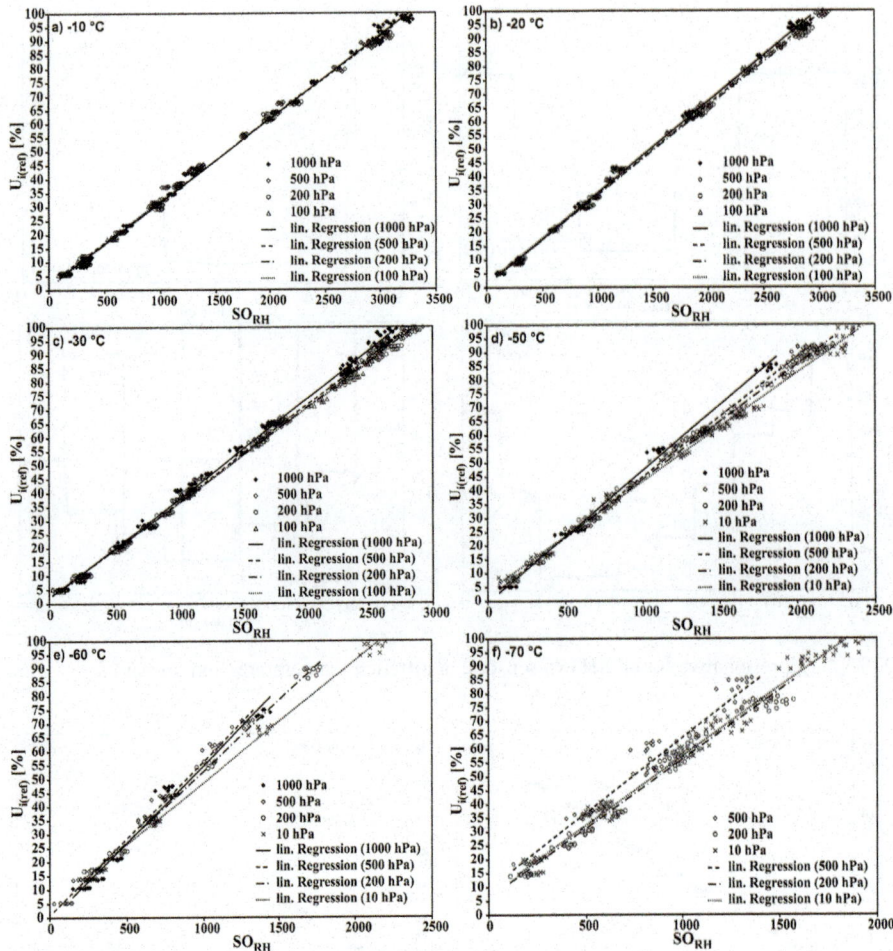

Figure 8. Pressure dependency of all SHT75 sensors at temperatures from -70 to $-10\,°C$.

Pt100 onto the SHT75 (see Sect. 3.1.3). A final determination of the reason for the error is not possible because an unaltered SHT75 was not tested.

3.4.2 Aging of the SHT75

Figure 7 illustrates the $U_{\mathrm{w(SHT75)}}$ values of all SHT75 at $10\,°C$ and $1000\,hPa$ at the start and end of the 136 day test time. The area marked in grey shows the expanded uncertainty from Sect. 3.4.1 but relative to the linear regression fit of the measurements rather than to the diagonal as in Fig. 6. Therefore, the systematic error (see Sect. 3.4.1) is included in the error analysis.

Even though the sensors were exposed to conditions beyond their specifications (see Sect. 3.1.1), for 74 days out of 136 days the $U_{\mathrm{w(SHT75)}}$ values in Fig. 7 are mainly within the expected uncertainty. A drift in the measurements was not observed.

3.4.3 Pressure dependency of the SHT75

In this section, the relation of the SO_{RH} values (see Sect. 3.1.1) and $U_{\mathrm{i(ref)}}$ values (see Sect. 3.4) is evaluated as a function of pressure.

We do not observe pressure dependence (Table 2) in the temperature range from -10 to $25\,°C$ (see Fig. 8a). In the range from -20 to $-50\,°C$ (Fig. 8b to d), an increase of the scatter in the measurements as a function of increasing pressure was detected. At $-60\,°C$ (Fig. 8e), the regression fits of $1000\,hPa$, $500\,hPa$ and $200\,hPa$ are in better mutual agreement again. Only the $10\,hPa$ fit differs clearly from the others. At $-70\,°C$ (Fig. 8f), a pressure dependence can no longer be observed. The deviation of the fit at $500\,hPa$ is a result of the strong variation of the measured values which are located below the fit as the humidity is decreased and above the fit as the humidity is increased. This could be an indication of a hysteresis of the sensors caused by sorption behavior of the polymer and not by instability of the polymer under these extreme conditions because there is no drift at the pressure of $10\,hPa$ observable (Fig. 8f) and no aging (see Sect. 3.4.2 and Fig. 7).

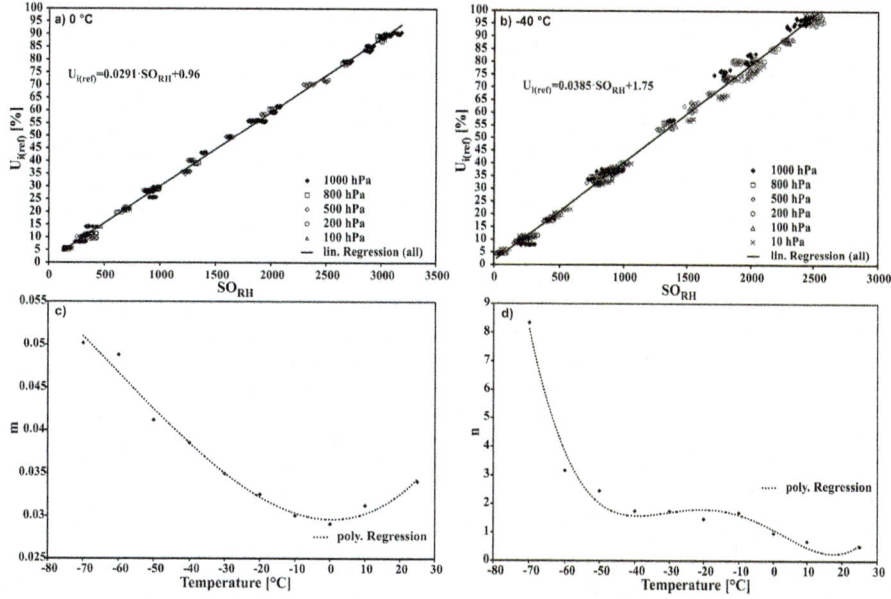

Figure 9. All SO_{RH} values of the nine SHT75 at **(a)** 0 °C and **(b)** −40 °C; **(c)** slope m and **(d)** offset n of the linear regression fit for all temperatures.

In general, the measurements showed a decrease in the equilibration times of the system with decreasing pressure and simultaneous increase of the sensitivity of the SHT75.

3.4.4 Temperature dependency of the SHT75

The SHT75 showed a strong temperature dependency of the SO_{RH} values which was also observed by the manufacturer (Sensirion AG, 2011). We perform a linear fit (see Fig. 9a, b)

$$U_{w,i(ref)} = mSO_{RH} + n \tag{6}$$

for the different temperatures under investigation (Fig. 9c, d). A polynomial fit performed for the data in Fig. 9c and d gives

$$m = 3.74557 \times 10^{-8}t^3 + 6.894 \times 10^{-6}t^2$$
$$- 6.42783 \times 10^{-6}t + 0.03 \tag{7}$$

$$n = 1.11357 \times 10^{-6}t^4 + 6.23536 \times 10^{-5}t^3$$
$$- 5.59115 \times 10^{-4}t^2 - 0.06308t + 1.09. \tag{8}$$

Figure 10 shows the $U_{w,i(SHT75/Pt100)}$ values calculated from Eqs. (6)–(8) based on all SO_{RH} values, measured at the temperatures and pressures listed in Table 2. In the calculation, the temperature values of the Pt100 glued onto the SHT75 (see Table 1, column 3) were used. The expanded uncertainty of $U_{w,i(SHT75/Pt100)}$ and $U_{w,i(ref)}$ is $U_{99} = 7 \%$ (grey area in Fig. 10). The statistical error analysis is based on Hässelbarth (2004) for a linear calibration of type A (GUM, 2008). The values outside the area marked in grey are mainly caused by the large scatter of the measured humidity values at −70 °C (see Fig. 8e).

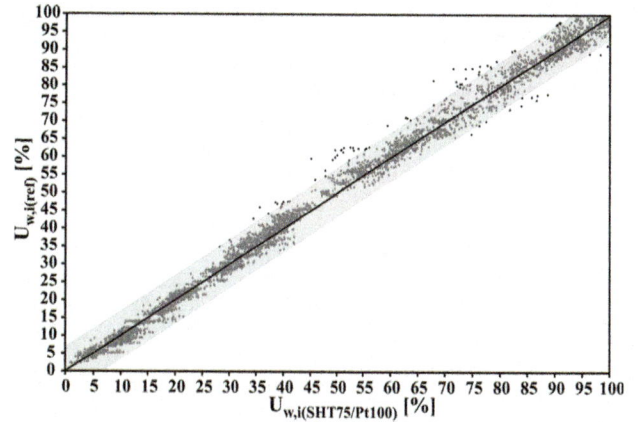

Figure 10. All measurements at all temperatures; the area marked in grey is the confidence interval (±7 %) of $U_{w,i(SHT75/Pt100)}$ calculated with Eqs. (6)–(9) and $U_{w,i(ref)}$ calculated with Eq. (5).

For fits with lower uncertainties, we recommend to calibrate the sensors for individual temperatures and/or pressures.

4 Conclusions

Pt100 and SHT75-sensors work reliably in the operating range given by manufacturer and the SHT75 even outside. The temperature measurement with SHT75 leads to significantly stronger variation of the measured values than with the Pt100 (Sect. 3.3). The comparison of the SHT75 humidity values with the reference humidity of the dew point hygrometer shows a difference to the manufacturers fit at relative humidities below 20 % (see Sect. 3.4.1). The reason for this deviation remains unclear. The SHT75 shows a strong temperature dependency (Sect. 3.4.4) and also a pressure dependency below $-10\,°C$ (Sect. 3.4.3). No aging effects were observed (Sect. 3.4.2). A characteristic curve for the Pt100–SHT75 combination (Sect. 3.1.3) was calculated with an U_{99} of 7 % of the measured values (Sect. 3.4.4).

The experiment has demonstrated that it is possible to perform reliable humidity measurements at low vacuum and temperatures down to $-70\,°C$ using off-the-shelf sensors. The use of additional Pt100 instead of the SHT75 internal band-gap sensor is recommended for the reduction of measurement uncertainties. Moreover, characteristic curves should be recorded for each pressure, especially at temperatures $< -20\,°C$ (Sects. 3.4.3 and 3.4.4).

Further measurements of cross-sensitivity to other gases (e.g., CO_2), cf. Koncz et al. (2010), are necessary in order to validate these sensors for humidity measurements under extreme conditions such as a simulated Martian atmosphere.

Acknowledgements. Funding from German Federal Ministry of Economics and Technology – BMWi (project no. SF11021A) is gratefully acknowledged. I would like to thank David Wolter and Hendrik Hansen-Goos for many valuable suggestions and corrections to this manuscript.

Edited by: R. Maeda
Reviewed by: two anonymous referees

References

Bögel, W.: Neue Näherungsformeln für den Sättigungsdampfdruck des Wassersdampfes und für die in der Meteorologie gebräuchlichen Luftfeuchte-Parameter, 1977.

ECCOBOND 286: Data sheet: ECCOBOND® 286 A/B Easy Mix Ratio, General Purpose Epoxy Adhesive, 2013.

GUM: Bureau international des poids et mesures, Organització Internacional per a la Normalització, International Electrotechnical Commission, International Organization of Legal Metrology, 2008, Guide to the expression of uncertainty in measurement, International Organization for Standardization, Genève, 2008.

Hässelbarth, W.: BAM-Leitfaden zur Ermittlung von Messunsicherheiten bei quantitativen Prüfergebnissen, 2004.

Hudson, S. R., Town, M. S., Walden, V. P., and Warren, S. G.: Temperature, Humidity, and Pressure Response of Radiosondes at Low Temperatures, J. Atmos. Ocean. Tech., 21, 825–836, doi:10.1175/1520-0426(2004)021<0825:THAPRO>2.0.CO;2, 2004.

IST AG: Platinum Temperature Sensors: 6W – Product Series Temperature Range: $-200\,°C + 600\,°C$, 2012.

Koncz, A.: Entwicklung und Schaffung eines in-situ Feuchtemessgerätes für den Mars im Zusammenhang mit der ESA Marsmission ExoMars, Universität Stuttgart, Stuttgart, 2012.

Koncz, A., Lorek, A., and Wernecke, R.: Characterisation of capacitive humidity sensors under Martian pressure and temperatures down to $-120\,°C$, in: Proceedings, MFPA Weimar, pp. 248–254, 2010.

Lorek, A. and Koncz, A.: Simulation and Measurement of Extraterrestrial Conditions for Experiments on Habitability with Respect to Mars, in: Habitability of Other Planets and Satellites, edited by: de Vera, J.-P. and Seckbach, J., Springer Netherlands, Dordrecht, 145–162, 2013.

Lorek, A., Koncz, A., and Wernecke, R.: Development of a gas flow independent coulometric trace humidity sensor for aerospace and industry, in: MFPA, MFPA Weimar, 289–296, 2010.

Miloshevich, L. M., Vömel, H., Paukkunen, A., Heymsfield, A. J., and Oltmans, S. J.: Characterization and Correction of Relative Humidity Measurements from Vaisala RS80-A Radiosondes at Cold Temperatures, J. Atmos. Ocean. Tech., 18, 135–156, doi:10.1175/1520-0426(2001)018<0135:CACORH>2.0.CO;2, 2001.

Sensirion AG: Datasheet SHT7x (SHT71, SHT75) Humidity and Temperature Sensor IC, 2011.

VDI/VDE 3514 Part 1: VDI/VDE 3514 Part 1: Measurement of humidity – Characteristics and symbols, 2007.

WMO: Technical Regulations Basic Documents No. 2 Volume I – General Meteorological Standards and Recommended Practices, World Meteorological Organization, Geneva, Switzerland, 2012.

Zent, A. P., Hecht, M. H., Cobos, D. R., Wood, S. E., Hudson, T. L., Milkovich, S. M., DeFlores, L. P., and Mellon, M. T.: Initial results from the thermal and electrical conductivity probe (TECP) on Phoenix, J. Geophys. Res., 115, 1–23, doi:10.1029/2009JE003420, 2010.

9

Electrophoretic deposition of Au NPs on CNT networks for sensitive NO$_2$ detection

E. Dilonardo[1], M. Penza[2], M. Alvisi[2], C. Di Franco[3], D. Suriano[2], R. Rossi[2], F. Palmisano[1], L. Torsi[1], and N. Cioffi[1]

[1]Department of Chemistry, Università degli Studi di Bari Aldo Moro, Bari, Italy
[2]ENEA, Italian National Agency for New Technologies, Energy and Sustainable Economic Development, Technical Unit for Materials Technologies – Brindisi Research Center, Brindisi, Italy
[3]CNR-IFN Bari, Bari, Italy

Correspondence to: E. Dilonardo (elena.dilonardo@uniba.it) and N. Cioffi (nicola.cioffi@uniba.it).

Abstract. In the present study, Au-surfactant core-shell colloidal nanoparticles (NPs) with controlled dimension and composition were synthesized by sacrificial anode electrolysis. Transmission electron microscopy (TEM) revealed that Au NPs core diameter is between 8 and 12 nm, as a function of the electrosynthesis conditions. Moreover, surface spectroscopic characterization by X-ray photoelectron spectroscopy (XPS) analysis confirmed the presence of nanosized gold phase. Controlled amounts of Au NPs were then deposited electrophoretically on carbon nanotube (CNT) networked films. The resulting hybrid materials were morphologically and chemically characterized using TEM, SEM (scanning electron microscopy) and XPS analyses, which revealed the presence of nanoscale gold, and its successful deposition on CNTs. Au NP/CNT networked films were tested as active layers in a two-pole resistive NO$_2$ sensor for sub-ppm detection in the temperature range of 100–200 °C. Au NP/CNT exhibited a p-type response with a decrease in the electrical resistance upon exposure to oxidizing NO$_2$ gas and an increase in resistance upon exposure to reducing gases (e.g. NH$_3$). It was also demonstrated that the sensitivity of the Au NP/CNT-based sensors depends on Au loading; therefore, the impact of the Au loading on gas sensing performance was investigated as a function of the working temperature, gas concentration and interfering gases.

1 Introduction

In recent years the use of carbon nanotubes (CNTs) has attracted great interest in gas sensing applications because of their reduced dimensionality, which means a high surface-to-volume ratio, together with an outstanding gas adsorption capability, and lower operating temperatures compared to the conventional metal oxide-based device; therefore, these properties make CNTs ideal candidates for environmental sensing applications. Presently, continued progress in CNT-based sensor development for gas detection has been achieved (Zhang et al., 2008; Bondavalli et al., 2009; Penza et al., 2014).

Semiconducting CNTs generally have a typical p-type electrical behaviour under specific ambient conditions (Martel et al., 1998): in the presence of adsorbing oxidizing gases, their resistance decreases; instead, the resistance variation is in the opposite direction in the presence of reducing gas molecules (Kong, 2000). The possible sensing mechanisms include electrostatic gating (Kong, 2000), interaction with pre-adsorbed oxygen species (Collins et al., 2000), charge transfer from adsorbed gas species to carbon nanotubes (Chang et al., 2001; Zhao et al., 2002), and alteration of the electrode work function which leads to a change in the carrier mobility due to formation or removal of the Schottky barrier (Peng et al., 2009).

Although significant progress has been made in understanding the sensing mechanisms of pristine CNTs towards gas molecules, the operation/sensing mode still remains ambiguous (Fan et al., 2005).

The need for air-quality monitoring necessitates the development of highly sensitive sensors that are selective for the detection of individual pollutant gases, especially NO_2 which is a very toxic air pollutant to be detected at sub-ppm level with high sensitivity and selectivity.

The sensing response of gas sensors based on unmodified CNTs is weak and scarcely selective since the ideal carbon hexagonal network is held together by strong sp^2 bonds characterized by a low chemical reactivity with the molecular environment (Peng and Cho, 2003). Consequently, the functionalization of the CNT sidewalls is mandatory to improve both the sensitivity and the selectivity of the CNT-based gas sensors.

In general, molecular sensing requires strong interactions between sensor material and target molecules; this is also the case for nanotubes. It has been found that nanotubes are scarcely sensitive to many types of molecules (e.g. hydrogen and carbon monoxide), indicating an apparent lack of specific interactions between nanotubes and these molecules; therefore, nanotube sensors with molecular selectivity can be obtained through rational chemical and/or physical modification of the nanotube's surface, involving simple deposition of functional materials on the nanotubes (Kong et al., 2001).

Moreover, it has been reported that the sensitivity of nanotube gas sensors could be enhanced through functionalization and defect generation, because single-walled carbon nanotube (SWNT) defect sites of various nature, such as topological, rehybridization, incomplete bonding defects and doping with elements other than carbon (Charlier, 2002), are more reactive than the pristine sp^2-bonded lattice. Appropriate deposition procedures can selectively decorate CNT defect sites with catalytic nanophases (Fan et al., 2005). As reported in Robinson et al. (2006), CNT defect sites play an important role in the electrical response for a broad spectrum of chemical vapours, and the controlled introduction of defects can be used to increase the sensitivity and chemical selectivity of both the conductance and capacitance responses. The defects form low-energy adsorption sites that also serve as nucleation sites for analyte condensation at high vapour concentrations; moreover, the chemical sensitivity of SWNTs can be increased significantly by introducing a controlled low density of defects along the nanotube sidewall. In addition, experimental studies (Lv et al., 2010) showed that defects of CNTs created nucleation sites for metal nanoparticles. In fact, the structural defects, such as topological defects and vacancies in CNTs, always existed in most of the CNTs. Studies (Kim et al., 2007; Pannopard et al., 2009) have indicated that the transition metal has a rich d-electron and empty orbit, and the small gas molecule can bond strongly to the metal when adsorbed on the surface. As previously reported (Kim et al., 2007; Zhang et al., 2014), when metal nanoparticles are deposited on the perfect surface of a nanotube or away from the surface defects, the interaction between the metal and the nanotube is weak; on the other hand, when the metal is adsorbed on the point defect site of the nanotube surface, the structure becomes stable.

Covalent and non-covalent methods have been employed to functionalize CNTs with various materials including polymers (Salavagione et al., 2014), metal oxides (Zhang et al., 2013), metals (Penza et al., 2014) and organometallic complex (Brunet et al., 2012). In particular, the functionalization with metal nanoparticles (NPs) can lead to highly sensitive and selective gas sensors thanks to the extraordinary catalytic properties of the metal NPs (Feldheim and Foss, 2002), as already suggested by several experimental (Khalap et al., 2010), theoretical (Pannopard et al., 2009) and combined (Kauffman et al., 2010) works.

Leghrib et al. (2010) reported on multi-walled carbon nanotubes (MWCNTs) decorated with different metal NPs, e.g. Rh, Pd, Au and Ni, to tailor gas recognition of benzene vapours at concentrations lower than 50 ppb with high sensitivity and selectivity at room temperature. Au- or Ag-decorated CNT films have been used to detect NO_2 at up to 500 ppb at room temperature (Espinosa et al., 2007). Star et al. (2006) and also Lu et al. (2004) decorated single-walled CNTs with Pt, Pd, Au or Rh for detecting a large variety of gases such as CO, NO_2, CH_4, H_2S, NH_3 and H_2, by discriminating from interfering gases. Moreover, recent works (Colindres et al., 2014; Doroodmand et al., 2013) concern the functionalization of CNTs with Pd and the electrodeposition of FeOOH nanostructures on CNTs for O_3 gas sensing, with a detection limit in the range of ppb and a high selectivity.

Among various metal NPs, Au-functionalized CNT-based resistive gas sensors show the best sensitivity and selectivity towards NO_2 gas detection (Penza et al., 2007).

Several methods have been used to functionalize CNTs by metal NPs, such as thermal evaporation (Scarselli et al., 2012), sputtering (Penza et al., 2011) and electrochemical deposition (Mubeen et al., 2011).

In this study, we report on the functionalization of the CNT-based gas sensor by Au NPs with defined dimension and controlled loading using an electrophoretic deposition, as an easier process for the mass production of metal-functionalized CNT-based gas sensor devices. The effects of the operating sensor temperature (range 100–200 °C) and of the metal loading on sensor performance towards NO_2 detection were reported. Moreover, the Au NP-functionalized CNT-based gas sensor response in the presence of interfering gases (e.g. NO, NH_3) was also evaluated.

2   Experimental

2.1   Electrochemical synthesis of Au NPs

The core-shell gold nanoparticle (Au NP) solution, used to metal directly decorate CNT-based gas sensors, was produced by the sacrificial anode electrolysis (SAE), in the presence of 0.05 M tetraoctylammonium chloride (TOAC), with the simultaneous function of electrolyte and Au NP

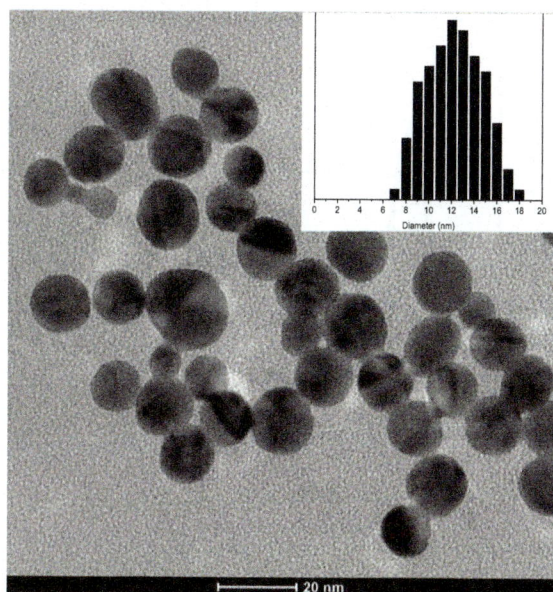

Figure 1. TEM image and dimensional dispersion histogram (inset) of electrochemically synthesized core-shell Au NPs/TOAC.

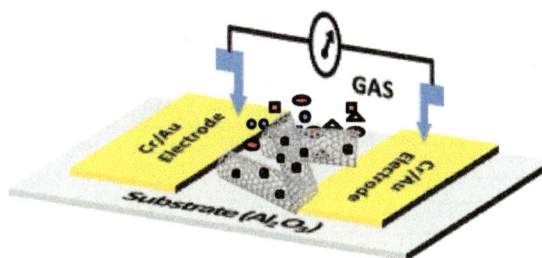

Figure 2. Scheme of Au-modified CNT-based gas sensor device.

stabilizer, dissolved in an anhydrous tetrahydrofuran (THF) and acetonitrile (ACN) solution, mixed in 3 : 1 ratio (Cioffi et al., 2011). A three-electrode cell was used, equipped with an Ag / AgNO$_3$ (0.1 M in ACN) reference electrode, a gold anode and a platinum cathode. During the process, the cell was kept under a nitrogen atmosphere.

Since in these systems the NPs shell thickness is roughly proportional to the chain length of the surfactant (Reetz and Helbig, 1994; Reetz et al., 1995), while the core size is mainly influenced by the electrochemical parameters, Au NPs with a shell thickness of about 1.2 nm were synthesized using TOAC. The working potential was set at +1 V corrosion voltages and the electrolysis charge at 300 mC. As discussed in details in previous works (Cioffi et al., 2011), and reported in Fig. 1, the electrochemically synthesized Au NPs have a uniform dispersion with diameter of 12 nm. Moreover, since the Au NPs are stabilized by the tetraoctylammonium surfactant with a positive charge, as revealed in previous work of our group (Cioffi et al., 2000; Ieva et al., 2008), the net surface charge of metal NPs is positive and it was used for the electrophoretic deposition to surface decorate CNT devices.

2.2 Growth of carbon nanotubes and sensor fabrication

MWCNT layers were prepared by a radio-frequency plasma-enhanced chemical vapour deposition (RF-PECVD) system, at a reasonably low growth temperature of 450 °C, onto low-cost alumina substrates (10 mm width × 10 mm length × 0.6 mm thickness). The experimental conditions used, as well as details about MWCNT composition,

structure, and abundance of amorphous matter, are reported elsewhere (Penza et al., 2008).

Modification of CNT-based devices was performed using an electrophoretic process consisting of a three-cell electrode in which the anode was a Pt foil, the cathode was the CNT device and Ag / AgNO$_3$ was the reference electrode. The distance between the anode and the cathode was set at 2 cm. The electrolytic solution was the Au NP colloidal solution. The deposition process is a cathodic process in which an applied working potential a little bit more negative than the open circuit potential was used to induce the migration of the colloidal Au NPs towards the negative cathode (CNT device), the reduction of the residual gold ions in the positive shell followed by its removal from the metal NPs surface, with the final result of the CNT surface decorated with Au NPs. The excess of the surfactant from the functionalized device was completely removed by washing it with ACN for three times.

The process was performed using two different deposition times, 90 and 600 s, to deposit different metal loadings. The scheme of a Au/CNT resistive gas sensor device is reported in Fig. 2.

2.3 Material characterization set-up

Surface chemical characterization was performed by means of a Thermo VG Theta Probe XPS spectrometer equipped with a μ spot monochromatic Al Kα source. Both survey and high-resolution spectra were acquired in fixed analyser transmission mode with pass energies of 150 and 100 eV, respectively.

The dimension of synthesized Au NPs was evaluated using a FEI TECNAI T12 transmission electron microscopy (TEM) instrument.

The morphology of pristine and Au-decorated CNTs was analysed by scanning electron microscopy (SEM), using a Field Emission Zeiss ΣIGMA instrument at 5–10 KV, 10 μm aperture, directly on the sensor devices.

2.4 Gas sensing measurements

The experimental set-up used for gas sensing measurements is reported elsewhere (Penza et al., 2010). The CNT-based gas sensors were placed in a sealed stainless test cell (500 mL

Table 1. XPS surface chemical composition of pristine and Au-functionalized CNTs.

Sample	C (%)	Au (%)	O (%)
Pristine CNTs	95.0 ± 0.5	–	5.0 ± 0.5
Au NPs/CNTs t: 90 s	94.4 ± 0.5	0.3 ± 0.2	5.3 ± 0.5
Au NPs/CNTs t: 600 s	91.2 ± 0.5	1.1 ± 0.2	7.8 ± 0.5

volume) for gas exposure measurements. They were in thermal contact with a home-made heater sink, powered by a DC power supply system (Agilent 6644A, 0–60 V/0–3.5 A), to control the desired set-point operating temperature. The DC electrical conductance of the metal oxide-based gas sensors was measured by the volt-amperometric technique in the two-pole format by a multi-metre (Agilent, 34401A). The chemiresistors in an array of four sensing elements were scanned with automated control by a multiplexing switch system (Keithley, 7001) equipped by a low-current scanner card (Keithley, 7158). All data were acquired and stored for further analysis in a PC-based workstation, equipped with software compiled in Agilent-VEE.

Dry air was used as the reference gas and diluting gas to air condition the sensors. The gas flow rate was controlled by distinct mass flow metres (MFC) with different full scales and controlled by GMIX software. The total flow rate per exposure was kept constant at 1000 sccm. The gas sensing experiments were performed by measuring the resistance change of deposited films under the exposure to the target gas at different concentrations. The sensing experiments were conducted at three different sensor temperatures (100, 150 and 200 °C), to evaluate the temperature effect on gas sensing performance.

The gas sensing cycle consisted of a period (at least 60 min) of stabilization of the sensor signals with dry air (analysis certificate on the gas bottle given by the company providing gases, AirLiquide) flowing, an exposure time of 10 min to various targeted gas concentrations at increasing steps and finally a recovery time (at least 60 min) to restore the sensor signals with dry air flowing to clean the test cell and sensor surface.

The sensor response to a given gas concentration was defined as the resistance change, ΔR, where ΔR is the change in resistance between the values of steady state of the electrical resistance, R_f and R_i, of the sensor upon a target gas and in air, respectively. The mean gas sensitivity ($\%\,\mathrm{ppm}^{-1}$) (S_m) is defined as weighted mean of the ratio between percentage relative resistance change (%) over gas concentration unit (ppm); it can be calculated with Eq. (1):

$$S_\mathrm{m} = \frac{1}{n} \sum_{j=1}^{n} \frac{|\frac{\Delta R}{R_i}|_j}{c_j} \; (\%\,\mathrm{ppm}^{-1}), \tag{1}$$

where c_j is a defined gas concentration which corresponds to the $[\Delta R / R_i]_j$ response.

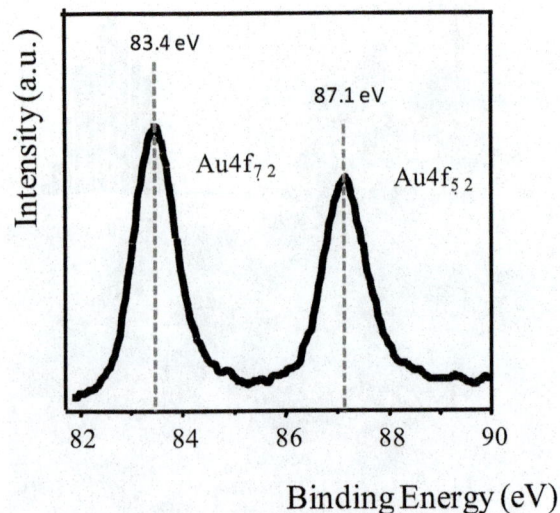

Figure 3. XPS Au4f spectra of Au-decorated CNTs.

3 Results

3.1 Material characterization

The surface of pristine CNTs and Au-decorated CNTs at different process times was chemically characterized by XPS. The detailed chemical quantification of the structures showed that by increasing the deposition time, as expected, the gold loading on CNTs increases, as reported in Table 1. Therefore, by controlling the process time it is possible to tune the deposited metal concentration.

In Fig. 3 the XPS Au4f spectrum of electrophoretically Au-decorated CNTs is reported. It is composed by a single doublet, attributed to Au in the elemental oxidation state. The position of the Au4f$_{7/2}$ peak at 83.7 eV $\pm$ 0.2 eV was significantly lower than that expected for bulk metallic Au at 84.0 eV (Moulder et al., 1992). This phenomenon is reasonably due to initial state size effects, highlighted for small gold particles (Radnik et al., 2002; Cioffi et at., 2011). Therefore, the effectiveness functionalization of CNTs with nano-size gold particles has been demonstrated. To corroborate these results, morphological analysis was performed by SEM.

Figure 4 displays SEM images of un-functionalized and surface-modified CNT films with of 0.3 and 1.1 at. % of Au loading. The morphological analysis of CNTs reveals a pronounced tubular structure, consisting of tangled nets of densely distributed chains.

Past Raman spectroscopy on pristine CNTs revealed the presence of some amount of amorphous carbon in the CNTs network (Penza et al., 2008, 2010); specifically, the high intensity ratio ($I(D)/I(G)$) of D peak, associated with the disordered graphitic layer, to G peak, due to the tangential modes of the graphitic planes, revealed qualitatively the presence of a large amount of disorder and defects. On the surface-functionalized CNTs, Au NPs effectively

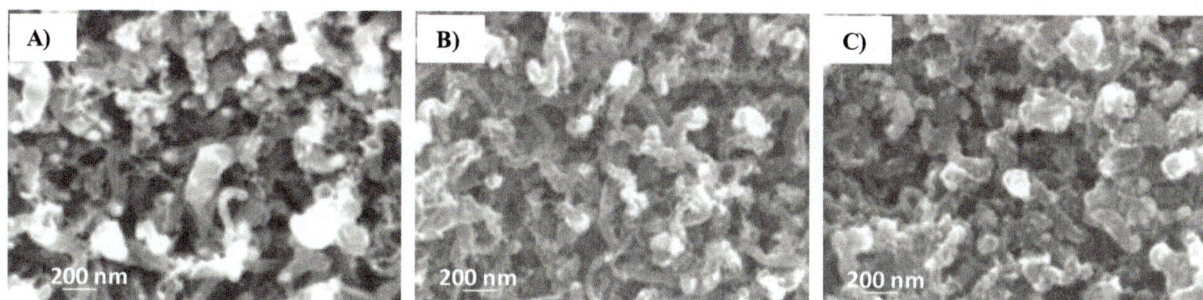

Figure 4. SEM images of (**a**) pristine CNTs, and CNTs with (**b**) 0.3 at. % and (**c**) 1.1 at. % Au loading.

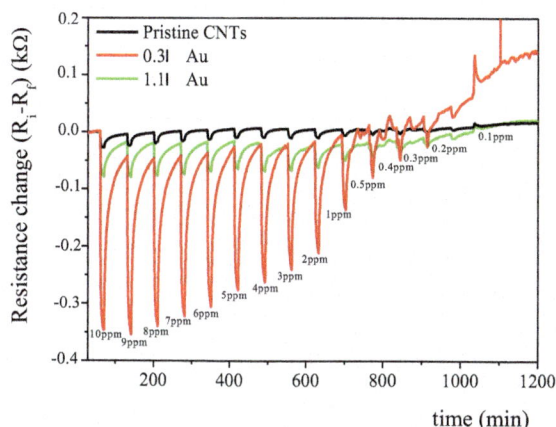

Figure 5. Time response of the NO$_2$ gas chemiresistor based on pristine CNT films, and functionalized with Au-loading of 0.3 and 1.1 at. %, at a working temperature of 150 °C.

decorate MWCNT sidewalls, forming isolated nanoclusters, with a density increase upon increasing deposited Au content. Specifically, a homogeneous distribution of Au NPs is found on CNT sidewalls in the top layers of films; even the presence of Au NPs on CNT sidewalls in the inner layers should not be excluded due to the film porosity. The tuned surface modification enables CNTs with controlled surface metal loading for the fabrication of selective gas sensors. The catalytic covering of nanotubes should strongly affect their properties concerning the gas adsorption and reactivity, hence, the tailoring of gas sensitivity, as discussed in the next paragraph.

3.2   Gas sensing characterization

Figure 5 reports the time response of the CNT chemiresistors, pristine and surface modified with a different Au NPs loading of 0.3 and 1.1 at. %, exposed to 10 min pulses of decreasing spot concentrations of NO$_2$ gas in the low range from 10 to 0.1 ppm. The working sensor temperature was 150 °C. The electrical resistance of Au-loaded and pristine CNT devices decreases rapidly when exposed to NO$_2$ gas; thus the p-type characteristic is maintained also after functionalization of the

CNTs with Au nanoclusters. In particular, the NO$_2$ sensing response for all sensors increases upon increasing the analyte gas concentration, as reported in the calibration curves in Fig. 6a for all gas sensors. As reported in Fig. 6b in which the NO$_2$ concentration range is magnified from 0 up to 1.0 ppm, the trend of the sensor response to NO$_2$ concentration is quite linear up to 1 ppm of NO$_2$; instead, after this value, at higher gas concentrations, the response variation is lower, probably due to the sensor saturation. Moreover, the NO$_2$ response at the operating sensor temperature of 150 °C is higher for Au-modified CNTs compared to the pristine CNT film; in particular, the CNT sensor containing 0.3 at. % of Au shows the better response.

The baseline resistance is not stable for all measured functionalized CNT-based gas sensors, due to the thermal drift induced by operating sensor temperature in CNT material, and probably also to the instability of the catalyst particles onto CNT sidewalls, activated by thermal energy (Penza et al., 2008); moreover, the baseline drift is more evident at low gas concentration because of the lower ratio between signal and noise.

Considering the sensor operating temperature effect on sensing properties of all gas sensor devices at all investigated temperatures in the range 100–200 °C, the mean sensitivity towards NO$_2$ of Au-doped CNT films is always higher than the un-doped, as reported in Fig. 7; specifically, the highest values are obtained for CNTs containing the lower Au loading of 0.3 at. %, especially at lowest tested temperatures. Moreover, for all gas sensor devices the maximum of mean sensitivity is obtained at the intermediate temperature of 150 °C at which both systems, Au NPs and CNTs, are stable, since at higher temperature CNTs could decompose and Au NPs could agglomerate, decreasing their catalytic properties. On the contrary, a lower temperature could not be enough to activate chemically effective interactions at the interfaces between the two structures, probably responsible of the dominant sensing mechanisms. The higher sensor response towards NO$_2$ of CNTs containing the lower loading of Au NPs is probably due to the fact that in this condition the Au NPs agglomeration in nanoclusters of higher dimension is prevented, since the surface density of Au NPs

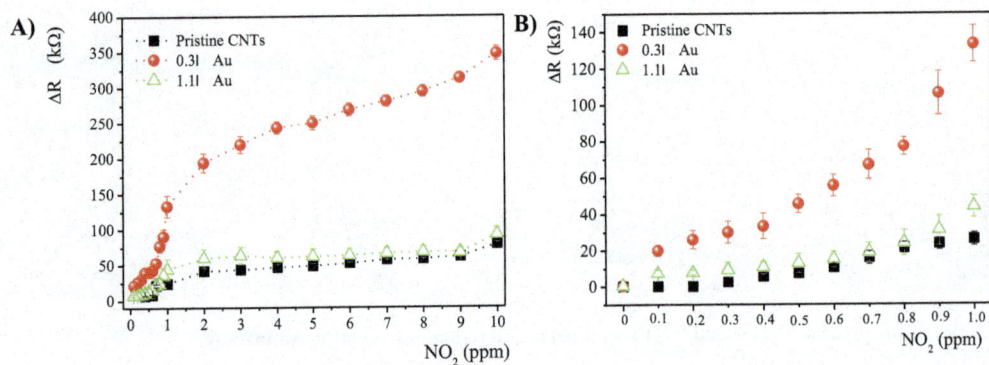

Figure 6. Calibration curves for pristine and Au-decorated CNT-based sensors toward NO_2 gas at 150 °C **(a)** in the concentration range of 0.1–10 ppm and **(b)** the magnification of 0.1–1.0 ppm concentration range.

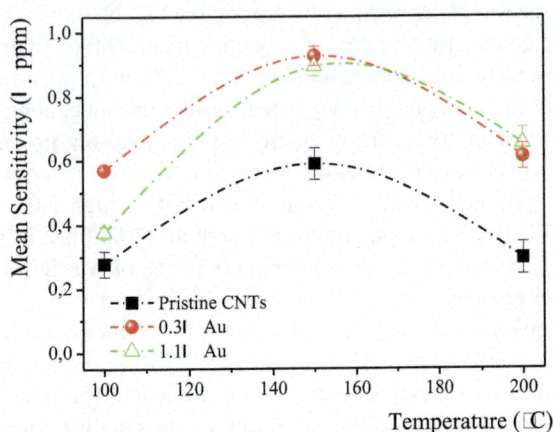

Figure 7. Mean sensitivity of pristine and Au-decorated CNT-based sensors toward NO_2 gas at different sensor operating temperatures in the range 100–200 °C.

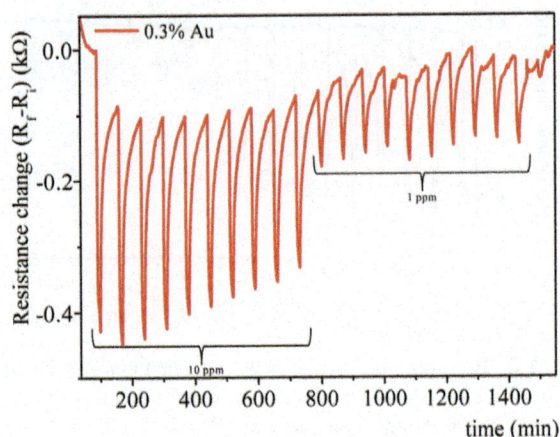

Figure 8. Repeatability test based on time response of functionalized CNTs with Au loading of 0.3 at. % at a working temperature of 150 °C and at 10 and 1 ppm of NO_2, repeating 10 times the exposure to each gas concentration.

on the sidewall of CNT is low, as revealed by SEM images; therefore, a high surface catalytic activity of nanosized Au is guaranteed. Instead, a high Au loading causes a high surface density of the Au NPs that could agglomerate, reducing their catalytic properties. As confirming this assumption, Hvolbaek et al. (2007) have demonstrated by density functional calculation that the fraction of low coordinated Au atoms scales approximately with the catalytic activity, suggesting that atoms on the corners and edges of the Au NPs are the active sites.

Moreover, excellent short-term repeatability of the response to the selected targeted gases has been also measured for the Au-modified CNT-based sensors, as reported in Fig. 8 in which the repeatability test has been evaluated exposing functionalized CNTs with Au-loading of 0.3 at. % at 10 and 1 ppm of NO_2 repeating 10 times the exposure to each gas concentration, at a working temperature of 150 °C.

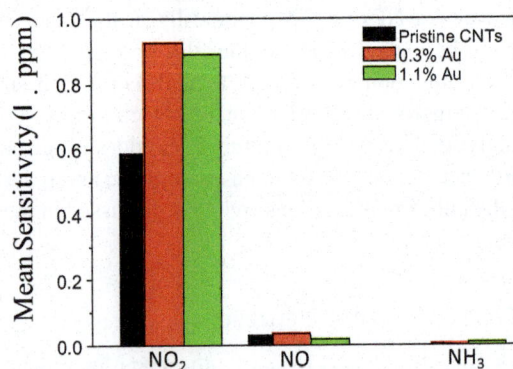

Figure 9. Mean sensitivity of CNT-based sensors towards NO_2, NO and NH_3 gases at a sensor temperature of 150 °C.

The cross-sensitivity of CNT sensors, un-modified and Au loaded, has been measured towards NH_3 and NO gases at a sensor operating temperature of 150 °C and, as reported in Fig. 9, all CNT-based devices are selective towards NO_2 detection with a higher sensitivity at about 1 order of magnitude. Finally, Au-modified CNT gas sensors exhibit a negligible response towards CO, CH_4 and SO_2 gases, as measured repeatedly (not shown).

4 Conclusions

In conclusion, CNT-based sensor devices have been directly decorated with Au NPs by means of an electrophoretic deposition method, for NO_2 gas detection at operating temperature in the range of 100–200 °C. The surface modification of the CNT networked films with size- and loading-controlled Au nanoclusters enhances the NO_2 gas sensitivity up to the detection of sub-ppm level of great interest for selective environmental NO_2 air monitoring. The effects of tailored Au loading onto the CNT surface on NO_2 gas sensitivity depend on nanoclusters density and sensor working temperature. An excellent short-term repeatability of the response to the selected targeted gases has been also measured for the Au-modified CNT-based sensors. A continuous gas monitoring at ppb level of the NO_2 gas has been effectively performed with CNT chemiresistors modified with a 0.3 at. % Au loading, at sensor temperature of 150 °C. The p-type character of the Au-modified CNT sensors has been also confirmed. Moreover, no cross-sensitivity has been revealed for the investigated gases, since all CNT-based devices show a higher selectivity for NO_2 gas.

Finally, the proposed electrochemical functionalization process of the CNTs seems to be easily applicable for low-cost mass production of modified CNT-based gas sensors.

Future work on different metal surface-modifications of the carbon nanotube networked films is planned for specific gas detection in sensor arrays concerning environmental monitoring applications.

Acknowledgements. The authors thank Gennaro Cassano for technical assistance during sensing experiments in ENEA. Authors acknowledge the Ministry of University and Scientific Research (MIUR) of Italy, PON program 2007–2013 for financial support. Finally, the authors are indebted to the COST Action TD1105 EuNetAir for the international networking activities in the field of the sensor materials for air pollution monitoring.

Edited by: A. Lloyd Spetz
Reviewed by: two anonymous referees

References

Bondavalli, P., Legagneux, P., and Pribat, D.: Carbon nanotubes basedtransistors as gas sensors: State of the art and critical review, Sensor Actuat. B-Chem., 140, 304–18, 2009.

Brunet, J., Dubois, M., Pauly, A., Spinelle, L., Ndiaye, A., Guérin, K., Varenne, C., Lauron, B.: An innovative gas sensor system designed from a sensitive organic semiconductor downstream a nanocarbonaceous chemical filter for the selective detection of NO_2 in an environmental context Part I: Development of a nanocarbon filter for the removal of ozone, Sensors Actuat. B-Chem., 173, 659–667, 2012.

Chang, H., Lee, J. D., Lee, S. M., and Lee, Y. H.: Adsorption of NH_3 and NO_2 molecules on carbon nanotubes, Appl. Phys. Lett. 79, 3863–3865, 2001.

Charlier, J.-C.: Defects in Carbon Nanotubes, Acc. Chem. Res., 35, 1063–1069, 2002.

Cioffi, N., Torsi, L., Sabbatini, L., Zambonin, P. G., and Bleve-Zacheo, T.: Electrosynthesis and characterisation of nanostructured palladium-polypyrrole composites, J. Electroanal. Chem., 488, 42–47, 2000.

Cioffi, N., Torsi, L., Losito, I., Sabbatini, L., Zambonin, P. G., and Bleve-Zacheo, T.: Nanostructured palladium–polypyrrole composites electrosynthesised from organic solvents, Electrochim. Acta, 46, 4205–4211, 2001.

Colindres, S. C., Aguir, K., Sodi, F. C., Vargas, L. V., Salazar, J. A. M., and Febles, V. G.: Ozone Sensing Based on Palladium Decorated Carbon Nanotubes, Sensors, 14, 6806–6818, 2014.

Collins, P. G., Bradley, K., Ishigami, M., and Zettl, A.: Extreme Oxygen Sensitivity of Electronic Properties of Carbon Nanotubes, Science, 287, 1801–1804, 2000.

Doroodmand, M. M., Nasresfahani, S., and Sheikhi, M. H.: Fabrication of ozone gas sensor based on FeOOH/single walled carbon nanotube-modified field effect transistor, Int. J. Environ. Anal. Chem., 93, 946–958, 2013.

Espinosa, E. H., Ionescu, R., Bittencourt, C., Felten, A., Erni, R., Van Tenderloo, G., Pireaux, J.-J., and Llobet, E.: Metal-decorated multi-wall carbon nanotubes for low temperature gas sensing, Thin Solid Films, 515, 8322–8327, 2007.

Fan, Y., Goldsmith, B. R. and Collins, P. J.: Identifying and counting point defects in carbon nanotubes, Nature Mater., 4, 906–911, 2005.

Feldheim, D. L. and Foss, C. A. (Eds.): Metal Nanoparticles: Synthesis, Characterization and Applications, New York: Marcel Dekke, 2002.

Hvolbæk, B., Janssens, T. V. W., Clausen, B. S., Falsig, H., Christensen, C. H., and Nørskov, J. K.: Catalytic activity of Au nanoparticles, NanoToday, 2, 14–18, 2007.

Ieva, E., Buchholt, K., Colaianni, L., Cioffi, N., Sabbatini, L., Capitani, G. C., LloydSpetz, A., Kall, P. O., and Torsi, L.: Au Nanoparticles as Gate Material for NOx Field Effect Capacitive Sensors, Sens. Lett., 6, 1–8, 2008.

Kauffman, D. R., Sorescu, D. C., Schofield, D. P., Allen, B. L., Jordan, K. D., and Star, A.: Understanding the Sensor Response of Metal-Decorated Carbon Nanotubes, Nano Lett. 10, 958–963, 2010.

Khalap, V. R., Sheps, T.ì Kane, A. A., and Collins, P. G.: Hydrogen Sensing and Sensitivity of Palladium-Decorated Single- Walled Carbon Nanotubes with Defects, Nano Lett., 10, 896–901, 2010.

Kim, S. J., Park, Y. J., Ra, E. J., Kim, K. K., An, K. H., Lee, Y. H., Choi, J. Y., Park, J. C., Doo, S. K., Park, M. H., and Yang, C. W.: Defect-induced loading of Pt nanoparticles on carbon nanotubes, Appl. Phys. Lett., 90, 023114–023116, 2007.

Kong, J.: Nanotube molecular wires as chemical sensors, Science, 287, 622–625, 2000.

Kong, J., Chapline, M. G., and Dai, H.: Functionalized Carbon Nanotubes for Molecular Hydrogen Sensors, Adv. Mater., 13, 1384–1386, 2001.

Leghrib, R., Felten, A., Demoisson, F., Reniers, F., Pireaux, J.-J., and Llobet, E.: Room-tempeature, selective detection of benzene at trace levels using plasma-treated metal-decorated multiwalled carbon nanotubes, Carbon 48, 3477–3484, 2010.

Lu, Y., Li, J., Han, J., Ng, H.-T., Binder, C., Partridge C., and Meyyappan, M.: Room temperature methane detection using palladium loaded single-walled carbon nanotube sensors, Chem. Phys. Lett., 391, 344–348, 2004.

Lv, J.-A., Cui, J.-H., Li , X.-N., Song, X.-Z., Wang, J.-G., and Dong, M.: The point-defect of carbon nanotubes anchoring Au nanoparticles, Physica E, 42, 1746–1750, 2010.

Martel, R., Schmidt, T., Shea, H. R., and Avouris, P.: Single- and multi-wall carbon nanotube field-effect transistors, Appl. Phys. Lett., 73, 2447–2449, 1998.

Metal Nanoparticles: Synthesis, Characterization and Applications, edited by: Foss, C. J. and Feldheim, D., Marcel Dekker, New York, 2002.

Moulder, J. F., Stickle, W. F., Sobol, P. E., and Bomben, K. D.: Handbook of X-Ray Photoelectron Spectroscopy, Perkin-Elmer Corporation, Eden Praire, Minnesota, 1992.

Mubeen, S., Lim, J.-H., Srirangarajan, A., Mulchandani, A., Deshusses, M. A., and Myung N. V.: Gas sensing mechanism of gold nanoparticles decorated single-walled carbon nanotubes, Electroanalysis, 23, 2687–92, 2011.

Pannopard, P., Khongpracha, P., Probst, M., and Limtrakul, J.: Gas sensing properties of platinum derivatives of singlewalled carbon nanotubes: a DFT analysis, J. Mol. Graph. Modelling, 28, 62–69, 2009.

Peng, N., Zhang, Q., Chow, C. L., Tan, O. K., and Marzari, N.: Sensing mechanisms for carbon nanotube based NH_3 gas detection, Nano. Lett., 9, 1626–1630, 2009.

Peng, S. and Cho, K.: Ab Initio Study of Doped Carbon Nanotube Sensors, Nano Lett., 3, 513–517, 2003.

Penza, M., Cassano, G., Rossi, R., Alvisi, M., Rizzo, A., Signore, M. A., Dikonimos, Th., Serra, E., and Giorgi, R.: Enhancement of sensitivity in gas chemiresistors based on carbon nanotube surface functionalized with noble metal (Au, Pt) nanoclusters, Appl. Phys. Lett., 90, 173123–173125, 2007.

Penza, M., Rossi, R., Alvisi, M., Cassano, G., Signore, M. A., Serra, E., and Giorgi, R.: Surface Modification of Carbon Nanotube Networked Films with Au Nanoclusters for Enhanced NO2 Gas Sensing Applications, J. Sensors, 2008, 107057, doi:10.1155/2008/107057, 2008.

Penza, M., Rossi, R., Alvisi, M., and Serra, E., Metal-modified and vertically aligned carbon nanotube sensors array for landfill gas monitoring applications, Nanotechnology, 21, 105501–1055014, 2010.

Penza, M., Rossi, R., Alvisi, M., Suriano, D., and Serra, E.: Pt-modified carbon nanotube networked layers for enhanced gas microsensors, Thin Solid Films, 520, 959–65, 2011.

Penza, M., Martin, P., and Yeow, J.: Carbon Nanotube Gas Sensors, in: Gas Sensing Fundamentals, edited by: Kohl, C.-D. and Wagner, T., Springer Series on Chemical Sensors and Biosensors, 15, 109–114, 2014.

Radnik, J., Mohr, C., and Claus, P.: The origin of binding energy shifts of core levels of supported gold nanoparticles and dependence of pretreatment and material synthesis, Phys. Chem. Chem. Phys., 5, 172–177, 2003.

Reetz, M. T. and Helbig, W.: Size-Selective Synthesis of Nanostructured Transition Metal Clusters, J. Am. Chem. Soc., 116, 7401–7402, 1994.

Reetz, M. T., Helbig, W., Quaiser S.A., Stimming U., Breuer, N., and Vogel R.: Visualization of surfactants on nanostructured palladium clusters by a combination of STM and high-resolution TEM, Science, 267, 367–369, doi:10.1126/science.267.5196.367, 1995.

Robinson, J. A., Snow, E. S., Badescu, S. C, Reinecke, T. L., and Perkins, F. K.: Role of Defects in Single-Walled Carbon Nanotube Chemical Sensors, Nano Lett., 6, 1747–1751, 2006.

Salavagione, H. J., Dìez-Pascual, A. M., Lazaro, E., Vera, S., and Gomez-Fatou, M. A.: Chemical sensors based on polymer composites with carbon nanotubes and graphene: the role of the polymer, J. Mater. Chem. A, 2, 14289–14328, doi:10.1039/C4TA02159B, 2014.

Scarselli, M., Camilli, L., Castrucci, P., Nanni, F., Del Gobbo, S., Gautron, E., Lefrant, S., and De Crescenzi, M.: In situ formation of noble metal nanoparticles on multiwalled carbon nanotubes and its implication in metal–nanotube interactions, Carbon, 50, 875–84, 2012.

Star, A., Joshi, V., Skarupo, S., Thomas, D., and Gabriel, J.-C. P.: Gas sensor array based on metal-decorated carbon nanotubes, J. Phys. Chem. B, 110, 21014–21020, 2006.

Zhang, T., Mubeen, S., Myung, N. V., and Deshusses, M. A.: Recent progress in carbon nanotube-based gas sensors, Nanotechnology, 19, 332001–332014, 2008.

Zhang, X., Dai, Z., Chen, Q., and Tang, J.: A DFT study of SO_2 and H_2S gas adsorption on Au-doped single-walled carbon nanotubes, Phys. Scr., 89, 065803–065809, 2014.

Zhang, Y., Cui, S., Chang, J., Ocola, L. E., and Chen, J.: Highly sensitive room temperature carbon monoxide detection using SnO_2 nanoparticle-decorated semiconducting single-walled carbon nanotubes, Nanotechnology, 24, 025503–025512, 2013.

Zhao, J., Buldum, A., Han, J., and Lu, J. P.: Gas molecule adsorption in carbon nanotubes and nanotube bundles, Nanotechnology, 13, 195–200, 2002.

10

Devices based on series-connected Schottky junctions and β-Ga$_2$O$_3$/SiC heterojunctions characterized as hydrogen sensors

S. Nakagomi, K. Yokoyama , and Y. Kokubun

Faculty of Science and Engineering, Ishinomaki Senshu University Minamisakai, Ishinomaki, Miyagi 986-8580, Japan

Correspondence to: S. Nakagomi (nakagomi@isenshu-u.ac.jp)

Abstract. Field-effect hydrogen gas sensor devices were fabricated with the structure of a series connection between Schottky junctions and β-Ga$_2$O$_3$/6H-SiC heterojunctions. β-Ga$_2$O$_3$ thin films were deposited on n-type and p-type 6H-SiC substrates by gallium evaporation in oxygen plasma. These devices have rectifying properties and were characterized as hydrogen sensors by a Pt electrode. The hydrogen-sensing properties of both devices were measured in the range of 300–500 °C. The Pt/Ga$_2$O$_3$/n-SiC device revealed hydrogen-sensing properties as conventional Schottky diode-type devices. The forward current of the Pt/Ga$_2$O$_3$/p-SiC device was significantly increased under exposure to hydrogen. The behaviors of hydrogen sensing of the devices were explained using band diagrams of the Pt/Ga$_2$O$_3$/SiC structure biased in the forward and reverse directions.

1 Introduction

Hydrogen gas has been expected to be employed as a clean energy source for various applications. For example, fuel cells will be used to power vehicles and as domestic and industrial energy systems in the near future. However, because the explosion limit of hydrogen gas is 4 % in air, quick detection of low hydrogen concentration is required to avoid a dangerous situation. Therefore, small, inexpensive hydrogen sensors would be required to ensure safety standards.

Many researchers have studied several types of hydrogen sensor devices with sintered metal oxide thin films formed by means of various methods that are based on the field effect.

β-Ga$_2$O$_3$ is a semiconductor material with a wide band gap of 4.9 eV and has been in focus as a new material for solar-blind deep UV detectors (Kokubun et al., 2007; Suzuki et al., 2009; Nakagomi et al., 2013b) and power devices (Higashiwaki et al., 2012).

Many Ga$_2$O$_3$-based gas sensors have also been investigated. Fleischer and Meixner (1991) studied an oxygen sensor with a Ga$_2$O$_3$ film prepared by sputtering that could be applied at temperatures over 500 °C. The sensor was also sensitive to reducing gases such as hydrogen (Fleischer and Meixner, 1992; Fleischer et al., 1992). Ogita et al. (1999) also fabricated an oxygen gas sensor based on Ga$_2$O$_3$ prepared by sputtering that exhibited dependence on the oxygen pressure used during the sputtering process. Trinchi et al. (2004) proposed a sensor device with a Pt/Ga$_2$O$_3$/SiC structure for the first time. A Ga$_2$O$_3$ layer was formed on an n-type SiC substrate using a sol-gel method. They demonstrated hydrogen-sensing properties due to a change in the Schottky barrier height. However, the sensor characteristics were measured only under forward bias conditions, and the current–voltage (I–V) characteristics were not examined in detail. The authors have inferred from the I–V characteristics presented by Trinchi et al. (2004) that the Ga$_2$O$_3$ layer acts as an electrical resistance layer because the change in the I–V characteristics caused by a change in the atmosphere is not just a simple shift in the turn-on voltage but in fact includes a change in slope. This indicates a change in the resistance component of the Ga$_2$O$_3$ layer.

We have previously reported the first hydrogen sensor with a Schottky diode structure based on a Ga$_2$O$_3$ single crystal (Nakagomi et al., 2011a, b). In addition, we have also reported field-effect hydrogen gas sensor devices based on β-Ga$_2$O$_3$ thin film formed on a sapphire substrate (Nakagomi et al., 2013a).

6H-SiC and 4H-SiC are well-known semiconductor materials for high-power applications with wide band gaps of 3.02 and 3.26 eV, respectively. Gas sensor devices based on SiC that can be operated at higher temperatures have also been reported (Spetz et al., 2004).

In this study, field-effect hydrogen gas sensor devices with a series connection between Schottky junctions and β-Ga$_2$O$_3$/6H-SiC heterojunctions were fabricated. β-Ga$_2$O$_3$ thin films were deposited on n-type and p-type 6H-SiC substrates, and the hydrogen sensitivities of the Pt/Ga$_2$O$_3$/n-type SiC and Pt/Ga$_2$O$_3$/p-type SiC structures were evaluated in detail. Both devices exhibited rectifying and hydrogen-sensing properties dependent on the bias conditions.

2 Experimental

2.1 Preparation of β-Ga$_2$O$_3$ thin film

The β-Ga$_2$O$_3$ layer was prepared using gallium evaporation in oxygen plasma. After the SiC substrate was heated to 800 °C, 4 mL min^{-1} of O$_2$ was supplied into the vacuum chamber to form oxygen plasma at 100 W RF power. The pressure of the chamber at that time was about 5×10^{-4} Torr. Gallium was thermally evaporated using a crucible in the chamber. Then the β-Ga$_2$O$_3$ thin film was formed on the SiC substrate. The preparation and crystal orientation of β-Ga$_2$O$_3$ layers on sapphire substrates with this method have been previously reported (Nakagomi and Kokubun, 2013). We prepared the Ga$_2$O$_3$ layer on a (001) c-plane 6H-SiC substrate using the same method as in the case of sapphire substrates and then evaluated the Ga$_2$O$_3$ layer. From the measurement of the X-ray diffraction pattern of the Ga$_2$O$_3$ layer, the estimated peak from β-Ga$_2$O$_3$ (111) was observed. This indicated that the Ga$_2$O$_3$ layer formed on the 6H-SiC substrate is β-Ga$_2$O$_3$. Additionally, several crystal grains were observed in a cross-sectional TEM image of the Ga$_2$O$_3$ layer. We think that the β-Ga$_2$O$_3$ layer included several crystal orientations.

2.2 Fabrication of device structures

Two types of sensor devices, shown in Fig. 1, were fabricated. One device consisted of a β-Ga$_2$O$_3$ layer deposited on an n-type 6H-SiC substrate with a resistivity of 0.09 Ωcm. As an ohmic electrode for n-type SiC, Ni (100 nm)/Ti (30 nm)/Pt (100 nm) layers were evaporated successively onto the substrate, followed by annealing at 1000 °C for 2 min in nitrogen. The other device consisted of a β-Ga$_2$O$_3$ layer deposited on a p-type 6H-SiC substrate with a resistivity of 2.2 Ωcm. As an ohmic electrode, Al (10 nm)/Ti (20 nm)/Al (100 nm)/Pt (100 nm) layers were evaporated successively onto the p-type 6H-SiC substrate, followed by annealing at 1000 °C for 2 min in nitrogen. A thin Pt layer (30 nm) with a diameter of ca. 1 mm was evaporated onto the β-Ga$_2$O$_3$ layer through a metal mask as a Schottky electrode. In this study,

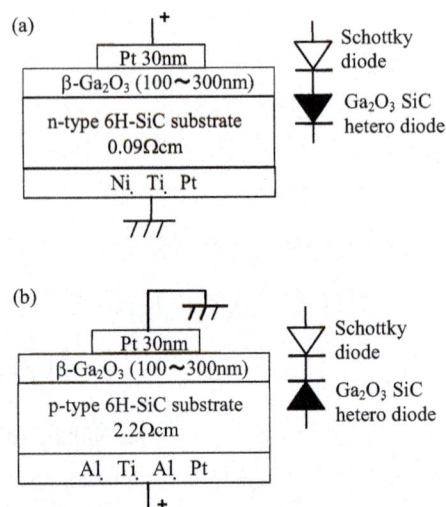

Figure 1. (a) Schematic diagram of a hydrogen gas sensor device based on the Pt/Ga$_2$O$_3$/n-SiC structure. A positive Pt electrode is referred to as forward bias. **(b)** Schematic diagram of a hydrogen gas sensor device based on the Pt/Ga$_2$O$_3$/p-SiC structure. A positive p-type SiC layer is referred to as forward bias.

forward bias is where the n-type SiC is biased negatively or when the p-type SiC is biased positively. The device based on the Pt/Ga$_2$O$_3$/n-SiC structure corresponds to a series connection between a Schottky diode and a Ga$_2$O$_3$/SiC heterojunction diode in the same direction. The device based on the Pt/Ga$_2$O$_3$/p-SiC structure corresponds to a series connection between a Schottky diode and a Ga$_2$O$_3$/SiC heterojunction diode in opposite directions to each other.

2.3 Device evaluation

A sensor device was placed in a quartz tube furnace with a thermocouple to monitor the temperature. The temperature of the device was controlled for measurements at 300, 400 and 500 °C. SiC and β-Ga$_2$O$_3$ are well known as wide band gap semiconductors; therefore, these devices based on β-Ga$_2$O$_3$ and SiC can operate at higher temperatures.

A mixed gas of N$_2$, O$_2$, H$_2$ and 1 % H$_2$/N$_2$ set using mass-flow controllers was supplied into the 25 mm diameter quartz tube. Oxygen and hydrogen concentrations were determined by means of a control flow rate for each gas. The total flow rate was maintained at 500 mL min^{-1} for all measurements. The I–V characteristics were measured in 20 % O$_2$/N$_2$ and 200 ppm H$_2$/N$_2$ atmospheres at 300, 400 and 500 °C using a source meter (Keithley 2400). A constant current source and digital recorder were used to measure hydrogen response curves. The hydrogen concentration was intermittently increased from 40 to 10 000 ppm in the 20 % O$_2$/N$_2$ atmosphere at 5 min intervals.

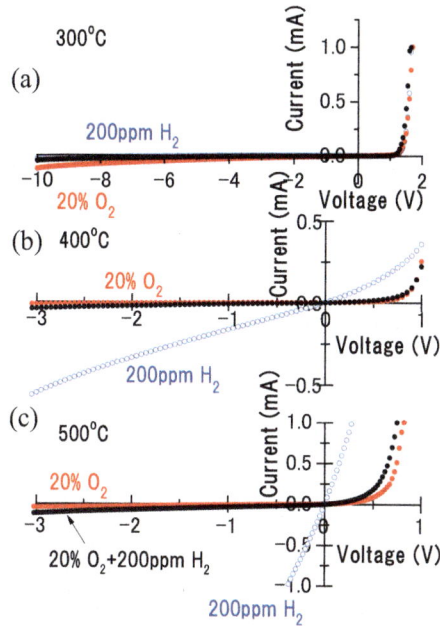

Figure 2. I–V characteristics of Pt/Ga$_2$O$_3$/n-SiC device in 20 % O$_2$ / N$_2$, 200 ppm H$_2$ / N$_2$ and 20 % O$_2$+ 200 ppm H$_2$ in N$_2$ measured at **(a)** 300, **(b)** 400 and **(c)** 500 °C.

Figure 3. Dependence of I–V characteristics of a Pt/Ga$_2$O$_3$/n-SiC device on H$_2$ concentration under 20 % O$_2$ atmosphere at 500 °C. I–V characteristics in 200ppm H$_2$ / N$_2$ are also shown.

3   Results and discussion

3.1 I–V characteristics

3.1.1   Pt/Ga$_2$O$_3$/n-type SiC heterojunction device

The I–V characteristics of the Pt/Ga$_2$O$_3$/n-SiC device measured in 20 % O$_2$ / N$_2$, 200 ppm H$_2$ / N$_2$ and 20 % O$_2$+ 200 ppm H$_2$ in N$_2$ at 300, 400 and 500 °C are shown in Fig. 2a–c. The I–V characteristics indicate rectifying properties at 300 °C. The current of the Pt/Ga$_2$O$_3$/n-SiC device increases when the n-type SiC is negatively biased. There was little difference in the I–V characteristics measured for three atmospheres at 300 °C. However, the current was increased in 200 ppm H$_2$ / N$_2$ compared with that in 20 % O$_2$/N$_2$ at 400 and 500 °C. In particular, the I–V characteristics in 200 ppm H$_2$ / N$_2$ were almost linear at 500 °C, which indicated a resistance without rectifying properties. A decrease in turn-on voltage in forward bias region and an increase in reverse current were observed at 500 °C.

Figure 3 shows dependence of the I–V characteristics of the Pt/Ga$_2$O$_3$/n-SiC device on hydrogen concentration under 20 % O$_2$ in N$_2$ at 500 °C. I–V characteristics in 200 ppm H$_2$ / N$_2$ are also shown. With an increase in H$_2$ concentration, the turn-on voltage decreased in the forward bias region and the current was increased in the reverse bias region. The dependence of the I–V characteristics on hydrogen concentration indicates that the Pt/Ga$_2$O$_3$/n-SiC device structure can be used as a hydrogen sensor both under reverse and forward bias conditions. The I–V characteristics

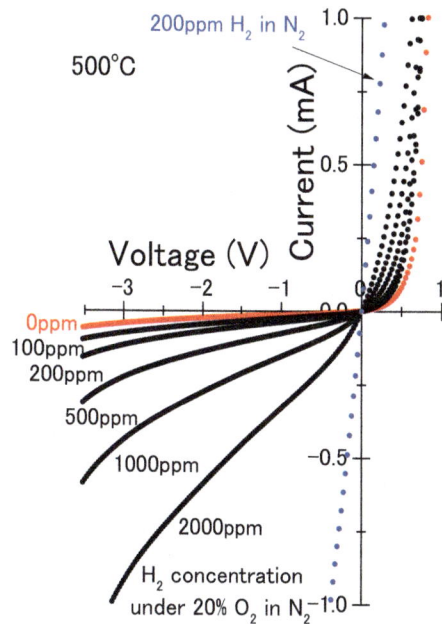

are similar to the behavior in a conventional gas sensor device with a Schottky diode structure. We have reported on a hydrogen gas sensor with a Schottky diode structure fabricated using a β-Ga$_2$O$_3$ single-crystal substrate (Nakagomi et al., 2011b). The dependence of I–V characteristics on hydrogen concentration is similar to the characteristics of the Pt/Ga$_2$O$_3$/n-SiC device shown in Fig. 3. In other words, the Pt/Ga$_2$O$_3$/n-SiC device reveals similar properties to conventional Schottky diode-type devices.

3.1.2   Pt/Ga$_2$O$_3$/p-type SiC heterojunction device

The I–V characteristics of the Pt/Ga$_2$O$_3$/p-SiC device measured in 20 % O$_2$/N$_2$, 200 ppm H$_2$ / N$_2$ and 20 % O$_2$+ 200 ppm H$_2$ in N$_2$ at 300, 400 and 500 °C are shown in Fig. 4a–c. The I–V characteristics show that good rectifying properties were maintained at all measurement temperatures. The current of the Pt/Ga$_2$O$_3$/p-SiC device device increased when the p-SiC substrate was positively biased. In contrast, little reverse current of the device flowed even at 500 °C. There was little difference in the I–V characteristics measured in the 20 % O$_2$ / N$_2$ and 200 ppm H$_2$ / N$_2$ atmospheres under reverse bias conditions. In contrast, the current under forward bias was changed according to the 200 ppm H$_2$ / N$_2$, 20 % O$_2$+ 200 ppm H$_2$ in N$_2$ or 20 % O$_2$ / N$_2$ atmosphere. The forward current in 200 ppm H$_2$ / N$_2$ was significantly increased with increasing temperature, and the change in the I–V characteristics caused by the change in atmosphere was increased. The voltage difference under constant current in the forward bias region can be used as a sensing signal; therefore,

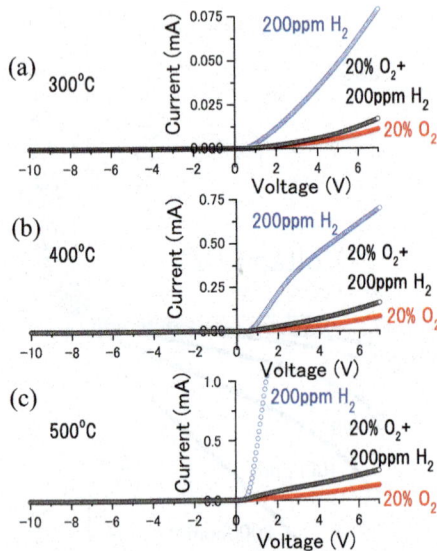

Figure 4. *I–V* characteristics of Pt/Ga$_2$O$_3$/p-SiC device in 20 % O$_2$/N$_2$, 200 ppm H$_2$ / N$_2$ and 20 % O$_2$+ 200 ppm H$_2$ in N$_2$ measured at **(a)** 300, **(b)** 400 and **(c)** 500 °C.

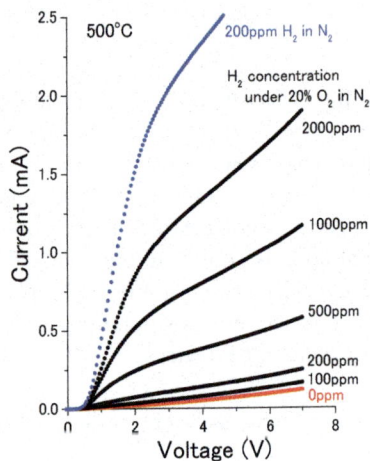

Figure 5. Dependence of *I–V* characteristics of Pt/Ga$_2$O$_3$/p-SiC device on H$_2$ concentration under 20 % O$_2$ atmosphere at 500 °C. *I–V* characteristics in 200ppm H$_2$ / N$_2$ are also shown.

the Pt/Ga$_2$O$_3$/p-SiC heterojunction structure is considered to be a feasible hydrogen sensor device with a large response under forward bias conditions.

Figure 5 shows dependence of the forward *I–V* characteristics of the Pt/Ga$_2$O$_3$/p-SiC device on hydrogen concentration under 20 % O$_2$ in N$_2$ at 500 °C. *I–V* characteristics in 200 ppm H$_2$ / N$_2$ are also shown. In 20 % O$_2$ atmosphere without hydrogen, the forward current was low. The forward current increased with increasing hydrogen concentration. The current was more increased in 200 ppm H$_2$ in a N$_2$ atmosphere without O$_2$. In the conditions of higher hydrogen concentration than 500 ppm, the rate of current increase

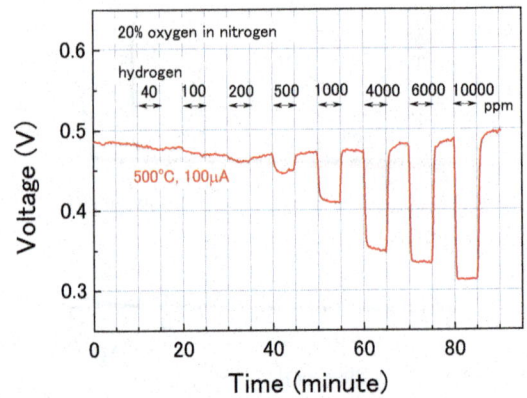

Figure 6. Response voltage curves for the Pt/Ga$_2$O$_3$/n-SiC gas sensor device with intermittent increases in the H$_2$ concentration in a 20 % O$_2$ / N$_2$ atmosphere at 500 °C. The forward bias current of the device was kept at 100 µA.

for voltage was decreased at the higher voltage than around 2–3 V. This suggests that the hydrogen-sensing mechanism of the gas sensor device with the Pt/Ga$_2$O$_3$/p-SiC structure does not proceed from a simple change in electrical resistance. The mechanism will be discussed in a later section (Sect. 4.3).

3.2   Response to hydrogen gas

3.2.1   Pt/Ga$_2$O$_3$/n-type SiC heterojunction device

Response curves of the sensor devices were measured to investigate the hydrogen gas sensitivity, response time and recovery time. Figure 6 shows voltage response curves for the Pt/Ga$_2$O$_3$/n-SiC structure device under constant current (100 µA) at 500 °C where the device was biased in the forward direction. The H$_2$ concentration was intermittently increased from 40 to 10 000 ppm in the 20 % O$_2$/N$_2$ atmosphere. The voltage changed with an increase in the H$_2$ concentration, except for hydrogen concentrations lower than 200 ppm. A sharp change in the voltage was observed in the region of 500–10 000 ppm H$_2$. The response and recovery were almost completed within a few seconds. The results demonstrate that the sensor device can detect almost 500 ppm H$_2$ gas in air. Despite the very small response voltage of less than 0.2 V and the higher sensing limit of 500 ppm, the Pt/Ga$_2$O$_3$/n-SiC device structure under forward bias demonstrated a quick response and recovery.

When the Pt/Ga$_2$O$_3$/n-SiC structure was biased in reverse, the device could detect lower hydrogen concentrations and give a larger voltage response than when forward biased. Figure 7 shows voltage response curves for the device under constant current (50 µA) at 300, 400 and 500 °C. The device was able to detect 40 ppm hydrogen gas in 20 % O$_2$. The behavior at 500 °C was remarkable. The change in voltage amounted to 5 V for 4000 ppm hydrogen, whereas the

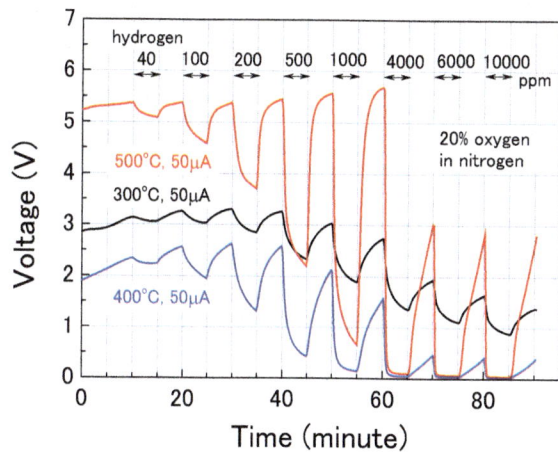

Figure 7. Response voltage curves for the Pt/Ga$_2$O$_3$/n-SiC gas sensor device with intermittent increases in the H$_2$ concentration in a 20 % O$_2$ / N$_2$ atmosphere at 300, 400 and 500 °C. The reverse bias current of the device was kept at 50 µA.

Figure 8. Response voltage curves for the Pt/Ga$_2$O$_3$/p-SiC gas sensor device with intermittent increases in the H$_2$ concentration in a 20 % O$_2$ / N$_2$ atmosphere at 300, 400 and 500 °C. The forward bias current of the device was kept at 50 µA.

voltage reached almost zero for the hydrogen concentrations higher than 4000 ppm. Although the response time for hydrogen concentrations higher than 4000 ppm was within a few seconds, the recovery time became slower than that when forward biased.

In the measurements of the response curve shown in Figs. 6 and 7, the temperature of the device rose and fell by around 1 °C for each temperature condition at intervals of about 10 min. These temperature fluctuations did not have a large influence on the response of the device.

3.2.2   Pt/Ga$_2$O$_3$/p-type SiC heterojunction device

The Pt/Ga$_2$O$_3$/p-SiC structure device is not very sensitive to hydrogen gas in the reverse bias condition; therefore, response curves were measured only for the forward bias condition. Figure 8 shows voltage response curves for the Pt/Ga$_2$O$_3$/p-SiC structure device under constant current (50 µA) at 300, 400, and 500 °C with forward bias. The device could detect 40 ppm hydrogen gas in 20 % O$_2$ / N$_2$ when forward biased. At 300 °C, the change in voltage due to the change in hydrogen concentration was small and gradual. However, an increase in temperature resulted in an increased voltage response that amounted to 8 V for 4000 ppm hydrogen in 20 % O$_2$ / N$_2$ at 500 °C. The voltage for hydrogen concentrations higher than 4000 ppm was almost 1 V. The sensor could detect 40 ppm hydrogen in 20 % O$_2$ / N$_2$ at 500 °C distinctly. Although the response time for hydrogen concentrations higher than 4000 ppm was within a few seconds, the recovery time became slow. This recovery behavior was similar to that for the Pt/Ga$_2$O$_3$/n-SiC structure device when biased in the reverse direction.

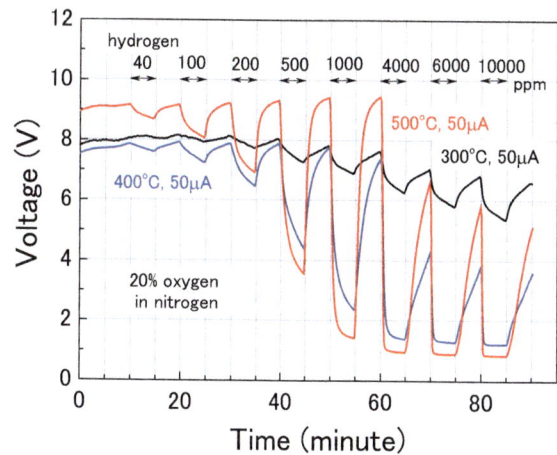

In the measurement of the response curve shown in Fig. 8, the temperature of the device rose and fell by around 1 °C for each temperature condition at intervals of about 10 min. These temperature fluctuations did not have a large influence on the response of the device.

Because the present sensor devices with Pt/Ga$_2$O$_3$/n-SiC or Pt/Ga$_2$O$_3$/p-SiC structure are fabricated from semiconductor materials, the sensor devices must be influenced by temperature fluctuation. If the sensors are used under the condition with large temperature fluctuation, the temperature compensation system should be contrived as we demonstrated in the previous report (Nakagomi et al., 2013a).

3.2.3   Comparison between Pt/Ga$_2$O$_3$/n-type SiC and Pt/Ga$_2$O$_3$/p-type SiC devices

Figure 9 shows the relationship between the sensor voltage output under a constant current of 50 µA and various hydrogen concentrations at 300, 400 and 500 °C. Figure 9 was constructed from the response curves shown in Figs. 7 and 8. Both the Pt/Ga$_2$O$_3$/n-type SiC device biased in reverse and the Pt/Ga$_2$O$_3$/p-type SiC device biased forward are included in Fig. 9. When the hydrogen concentration was increased in 20 % O$_2$ / N$_2$ at 300 °C, there was a baseline drift for both devices. However, the output voltage decreased largely in the region between 200 and 1000 ppm hydrogen at 400 and 500 °C. The change in output voltage is caused by the reaction between oxygen and hydrogen, which was noted in one of our previous works (Nakagomi et al., 2011a). For example, the voltage output decreases significantly for 200 ppm hydrogen in 20 % O$_2$ / N$_2$. Therefore, the present sensor device could detect hydrogen concentrations lower than 1/200 of the explosion limit of hydrogen gas in air. Thus the concentration ratio of H$_2$ / O$_2$ is 1/1000.

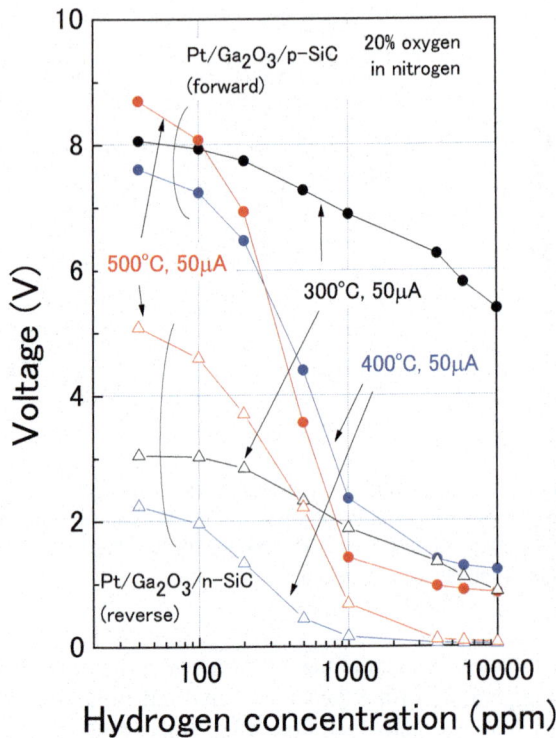

Figure 9. Relationship between the voltage response and hydrogen concentration for the Pt/Ga₂O₃/n-SiC and Pt/Ga₂O₃/p-SiC devices in a 20 % O₂ / N₂ atmosphere at 300, 400 and 500 °C.

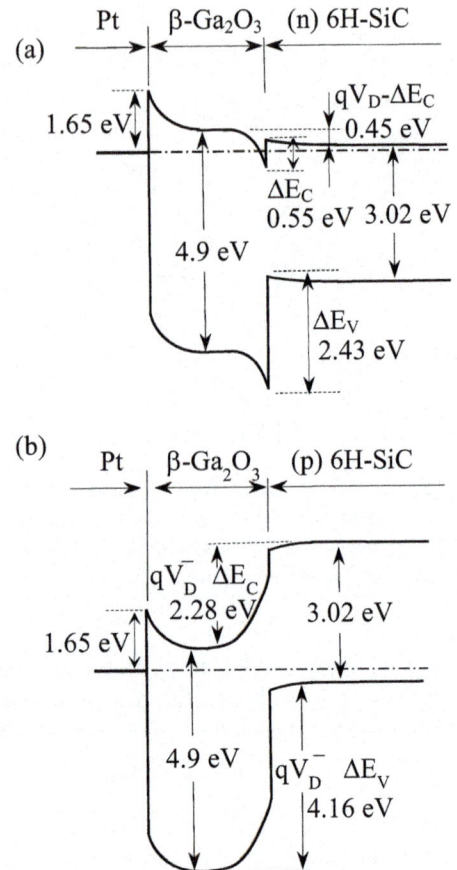

Figure 10. Energy band diagrams for hydrogen gas sensor devices based on the **(a)** Pt/Ga₂O₃/n-SiC and **(b)** Pt/Ga₂O₃/p-SiC structures.

4   Discussion

4.1   Energy band diagram

The expected band diagram for the hydrogen gas sensor device based on the Pt/Ga₂O₃/n-SiC and Pt/Ga₂O₃/p-SiC structures under zero bias is shown in Fig. 10a and b. β-Ga₂O₃ has a wider band gap than 6H-SiC, and the electron affinity of β-Ga₂O₃ and 6H-SiC has been reported as 4.0 and 3.45 eV, respectively (Mohamed et al., 2012; Davydov 2007). Therefore, the offset in the conduction band, ΔE_C of 0.55 eV, and the offset in the valence band, ΔE_V of 2.43 eV, should exist at the heterointerface in the ideal case. As a result, it is estimated that there was a barrier (qV_D-ΔE_C) of 0.45 eV for conduction electrons in n-type SiC of the Pt/Ga₂O₃/n-SiC structure. V_D is the sum of the built-in potential formed at both sides of heterointerface. For the Pt/Ga₂O₃/p-SiC structure device, the barrier ($qV_D + \Delta E_C$) for conduction electrons in the Ga₂O₃ layer is 2.28 eV. A Schottky barrier of 1.65 eV is also formed between Pt and the β-Ga₂O₃ layer if the work function of Pt is 5.65 eV. However, it is known that this Schottky barrier height is changed depending on the hydrogen concentration in the atmosphere.

It is important that holes in the SiC region not be able to flow into the β-Ga₂O₃ layer, because the energy barrier of 4.16 eV for holes from p-type 6H-SiC to β-Ga₂O₃ is

considerably higher than the barrier of 2.28 eV for conduction electrons from β-Ga₂O₃ to p-type 6H-SiC when p-type SiC is used as a substrate. Even when n-type SiC is used as substrate, the energy barrier for holes from n-type 6H-SiC to β-Ga₂O₃ of 2.43 eV is considerably higher than the barrier for conduction electrons from 6H-SiC to β-Ga₂O₃ in the conduction band. Therefore, only electrons can be considered as the charge carriers for both device structures.

4.2   Ga₂O₃/n-type SiC heterojunction device

We considered reasons why the device based on heterojunctions of Ga₂O₃/n-type SiC has hydrogen-sensing properties for each bias condition. Figure 11a and b show schematic band diagrams for the Pt/Ga₂O₃/n-type SiC structure in the forward and reverse bias directions, respectively. The applied voltage of the device V is distributed to the bias voltage V_1 applied to the Schottky junction and the bias voltage V_2 applied to the n–n heterojunction between Ga₂O₃ and n-type SiC.

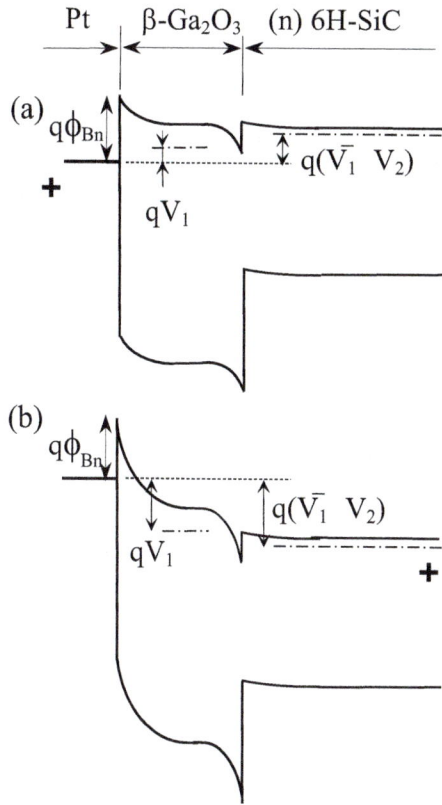

Figure 11. Energy band diagrams for a hydrogen gas sensor device based on the Pt/Ga$_2$O$_3$/n-SiC structure biased in the **(a)** forward and **(b)** reverse directions.

The current of the Schottky diode is given by

$$I_1 = SA^*T^2 e^{-\frac{q\varphi_{Bn}}{kT}}(e^{\frac{qV_1}{kT}} - 1), \tag{1}$$

where S is area of the device, A^* is the effective Richardson's constant, T is absolute temperature, q is the charge of an electron, and $q\varphi_{Bn}$ is the Schottky barrier height (Sze and Ng, 2007).

Only electron flow can be considered in the Pt/Ga$_2$O$_3$/n-type SiC structure device; therefore, the current of the n–n heterojunction between Ga$_2$O$_3$ and n-type SiC is given by

$$I_2 = SBe^{-\frac{qV_D - \Delta E_C}{kT}}(e^{\frac{qV_2}{kT}} - 1), \tag{2}$$

where S is the area of the device, B is constant including several parameters, ΔE_C is the offset in the conduction band between β-Ga$_2$O$_3$ and 6H-SiC, and V_D is the sum of the built-in potential formed at both sides of the heterointerface, i.e., $(qV_D - \Delta E_C)$ is the barrier height for electrons to migrate from n-type 6H-SiC to β-Ga$_2$O$_3$ (see Fig. 10a). In addition, the Schottky diode and heterojunction of Ga$_2$O$_3$/n-type SiC are connected in series; therefore

$$I_1 = I_2. \tag{3}$$

When the device with the Pt/Ga$_2$O$_3$/n-type SiC structure is forward biased, both the Schottky junction and the n–n heterojunction are biased in the forward direction. Therefore, both V_1 and V_2 are positive. The band diagram in Fig. 11a shows that electrons in n-type SiC flow into the Ga$_2$O$_3$ region by getting over the barrier at the interface and flow to the Pt electrode over the Schottky barrier. In this case, two barriers are lowered because both barriers are forward biased. When the barrier $q\varphi_{Bn}$ is lowered due to hydrogen gas exposure, the flow of electrons from Ga$_2$O$_3$ to Pt is increased. However, the influence of the barrier height change is small because the Schottky junction is already biased in the forward direction and the barrier height for conduction electrons from Ga$_2$O$_3$ to Pt is already lowered.

The barrier for electron transport from Ga$_2$O$_3$ to Pt is higher than the barrier of $qV_D - \Delta E_C$ from n-type SiC to Ga$_2$O$_3$; therefore, $V_1 \gg V_2$ and assume $V = V_1$, and so

$$V = \varphi_{Bn} + \text{const.} \tag{4}$$

This equation indicates that a change in φ_{Bn} is equivalent to a change in applied voltage, V. Therefore, when the φ_{Bn} is changed depending on the hydrogen concentration in the atmosphere, the variation appears directly in the applied voltage.

The hydrogen sensor with the Pt/Ga$_2$O$_3$/n-type SiC structure reported by Trinchi et al. (2004) was used under forward bias conditions (Trinchi et al., 2004). The sensor device corresponds to the same situation; therefore, the present results indicate similar $I–V$ characteristics to those reported by Trinchi et al. (2004).

When the Pt/Ga$_2$O$_3$/n-type SiC structure device is biased in reverse, both the Schottky junction and the n–n heterojunction are biased in the reverse direction. However, because $q\varphi_{Bn} \gg \Delta E_C$, almost all of the voltage is applied to the Schottky junction. Thus, the ΔE_C barrier does not act as an obstacle. Almost all of the electron flow is determined by electrons that can get over the barrier from the Pt electrode to β-Ga$_2$O$_3$ layer; therefore, the current is mainly determined by the Schottky barrier height, $q\varphi_{Bn}$. A change in φ_{Bn} thus leads to a change in the $I–V$ characteristics. This situation corresponds to the case of a single Schottky diode. Therefore, the current is given by

$$I = -SA^*T^2 e^{-\frac{q\varphi_{Bn}}{kT}}. \tag{5}$$

4.3 Ga$_2$O$_3$/p-type SiC heterojunction device

Figure 12a and b show schematic band diagrams for the Pt/Ga$_2$O$_3$/p-type SiC structure biased in the forward and reverse directions, respectively. V_1 and V_3 correspond to the bias voltage applied to the Schottky junction and to the n–p heterojunction between β-Ga$_2$O$_3$ and p-type SiC, respectively.

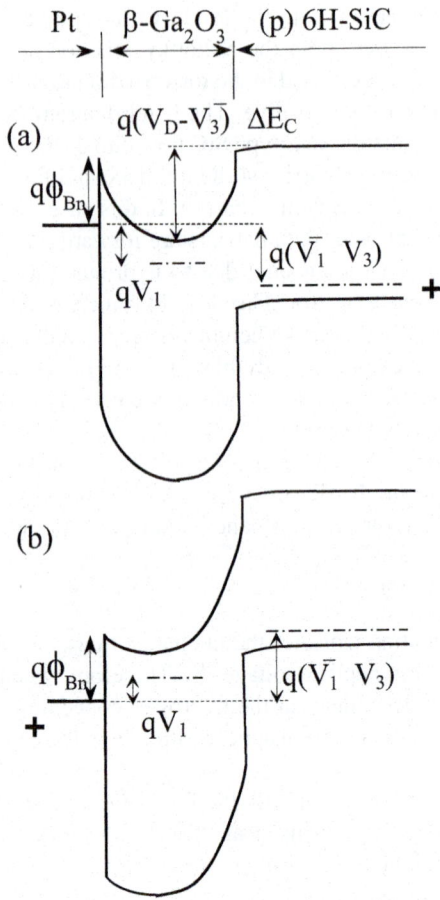

Figure 12. Energy band diagram for a hydrogen gas sensor device based on the Pt/Ga$_2$O$_3$/p-SiC structure biased in the **(a)** forward and **(b)** reverse directions.

Only the electrons are considered as charge carriers; therefore, the current of the n–p heterojunction is given by

$$I_3 = SBe^{-\frac{qV_D + \Delta E_C}{kT}}(e^{\frac{qV_4}{kT}} - 1). \tag{6}$$

When the device is biased in the forward direction, the Schottky junction between the Pt electrode and the β-Ga$_2$O$_3$ layer is biased in the reverse direction, while the n–p heterojunction is biased in the forward direction. The electrons in the β-Ga$_2$O$_3$ region flow to the p-type SiC region over the potential barrier in the conduction band. The barrier is lowered with increasing forward bias; however, the Schottky junction is biased in reverse; therefore, only those electrons that can get over the Schottky barrier flow into the β-Ga$_2$O$_3$ region and can also reach the SiC region. When $q\varphi_{Bn}$ is lowered due to hydrogen gas exposure, the electrons are increased and the current is thus increased. This increase in electrons getting over the Schottky barrier corresponds to a decrease in electrical resistance.

This process allows us to expect a saturation property in I–V characteristics observed in Fig. 6. Because the forward current does not quite increase with an increase in bias voltage due to electron flow limited by the Schottky barrier, the current is saturated. If the electron flow from the Schottky barrier is increased with an increase in hydrogen concentration, the saturation current level also rises.

The voltage applied to the device is shared by both the Schottky junction and heterojunction:

$$V_1 + V_3 = V. \tag{7}$$

Therefore, under constant applied voltage, the decrease in voltage shared at the Schottky junction V_1 due to a decrease in electrical resistance leads to an increase in the voltage shared at the heterojunction V_3. Thus, the flow of electrons is increased by this amplification effect. This is the mechanism under the forward bias condition.

When the device is biased in the forward direction, $e^{\frac{qV_1}{kT}}$ in Eq. (1) and 1 in Eq. (6) are omitted.

Using $-I_1 = I_3$,

$$(qV_D + \Delta E_C) - qV_3 = q\varphi_{Bn} - \text{const.} \tag{8}$$

When φ_{Bn} is decreased, V_3 must be increased, which leads to a lowering of the $q(V_D - V_3) + \Delta E_C$ barrier height and the increase in electron flow increases the forward current exponentially. We consider that this amplification effect leads to the large change in the forward current.

In contrast, when the device is biased in the reverse direction, 1 in Eq. (1) and $e^{\frac{qV_3}{kT}}$ in Eq. (6) are omitted to give

$$q(\varphi_{Bn} - V_1) = \Delta E_C + qV_D + \text{const.} \tag{9}$$

Therefore, if φ_{Bn} is decreased, V_1 will also be decreased. However, almost all of the applied voltage is applied to the reverse bias of the n–p heterojunction between Ga$_2$O$_3$ and p-type SiC, and because the Schottky junction is already biased in forward direction, the device has low hydrogen sensitivity.

5 Conclusions

Field-effect hydrogen gas sensor devices based on Pt/Ga$_2$O$_3$/SiC heterojunction structures were fabricated. Two types of hydrogen gas sensor devices were fabricated with Pt/Ga$_2$O$_3$/n-type SiC heterojunction and Pt/Ga$_2$O$_3$/p-type SiC heterojunction structures. The I–V characteristics were measured and the hydrogen response properties were evaluated with respect to the bias conditions. Both the sensor with the Pt/Ga$_2$O$_3$/n-type SiC structure biased in reverse and that with the Pt/Ga$_2$O$_3$/p-type SiC structure forward biased had large responses; when the Schottky junction between Pt and β-Ga$_2$O$_3$ is biased in reverse, the sensors have a large response. The sensors could detect 40 ppm hydrogen for certain under 20 % O$_2$ / N$_2$ at 500 °C under appropriate bias conditions. The behavior of the hydrogen sensing was explained using several band diagrams for bias in the forward and reverse directions.

Edited by: Y. X. Li
Reviewed by: two anonymous referees

References

Davydov, S. Y.: On the Electron Affinity of Silicon Carbide Poly-types, Semiconductors, 41, 696–698, 2007.

Fleischer, M. and Meixner, H.: Gallium Oxide Thin Films: a New Material for High-temperature Oxygen Sensors, Sens. Actuators B, 4, 437–441, 1991.

Fleischer, M. and Meixner, H.: Sensing reducing gases at high temperatures using long-term stable Ga_2O_3 thin films, Sens. Actuators B, 6, 257–261, 1992.

Fleischer, M., Giber, J., and Meixner, H.: H_2-Induced Changes in Electrical Conductance of β-Ga_2O_3 Thin-Film Systems, Appl. Phys., A54, 560–566, 1992.

Higashiwaki, M., Sasaki, K., Kuramata, A., Masui, T., and Yamakoshi, S.: Gallium oxide (Ga_2O_3) metal-semiconductor field-effect transistors on single-crystal β-Ga_2O_3 (010) substrates, Appl. Phys. Lett., 100, 013504, doi:10.1063/1.3674287, 2012.

Kokubun, Y., Miura, K., Endo, F., and Nakagomi, S.: Sol-gel prepared β-Ga_2O_3 thin films for ultraviolet photodetectors, Appl. Phys. Lett., 90, 031912, doi:10.1063/1.2432946, 2007.

Mohamed, M., Janowitz, C., Manzke, R., Irmscher, K., Galazka, Z., and Fornari, R.: Schottky barrier height of Au on the transparent semiconducting oxide β-Ga_2O_3, Appl. Phys. Lett. 101, 132106, doi:10.1063/1.4755770, 2012.

Nakagomi, S. and Kokubun, Y.: Crystal Orientation of β-Ga_2O_3 Thin Films Formed on c-Plane and a-Plane Sapphire Substrate, J. Cryst. Growth, 349, 12–18, 2013.

Nakagomi, S., Kaneko, M., and Kokubun, Y.: Hydrogen Sensitive Schottky Diode Based on β-Ga_2O_3 Single Crystal, Sensor Lett., 9, 35–39, 2011a.

Nakagomi, S., Ikeda, M., and Kokubun, Y.: Comparison of Hydrogen Sensing Properties of Schottky Diodes Based on SiC and β-Ga_2O_3 Single Crystal, Sensor Lett., 9, 616–620, 2011b.

Nakagomi, S., Sai, T., and Kokubun, Y.: Hydrogen Gas Sensor with Self Temperature Compensation Based on β-Ga_2O_3 Thin Film, Sens. Actuators B, 187, 413–419, 2013a.

Nakagomi, S., Momo, T., Takahashi, S., and Kokubun, Y.: Deep Ultraviolet Photodiodes Based on β-Ga_2O_3/SiC Heterojunction, Appl. Phys. Lett., 103, 072105, doi:10.1063/1.4818620, 2013b.

Ogita, M., Saika, N., Nakanishi, Y., and Hatanaka, Y.: Ga_2O_3 thin films for high-temperature gas sensors, Appl. Surf. Sci., 142, 188–191, 1999.

Spetz, A. L., Nakagomi, S., and Savage, S. S.: Advances in Silicon Carbide Processing and Application, edited by: Saddow, S. E. and Agarwal, A., Artech house, Boston, Chapter 2, 29–67, 2004.

Suzuki, R., Nakagomi, S., Kokubun, Y., Arai, N., and Ohira, S.: Enhancement of responsivity in solar-blind β-Ga_2O_3 photodiodes with a Au Schottky contact fabricated on single crystal substrates by annealing, Appl. Phys. Lett., 94, 222102, doi:10.1063/1.3147197, 2009.

Sze, S. M. and Ng, K. K.: Physics of Semiconductor Devices, John Wiley & Sons, Inc., 2007.

Trinchi, A., Wlodarski, W., and Li, Y. X.: Hydrogen sensitive GA2O3 Schottky diode sensor based on SiC, Sens. Actuators B, 100, 94–98, 2004.

11

A novel horizontal to vertical spectral ratio approach in a wired structural health monitoring system

F. P. Pentaris

Dept. of Electronic and Computer Engineering, Brunel University, London, UK

Correspondence to: F. P. Pentaris (fpentaris@gmail.com)

Abstract. This work studies the effect ambient seismic noise can have on building constructions, in comparison with the traditional study of strong seismic motion in buildings, for the purpose of structural health monitoring. Traditionally, engineers have observed the effect of earthquakes on buildings by usage of seismometers at various levels. A new approach is proposed in which acceleration recordings of ambient seismic noise are used and horizontal to vertical spectra ratio (HVSR) process is applied, in order to determine the resonance frequency of movement due to excitation of the building from a strong seismic event. The HVSR technique is widely used by geophysicists to study the resonance frequency of sediments over bedrock, while its usage inside buildings is limited. This study applies the recordings inside two university buildings attached to each other, but with different construction materials and different years of construction. Also there is HVSR application in another much older building, with visible cracks in its structure. Sensors have been installed on every floor of the two university buildings, and recordings have been acquired both of ambient seismic noise and earthquakes. Resonance frequencies for every floor of every building are calculated, from both noise and earthquake records, using the HVSR technique for the ambient noise data and the receiver function (RF) for the earthquake data. Differential acceleration drift for every building is also calculated, and there is correlation with the vulnerability of the buildings. Results indicate that HVSR process on acceleration data proves to be an easy, fast, economical method for estimation of fundamental frequency of structures as well as an assessment method for building vulnerability estimation. Comparison between HVSR and RF technique shows an agreement at the change of resonance frequency as we move to higher floors.

1 Introduction

Horizontal to vertical spectra ratio (HVSR) method was first proposed by Nogoshi and Igarashi (1971) and was subsequently widely spread by Nakamura (1989). HVSR has been applied by Dimitriu et al. (1999), who found the fundamental frequency of the ground. Yuncha and Luzon (2000) tested the HVSR technique in low and high impedance contrast between surface and bedrock. Their results show that the HVSR technique can give reliable results when the impedance contrast between surface layer and bedrock is high. The above authors introduce the problem of superposition of different incoming P-SV waves in HVSR. This problem is discussed in detail by Fäh et al. (2001), who apply the HVSR method using both classical and wavelet techniques, find stable recordings and use an inversion method of genetic

algorithm in order to present the S velocity without P wave effect on an ambient vibration measurement. They use an empirical model which may not be applicable at every site due to the need for training based on stable data values which may vary with geographical location.

The classic method of spectral ratio has been used by Parolai et al. (2004), who observed that fundamental frequency is stable in time but unstable in amplitude for a site. They conclude that HVSR should be verified with a lot of measurements, and they address that the spectrum analogy remains almost the same for seismic and environmental noise. They also conclude that high impedance contrast between surface sediment layer and bedrock can reveal the fundamental frequency when the higher harmonics are hidden. The combination of HVSR results combined with geological data is

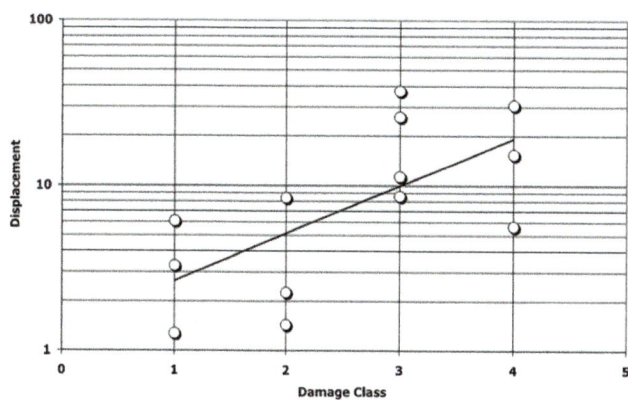

Figure 1. Damaged and calculated inter-storey displacement index correlation by Mucciarelli et al. (2001).

applied by Panou et al. (2005), in the large area of Thessaloniki, where they find similarities between HVSR results and the geological data and suggest that HVSR is a reliable method for site characterization. HVSR is applied by Lombardo and Rigano (2007), who use this method in order to characterize the terrain in an urban area, by recording measurements from different terrains and comparing the ambient noise with preliminary seismic recordings. They address the need for validation of the results with earthquake recordings. HVSR has also been applied by Sokolov et al. (2007) in earthquake recordings in which they study rock site amplifications in the large area of Taiwan. Wavelet analysis in HVSR has been applied by Carniel et al. (2008) using the key point advantage of wavelet analysis method (namely the ability to analyse data in both time and frequency domain) in order to improve the ability of HVSR for site effect estimation. The main disadvantage is the high complexity of the algorithm application. Another technique for improving HVSR is by self-organizing map (SOM) applied again by Carniel et al. (2009), who apply this neural network technique (SOM) for site characterization, but this method is computationally inefficient as it requires educating the neural network, for every site amplification analysis. Accelerometers and seismometers recording seismic and ambient noise activity have been used by Chávez-García et al. (2010), estimating spectral ratios of a site, presenting the local transfer function of case study buildings. Their main finding was the requirements for lower noise of accelerometers and seismometers in order to study efficiently the ambient vibrations.

2 Related work on HVSR method in structural health monitoring

Triwulan et al. (2010) claim that geological characteristics of a region, structural characteristics of a building and also the correlation of both geology and structure can be described by the application of HVSR data recorded on the ground of the

building and inside the building. They apply HVSR method in both ground and buildings in order to study the natural frequency, the index of vulnerability, the amplification factor of the ground as well as indexes of vulnerability of buildings and ground.

Ambient noise in concrete reinforced building, affected by subway trains, is also studied in Beijing by Luo et al. (2011), who present the resonance frequency of the building on each floor. They study the range of the fundamental frequency of the building (around 2.4 Hz), the frequency which is generated by the nearby traffic on the building (around 10 Hz) and the geological fundamental frequency of the region in which the building is located (around 2–3 Hz). They indicate that, although the amplification of the site is critical for the specific building, its damping ratio of 0.17 is very effective in structural integrity.

Mucciarelli et al. (2001) claim that study of microtremors and weak motion provides fast and reliable data for site amplification and structure vulnerability compared with other traditional methods (like geological study by drilling), and the correlation of damage and structural integrity is with physical parameters and not normalized adimensional indexes. Furthermore, they apply the technique (filtering of the signals and avoid wind, traffic and man-made disturbance effect) based on empirical decomposition method to estimate the structural vulnerability of buildings under seismic excitation. Figure 1 presents their proposed index of damage with inter-storey displacement.

Liu et al. (2014) applied ambient noise survey on a seven-floor building using seismometers in order to assess the site amplification in an urban area and to study the ability of HVSR to assess building vulnerability under strong motion excitation. The HVSR results are shown in Fig. 2. They consider that ambient noise could be used for earthquake and seismology engineering in urban areas, but they do not correlate these HVSR results with specific building vulnerability characteristics. Vertical red dashed lines present the resonance frequencies.

Literature reveals that HVSR application of ambient seismic vibration in structural health monitoring is limited. HVSR is used to evaluate the fundamental frequency of the surface layer over bedrock. By studying HVSR in buildings, there is the ability to study the fundamental frequency of the building excited by environmental noise of seismic activity, and compare it with the fundamental frequency of the surface around the building. The study of HVSR outside and inside the building potentially can provide two factors: initially, how close the fundamental frequency of the building is to the fundamental frequency of the surface layer, and also how high the amplitude of this frequency is.

2.1 Case study infrastructures

Case study includes two different university buildings, which are attached to each other on one side. They have different

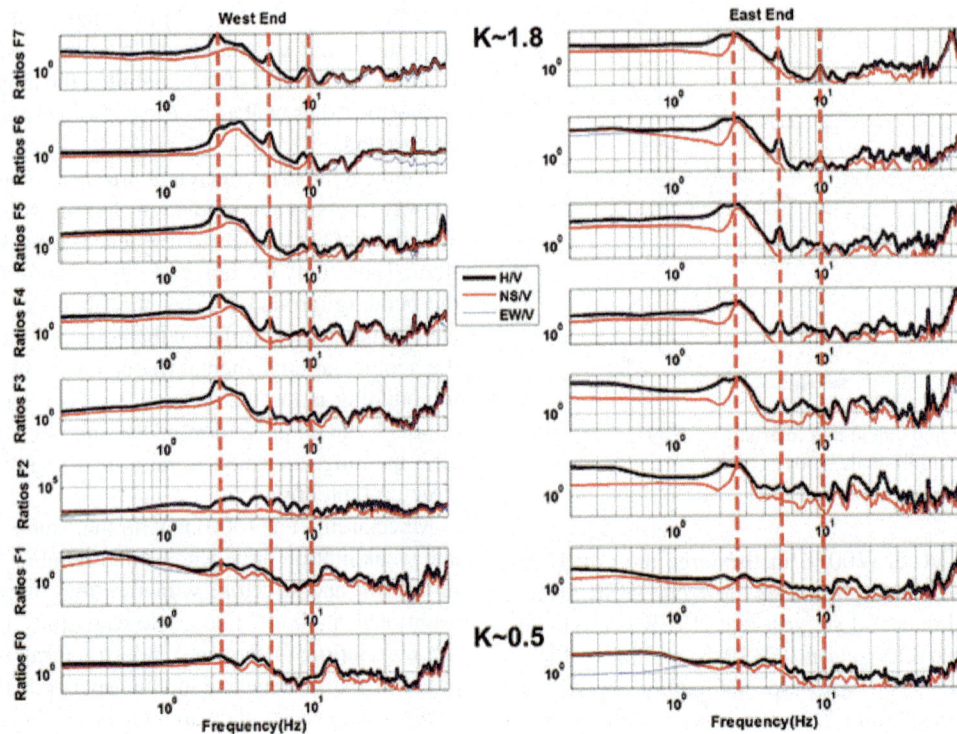

Figure 2. HVSR with black colour, N–S/V with red colour and E–W/V with blue colour on each floor for both sides (east and west) by Liu et al.(2014).

construction materials and different years of construction. In more detail, the two buildings host the Chania branch of the Technological Institute of Crete (TEI). Building (A) was built on the west side in 1995, with concrete, glass and steel, while building (B) was built on the east side in 2007 with concrete. Table 1 presents characteristics for both buildings, and Fig. 3 illustrates the schematic diagram of the two buildings, showing the sensor's location.

Sensors recorded data at various instances throughout a day, in order to study the effect of human activity during day and night.

HVSR method is applied for every floor of the buildings in order to discover the effect of floor amplification on each floor and the similarities which may exist under seismic and environmental noise. The approach of studying the HVSR on each floor in a building, under seismic activity, and ambient noise and correlate the increase of HVSR value with the building internal drift, is done for first time. In this context this work proposes an index which is correlated with HVSR disturbances from floor to floor in a building and could probably present the vulnerability of a building (under excitation) on every floor.

Figure 4 presents a map of the HVSR frequency for the large area of the city of Chania. Case study buildings are located in the geographical region where recorded HVSR frequency is in the range of 0.49–0.69 Hz (in the red square of

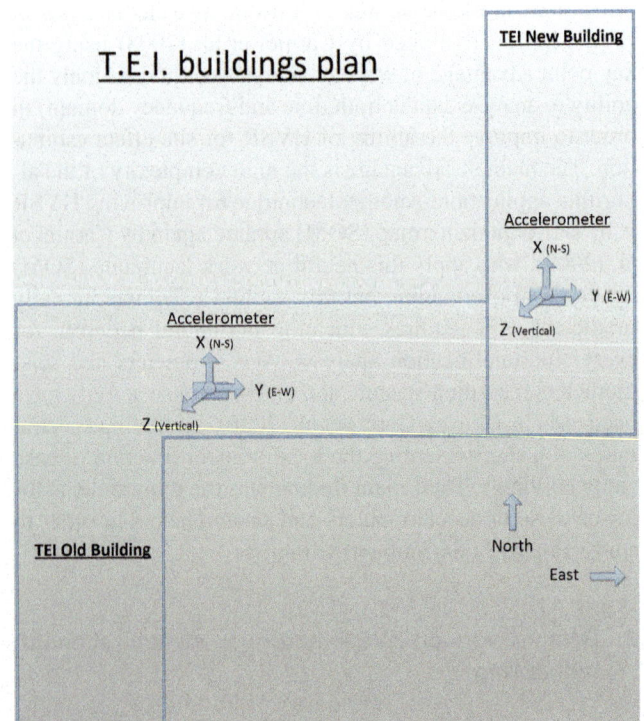

Figure 3. Schematic plan of the TEI building at Chania. The location of the sensors is presented for each building.

Table 1. Characteristics of the TEI buildings at Chania

Building code	Age (years)	Size (m)	Shape (direction)	Floor height	Number of rooms	Number of floors
A	19	30.62 × 18.03	Rectangle (N–S)	3.65	7 5	4 4
		11.29 × 43.30	Rectangle (E–W)	3.65		
B	7	51.55 × 21.85	Rectangle (N–S)	3.67	15	4

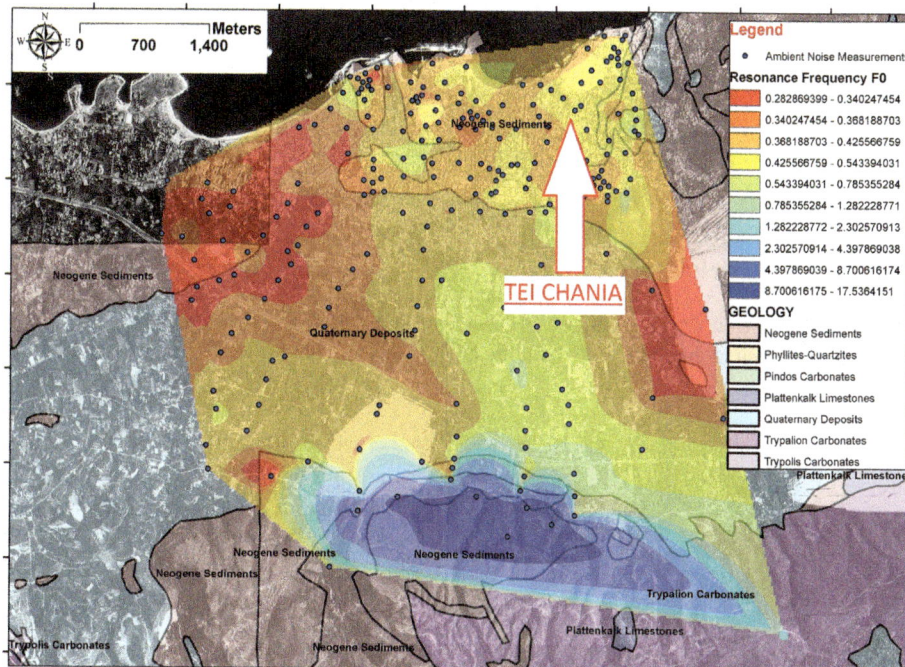

Figure 4. Case study building location is indicated. This map presents frequency of recorded HVSR for the broader area of Chania (Papadopoulos, 2013).

Fig. 4). The HVSR amplitude for the area close and on the case study university buildings is in the range of 2.40–2.54.

The instrumentation that is used in this work is from the wired structural health monitoring (SHM) system which is deployed in the Technological Educational Institute of Crete, Chania, with accelerometers of high sensitivity, sample rate at 125 Hz, and configured in triggered mode (Pentaris et al., 2013).

2.2 Instrumentation for HVSR recordings

In this study three kinds of sensors are used, in order to present the recording capabilities of each approach. Initially a Reftek accelerometer 130-SMA is used with a scale range of ±4 g, dynamic range of 112 dB at 1 Hz and a sensitivity of $1.6 \, \mathrm{V} \, (\mathrm{g} \, \mathrm{m}^{-1})^{-1}$. The frequency response is from flat DC up to 500 Hz. More information about Reftek accelerometer 130-SMA can be found at www.reftek.com (last access: 1 December 2013). Next a Guralp CMG-3ESP velocimeter is used, which is a broadband seismometer with response from 0.016 Hz until 50 Hz,

sensitivity $2 \times 1000 \, \mathrm{V} \, (\mathrm{m} \, \mathrm{s}^{-1})^{-1}$ and dynamic range higher than 140 dB. More details about Guralp Seismometer are on http://www.guralp.com (last access: 1 December 2013). Also HVSR is studied by seismometer Lennartz Le3D/5s. The sensitivity is $400 \, \mathrm{V} \, (\mathrm{m} \, \mathrm{s}^{-1})^{-1}$ with a frequency spectrum from 0.2 to 100 Hz. The sensitivity of the seismometer is flat between 0.4 and 100 Hz. Details about the ground velocity sensor Lennartz LeD/5s can be found at http://www.lennartz-electronic.de (last access: 10 November 2013).

2.3 Comparison of velocimeter and accelerometer sensors for HVSR method

HVSR is applied in seismometers and accelerometer, in order to study whether they have the same sensitivity, the same levels of electronic noise, and whether the recorded data have the same ability to reveal the HVSR data. The recording of the ambient noise will be divided in specific length segment and will be processed by Fourier transform, presenting the frequency spectrum with the resonance frequencies.

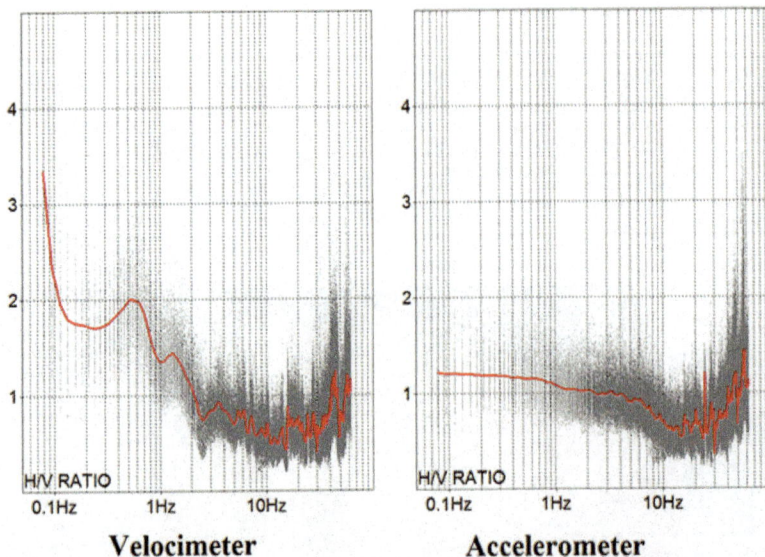

Figure 5. Data of Guralp velocimeter (left graph) and REFTEK accelerometer 130-SMA (right graph), analysed with HVSR technique for a recording of 60 min length on the ground floor of the old building of TEI Chania.

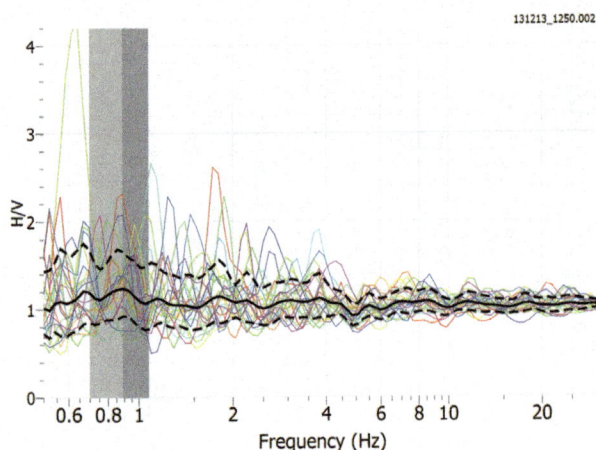

Figure 6. HVSR analysis of Lennartz LeD/5s velocimeter, for recording of 60 min length, on the ground floor of old building of TEI Chania.

Figures 5 and 6 agree with the findings of Chávez-García et al. (2010) in which both instrumentations present the same characteristics for frequencies higher than 2 Hz. Also the HVSR frequency and amplitude, from Figs. 5 and 6, recorded on the ground floor of old building of TEI at Chania, and for frequency spectrum higher than 2 Hz, are strongly related to the recorded HVSR frequency and amplitude of the surrounding area of TEI buildings (see Fig. 4, which presents the HVSR map of the large region and shows HVSR frequency close to 0.5–0.6. The lower HVSR amplitude (2.0) on the ground floor of TEI (related to the HVSR amplitude of 2.5 for the surrounding open field of TEI) is expected as building functions like a filter with specific transfer function, where

energy of ambient noise is slightly weakens, from the open field to the building infrastructure. In the case study experiment the fundamental frequency of both new and old building is much higher than 2 Hz (around 5.5 Hz). Fundamental frequency was computed from fast Fourier transform (FFT) of acceleration recordings at every floor of both buildings. HVSR method through specific accelerometers can function properly for the suggested approach of HVSR in SHM. The fundamental frequency of almost 5.5 Hz was recorded in every floor of the building. From the recordings of seismometers in both buildings, the FFT reveals that below 2 Hz the seismic acceleration has very low amplitude. As a result of the specific structures, the accelerometers that have an eigenfrequency of 1 Hz and are able to measure frequencies higher than 1 Hz are very efficient for studying HVSR method at these buildings. Other measurements that applied in the surround area outside the buildings of TEI Chania (with accelerometer) presented that HVSR resonance frequency is very close to the values of frequency spectral of velocimeter for frequencies higher than 2 Hz (Fig. 5).

2.4 Recordings of acceleration for both buildings

In this work, recordings from high-sensitivity accelerometers are going to be analysed with HVSR method. The recordings are from earthquakes that occurred in the wired area of the buildings and affected them, from midnight and noon recordings of ambient noise (each of 20 min recording duration). Each recording has been captured from every wired accelerometer (in every floor) of the new and the old building. The analysed data are discussed in detail in order to present similarities and differences. HVSR data are studied, from earthquakes, midnight ambient noise and midday

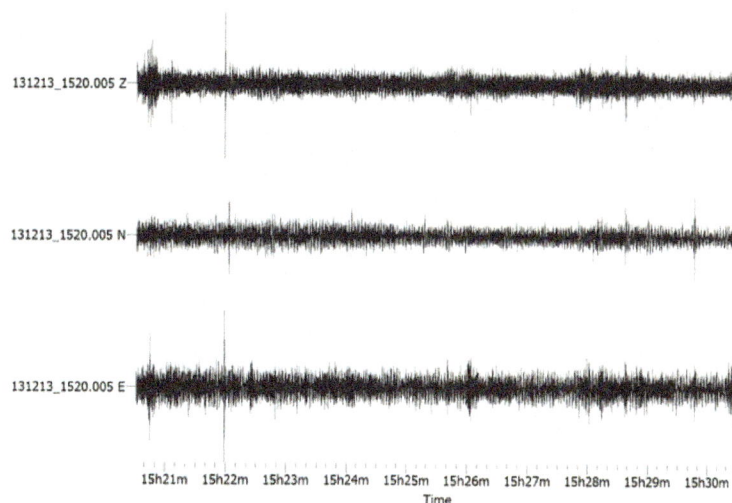

Figure 7. 10 min recording of ambient noise on the second floor of the new building of TEI Chania.

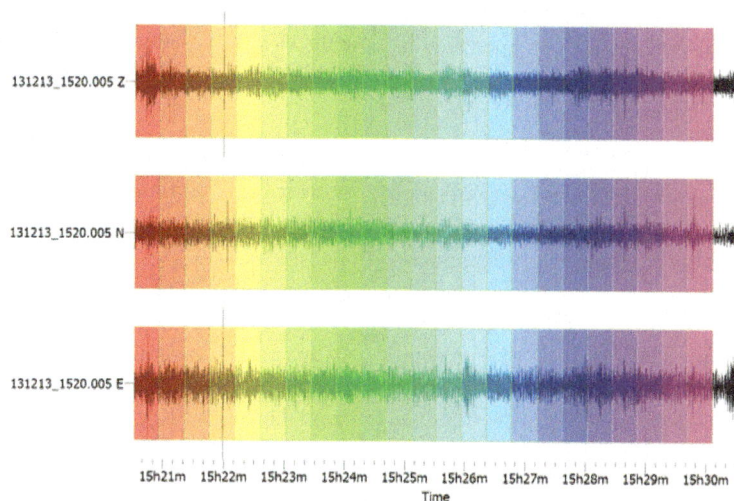

Figure 8. The time window (top) of 25 s for the previous recording with the HVSR result (bottom). Different colours corresponding in different time windows. Grey background colour presents the automatic recognition of fundamental frequency of the Geopsy program.

ambient noise, all recorded from the accelerometers of the wired structural health monitoring system of Technological Educational Institute in the city of Chania. One measurement of each kind for each sensor is presented. The analysis of all recorded data proves that there are very few and low-range disturbances in the measurements, and almost all the amplitudes and resonance frequencies are the same in HVSR graphs, for each kind of measurement.

2.5 Analysis of HVSR with specific processing software

Analysis of data with HVSR method is achieved by specific software (Geopsy) (www.geopsy.org, last access: 1 November 2013), developed in the frame of SESAME program (Site Effects assessment using Ambient Excitations). The

data process enquires specific procedures. Initially there is mean removal in the recordings. Then there is the specification of time. This section includes band pass filtering of the data in the frequency spectrum 0.5–30 Hz, range that it in the range of interest for in structural health monitoring (this frequency range includes the majority of resonance frequencies for structures and buildings), application of time window of 25 s according with the minimum frequency measured (0.5 Hz), no overlap of the windows and finally computation of frequency spectrum with fourier transform for the components north–south, east–west and Vertical and every time window. The next process part is referred in the parameters of smoothing of data. The Ohmachi and Konno smoothing method is applied with smoothing constant value at 40 and cosine taper width at 5 %. Finally there is computation of

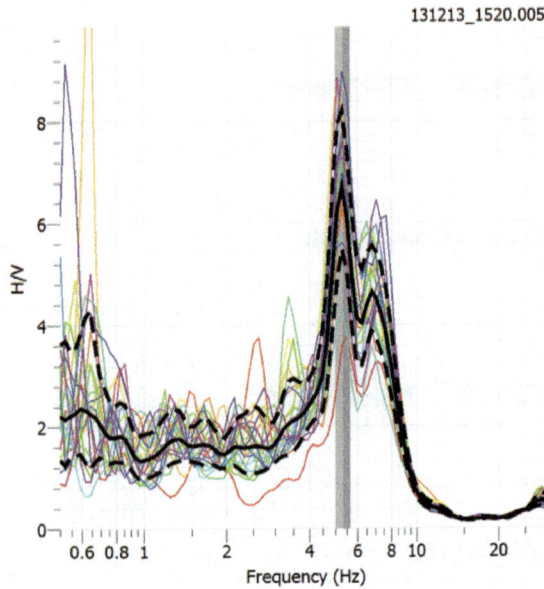

131213_1520.005

Figure 9. The time window of 25 s for the above recording (left) with the HVSR result (right). Different colours corresponding in different time windows. Grey background colour presents the automatic recognition of fundamental frequency of the Geopsy program.

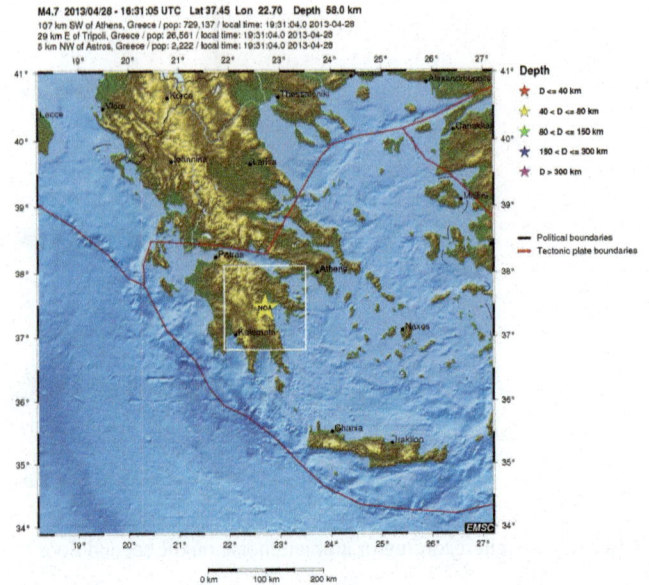

Figure 10. The 28 April 2013, 16:33 UMT time seismic event at 37.45 (latitude) 22.70 (longitude) 58.0 km (depth) by EMSC (http://www.emsc-csem.org).

horizontal to vertical ratio for every time window with log step of 100 number of samples and presentation of HVSR results. The horizontal component is computed by the geometric mean of N–S and E–W component for every time window with the relation $H(t) = \sqrt{\text{N–S}(t) \times \text{E–W}(t)}$. In case study recordings all time windows are kept (even if they present very high amplitudes) in order to reveal the effect that they have on the structures. There is interest also for microtremors (man-made excitation) and not only for microseismicity (environmental noise) because both affect the structural characteristics of a structure. Below, Fig. 7 presents the three components of a 10 min duration recording. (Vertical component is Z, north–south N and east–west E.) As the figure shows, some noises excitations are presented on three components and other only on two or one.

Figure 8 presents the same recording (with Fig. 7) of ambient noise acceleration on three components, windowed with specific length of 25 s (with different colour each time window), in order to separate specific time durations of the recording signal and analyse it with HVSR technique. Figure 9 presents the HVSR of the corresponding time window (for each colour). Such Fig. 9 presents the sum of HVSR plots of all the time windows of 25 s duration. The highest amplitude of HVSR plots is indicated with grey bold line on the background of graph in Fig. 9. The specific frequency is presented for the highest amplitude around 5.5 Hz. The corresponding amplitude reach up to 9. There is a dispersion on amplitudes because each time window (of Fig. 8) contains different amount of energy. Artificial noises that induces in case study recordings, contain more energy. Such in order

to approximate the real HVSR value there is a statistical approach of many time windows in order to indicate the containing artificial noises and minimize their effect.

3 Data analysis

In this part of study there is process analysis of an earthquake that occurred at 28 April 2013 at 16:31 UTC 250 km away from the buildings at TEI Chania, with a magnitude of 4.7 M. Data were recorded by 130-SMA instrumentation. On Fig. 10 is depicted the map epicentre of the case study earthquake.

Figure 11 reveals the HVSR analysis of the seismic event of Fig. 10, as it was recorded by the SMA accelerometers on TEI buildings at Chania (old and new building). The acceleration time series have been separated in three time duration zones (red, green and blue) with 25 s time duration, each time window. Red window contains the "primary wave" of the earthquake, green the "secondary wave" and blue the "after the secondary wave". Figure presents that the amplitude of HVSR plot is higher as the energy is getting higher. Such the HVSR amplitude plot of P wave is much lower than the S wave plot. Also the HVSR amplitude rises as the floor rises. Another four seismic tests will reveal if the HVSR remains stable due to different parameter seismic excitation on both case study buildings. Figure 12 presents broad region map of these four seismic events with code names 1 (top left), 2 (top right), 3 (bottom left) and 4 (bottom right).

Below Fig. 13 presents the computation of HVSR for seismic event with code names 1 (top left), 2 (top right) of Fig. 12. On Fig. 13, HVSR plot graph with label A3OB is the recording of old building third floor, HVSR plot graph

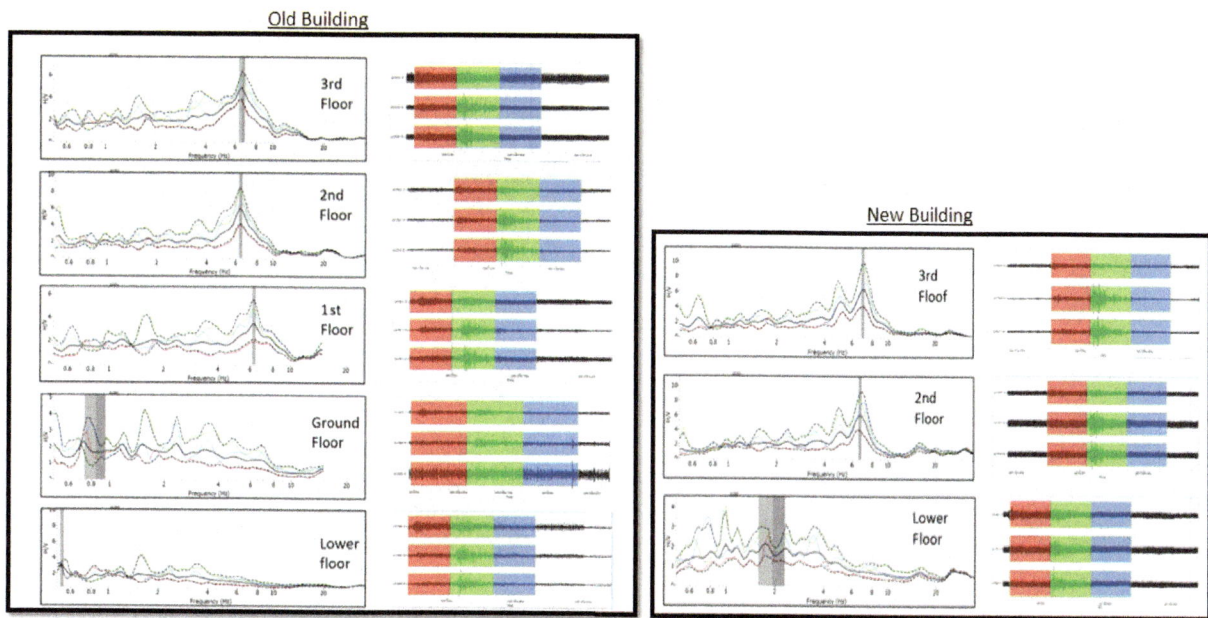

Figure 11. HVSR recordings on 28 April 2013, 16:33 UMT (28 April 2013, 19:33 LT) on the old (left) and new (right) building of TEI in Chania for the above seismic event. With red colour is 25 s window of P wave, with green the 25 s window of S wave and with blue the 25 s window of after S wave window.

with label B2OB is the recording of old building second floor, HVSR plot graph with label C1OB is the recording of old building first floor, HVSR plot graph with label D0OB is the recording of old building ground floor, HVSR plot graph with label ELOB is the recording of old building lower floor, HVSR plot graph with label F3OB is the recording of new building third floor, HVSR plot graph with label G2OB is the recording of new building second floor, and HVSR plot graph with label HLOB is the recording of new building lower floor.

On Fig. 14 are presented the computations of HVSR for seismic events with code names 3 (bottom left), 4 (bottom right) of Fig. 11.

3.1 One-hour time duration HVSR recordings of ambient noise during noon and midnight

It is going to present an HVSR recording of 30 min duration at noon (13:00 LT) in order to study the HVSR rise from floor to floor. Figures 15 and 16 present the HVSR recordings at noon and at night, on the old building of TEI with the presentation of, 25 s time duration windows, which are extended in the whole length of the recording.

On Fig. 17, is presented the HVSR analysis of ambient noise, five months later for the old and the new building of TEI at Chania. The label of the HVSR plot graphs, refers to the same floor and building, as with Figs. 13 and 14.

From Figs. 15, 16 and 17 we conclude that resonance frequencies of both buildings remain stable for each floor and also the amplitude remains the same for each corresponding

frequency. Furthermore, the increase of the amplitude for old and new building follows the same pattern. For the old building, on the third floor, the first resonance frequency is in the range of 6 Hz and second at 8 Hz. The same resonance frequencies are reveal on the second floor and first floor with lower HVSR amplitude, during noon and midnight. On the ground floor and lower floor of the old building the energy of the ambient noise is so low, that there is no effect from the transfer function of the building, and the ambient noise of the surround area is revealed on the lower floors of the building (the HVSR outside TEI is in the range of 0.55–0.65 Hz). For the new building on the third floor the first resonance frequency is presented in the range of 5.7 Hz and the second in the range of 8 Hz. On the second floor the value of HVSR lower. On the lower floor of the new building the energy of ambient noise is again so low, that it is not able to excite the resonance frequencies of the building and such the HVSR plot is almost a smooth curve, highlighting the HVSR value of the area outside TEI.

3.1.1 Measurements with Lennartz seismometer

In this section it is going to analyse HVSR measurements that were recorded in old and new building of TEI Chania and also in a public building located in the city of Chania Technical Chamber of Greece (TEE). Below Fig. 19 presents HVSR recordings for lower ground, first and second floor of the old building TEI. Length of time duration recording is on 10 min. The same time duration of acceleration recording is

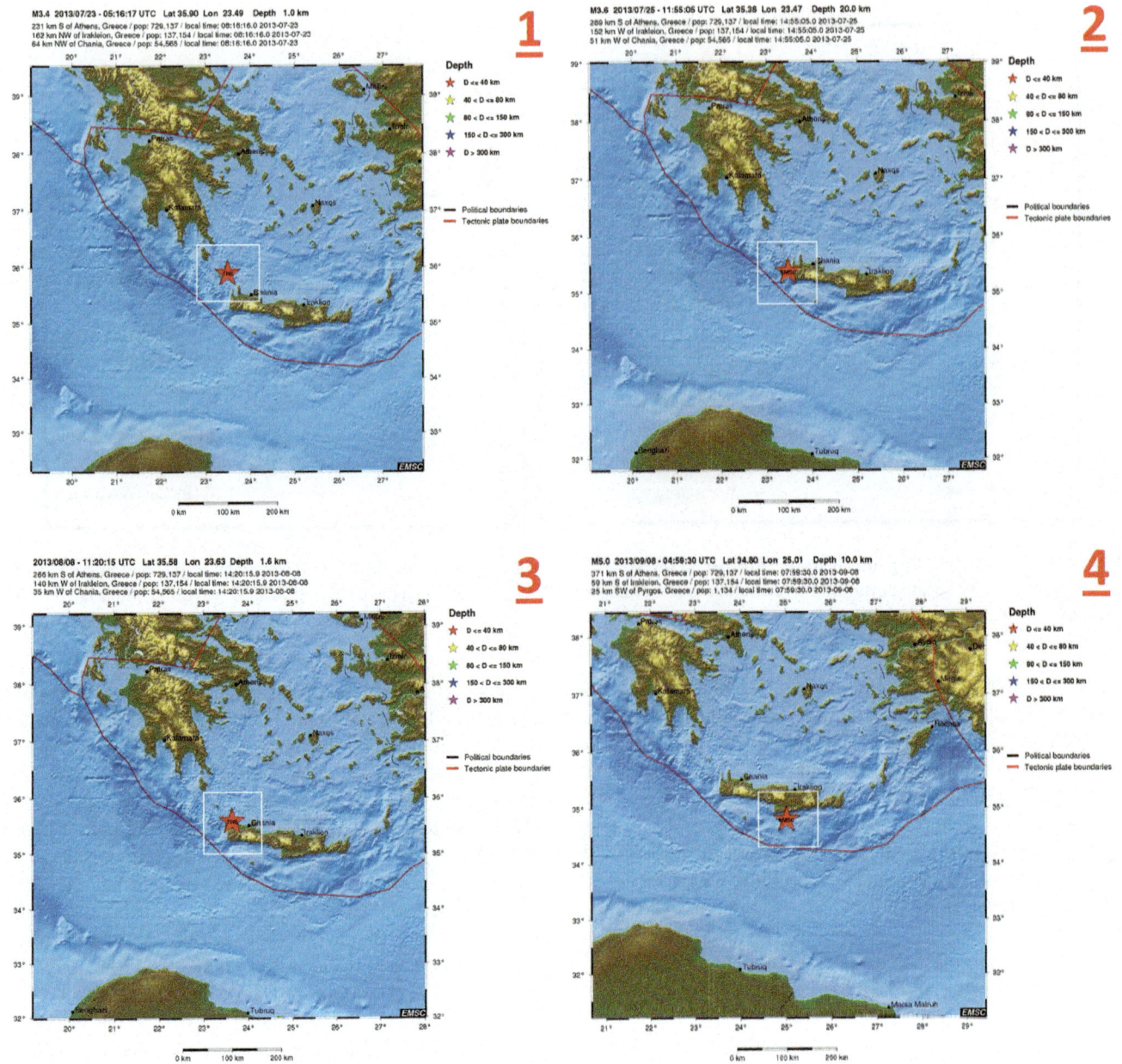

Figure 12. Four seismic events with code names 1 (top left), 2 (top right), 3 (bottom left) and 4 (bottom right) by EMSC (http://www.emsc-csem.org).

on Fig. 18 which presents HVSR recordings for outside and inside the building of TEI.

On Fig. 20 is presented HVSR recordings for lower, ground and first floor of the new building TEI. Length of time duration recording is on 10 min. Figure 21 presents HVSR recordings for second and third floor of the new building TEI.

On this part of the study will be presented the HVSR from acceleration recordings of the TEE building located in the city of Chania. Figure 22 presents the HVSR plots for the lower, ground and first floor of the case study building.

Every kind of HVSR measurement in the old and new building, reveals that HVSR rises with higher rate in the old building and with lower in the new building. Under earthquake excitation or under ambient noise, old building, which has been affected by much more load from seismic and man-made excitation but also from the time (as it is much older than the new building), presents higher HVSR rise, from floor to floor. Also the TEE building which is a very old building, with many visible cracks on its structure presents much higher HVSR rise than the two buildings of TEI. HVSR rise could function as an indicate from possible

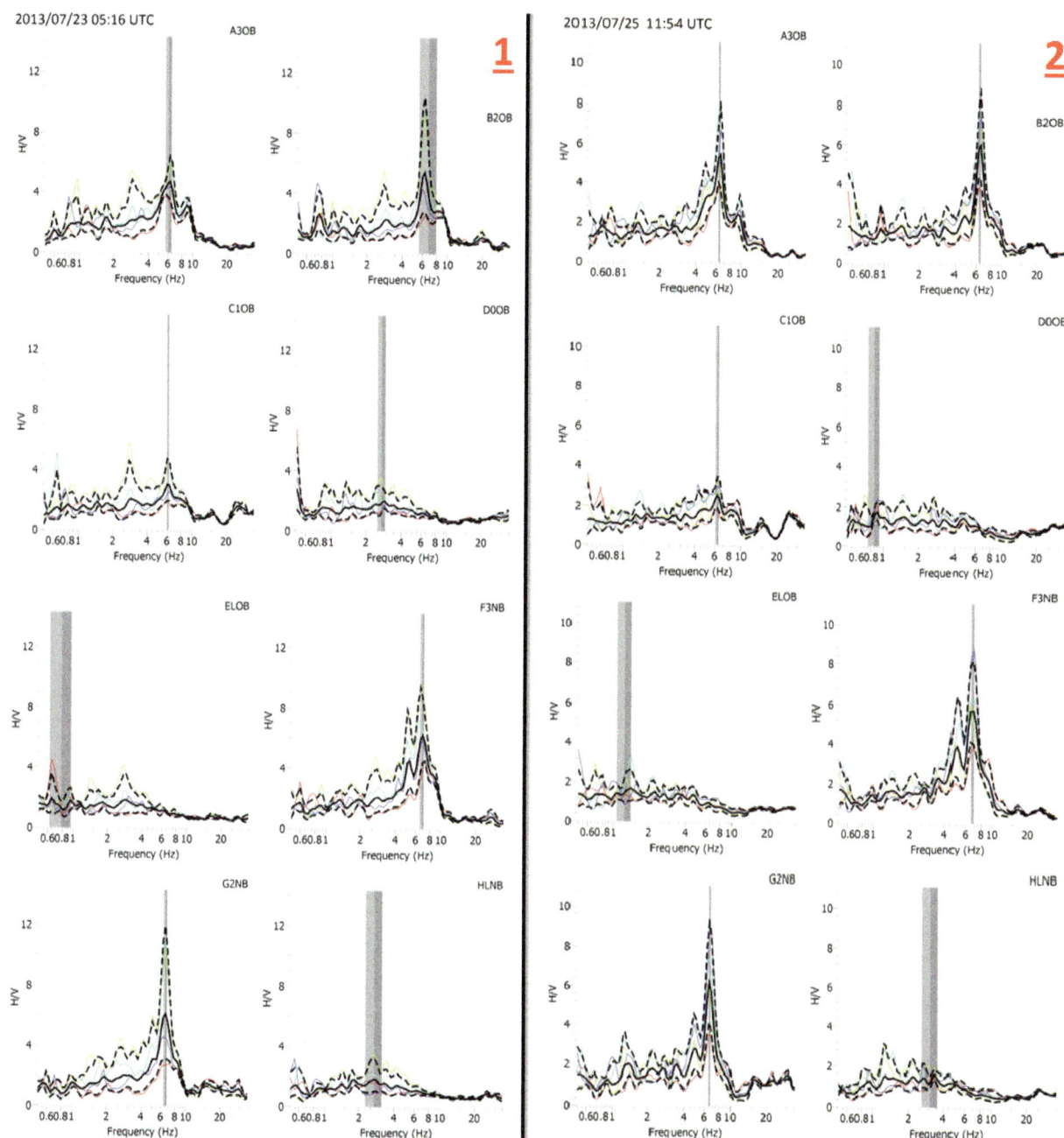

Figure 13. HVSR analysis of seismic events with code names 1 (left) and 2 (right).

high vulnerability of a structure. From the above recordings, it is observed that in both earthquakes the FR was almost the same for the old and the new building for each floor. As the floor goes higher the FR also goes higher. The windows that include S waves have much higher FR than the windows with P waves and the windows after S waves. In the programmed recordings of 30 min at noon and night the HVSR is much more lower at every floor but it also rises as the floor gets higher with a much more low rate that in seismic events.

4 Results

Accelerometers present lower HVSR than seismometers. This is expected as acceleration is the derivative of speed. The HVSR recordings of accelerometers are the derivative of HVSR of seismometers. The resolution of accelerometers, from frequency spectrum higher than 2 Hz is very high, compared with resolution of seismometers, presenting all the resonance frequencies of the structures with great detail. For buildings that fundamental frequency is higher than

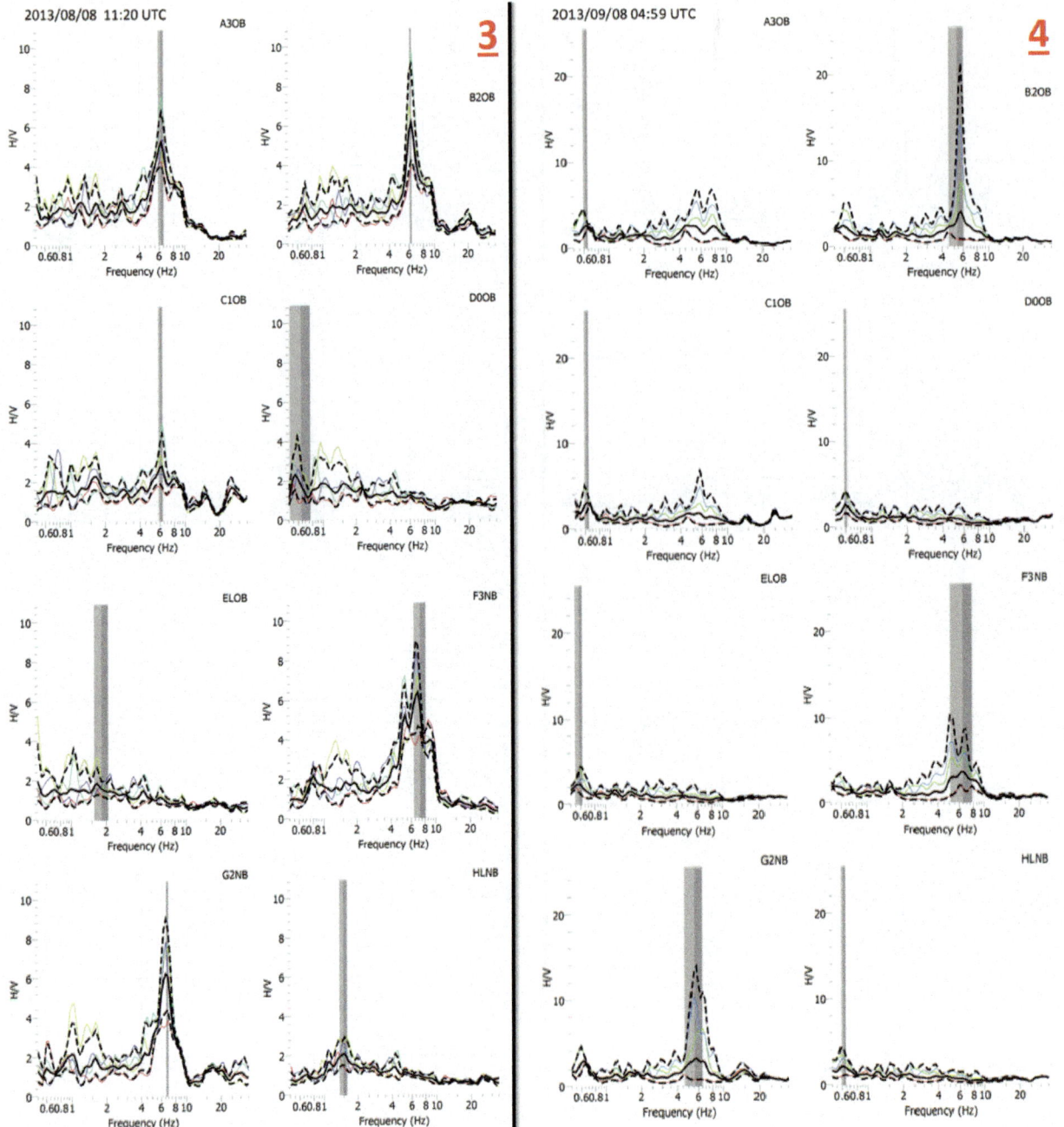

Figure 14. HVSR analysis of seismic events with code names 3 (left) and 4 (right).

2 Hz accelerometers could be used of HVSR measurements. As the floor gets higher the amplitude of the HVSR index rises in the range of resonance frequencies. This indicates that the differential acceleration from floor to floor increases and such increase the vulnerability of the structures as the getting higher. (There is specific threshold of differential acceleration from floor to floor that indicates damage when the value overpasses this threshold).

HVSR plots of ambient noise of old and new TEI buildings, for each floor reveal that:

- Analysis of acceleration recordings under seismic activity present the same frequency spectrum under FFT analysis and under FR analysis.

- Processing of ambient data with HVSR method and earthquake data with FR method, present almost the same analogy of amplitudes increase, for the same frequencies.

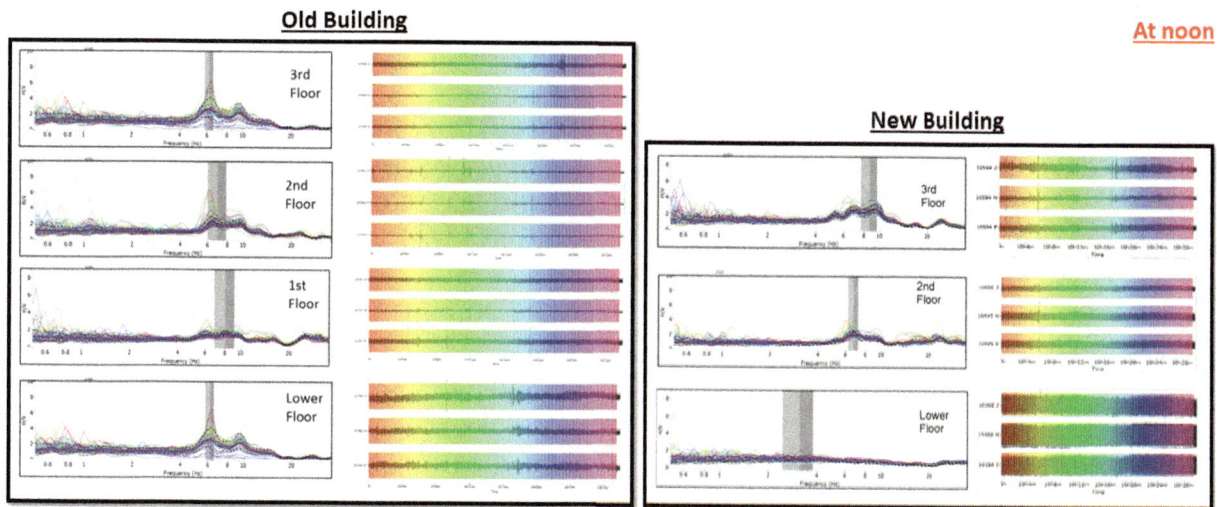

Figure 15. Programmed 30 min HVSR recordings on 8 May 2013, 10:00 UMT (8 May 2013, 13:00 LT) on the old building of TEI. On the right is the presentation of 25 s time duration windows which are extended in the whole length of the recording.

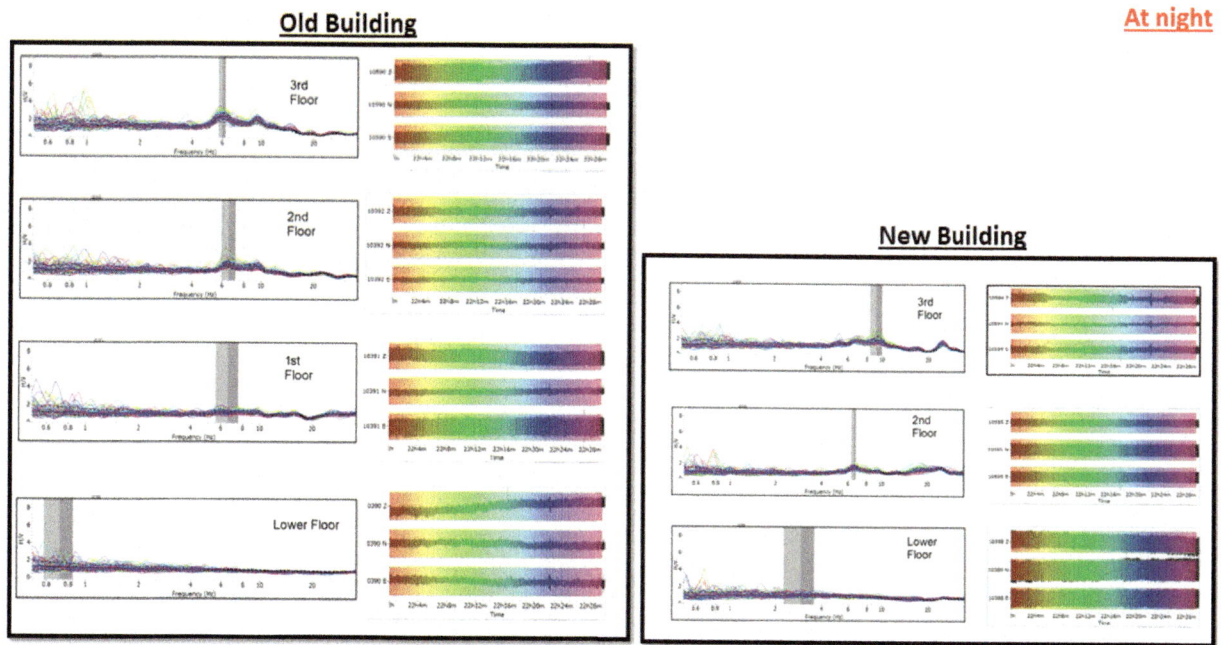

Figure 16. Programmed 30 min HVSR recordings on 7 May 2013, 22:00 UMT (8 May 2013, 01:00 LT) on the old (left) and new (right) building of TEI. On the right is the presentation of 25 s time duration windows which are extended in the whole length of the recording.

– Old building of TEI, is revealed to have higher receiver function (RF) under earthquake excitation, but also higher HVSR under night and day ambient noise.

Conclusively ambient noise analysed with HVSR method could present the amplitudes that affect each floor of a structure and also present an index (Figs. 23 and 24) of effect of seismic activity on each floor of a building. This index is the tilt of the graph from floor to floor in each building for each kind of measurement and presents how the HVSR rises as the floor gets higher in each condition (under earthquake excitation, and/or under ambient noise during night and noon).

5 Description of the proposed index

The specific index presents the change of HVSR of the fundamental frequency of the structure from one floor to another. In ideal conditions this index should be stable. As the HVSR increase the influence of site amplification in the specific structure increases. And if the HVSR increase on

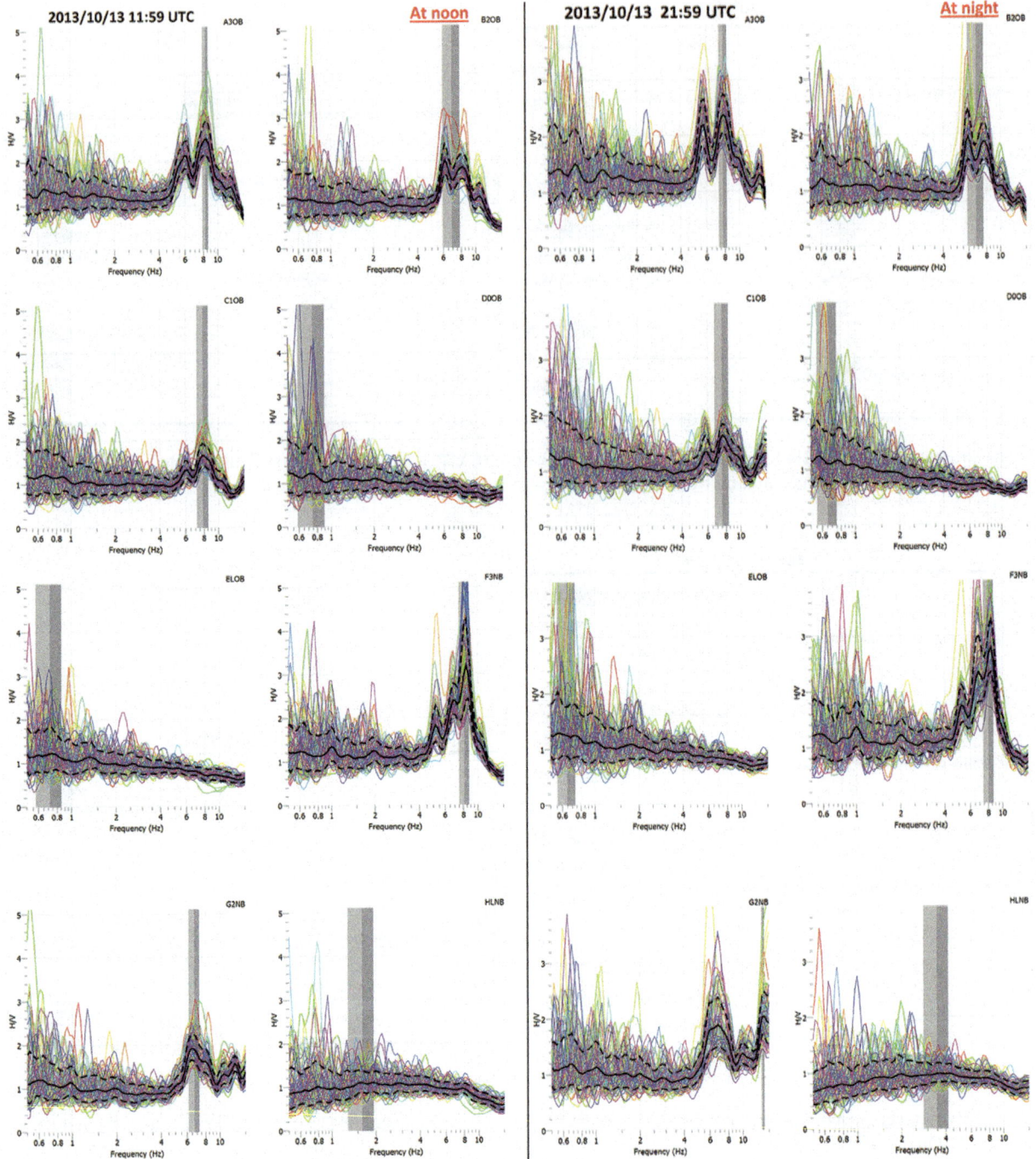

Figure 17. HVSR analysis of ambient noise recording with time duration of 1 h.

higher floors it indicates that there is higher vulnerability of the structure in higher floors. The fundamental frequency of HVSR recordings at each floor is being measured and there is comparison of the increase of the value of HVSR, and it is correlated this rise with the increase of probably vulnerability of the structure. On Figs. 23 and 24, is presented the rise of HVSR maximum amplitude value, for the three case scenarios (night, noon and under seismic excitation) for the old and the new TEI building respectively.

There is effort to approach this index from the field of digital signal processing rather than the civil engineering, and correlate the increase of HVSR with the increase of the amplitude of structure acceleration of the building. This approach is instead of a simple value of horizontal to vertical spectral rations to study the rate of increase or decrease of

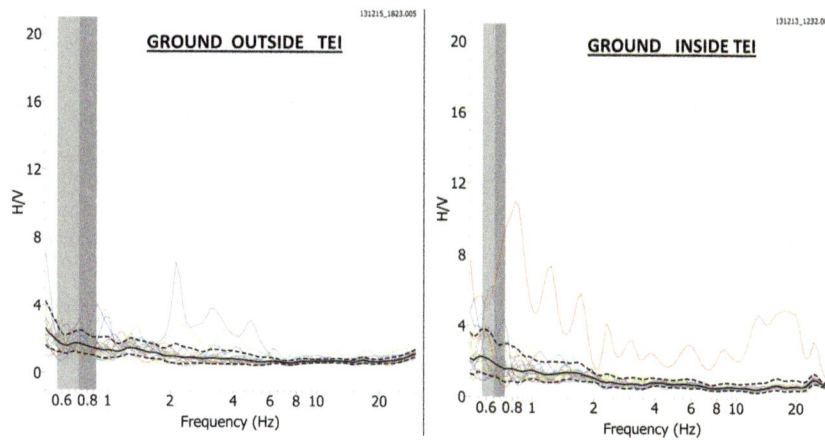

Figure 18. HVSR plots of outside area (left figure) and inside building of TEI on the lower floor (right figure).

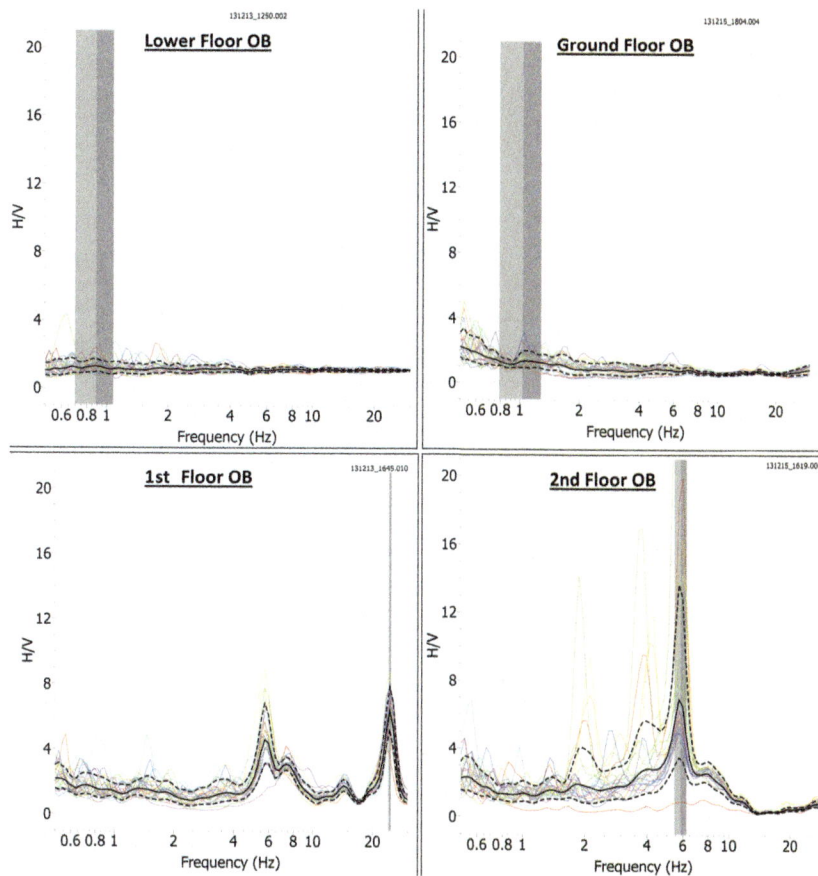

Figure 19. HVSR plots of acceleration recordings for lower floor old building (top left figure), ground floor old building (top right figure), first floor old building (bottom left figure) and second floor old building (bottom right figure).

this value as the floor gets higher or lower. The RF under seismic excitation, as well as the HVSR under night and day ambient noise, is higher for the old building in relation with the new one.

6 Laboratory validation

The proposed approach is verified by results of laboratory model. On a metallic model (Dexion) of dimensions 2 m (height) 1.2 m (length) 50 cm (width) with four rack levels have been placed four accelerometers Reftek 130-SMA, one

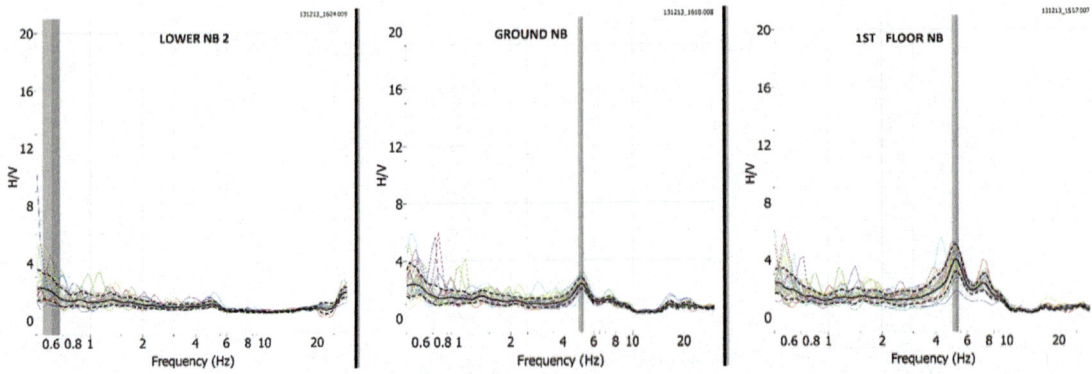

Figure 20. HVSR plots of acceleration recordings for lower floor new building (left figure), ground floor new building (middle figure) and first floor old building (right figure).

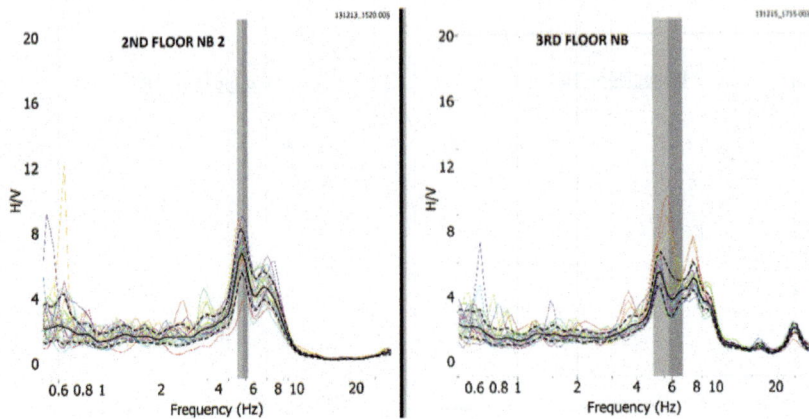

Figure 21. HVSR plots of acceleration recordings for second floor new building (left figure), and third floor new building (right figure).

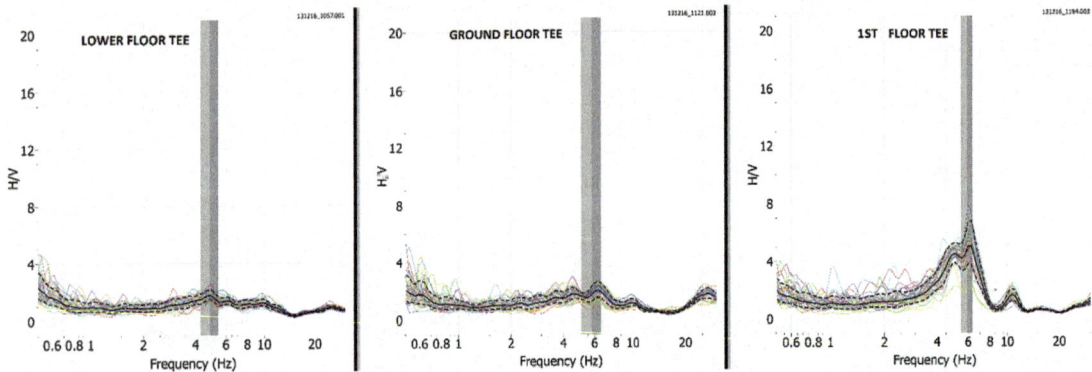

Figure 22. HVSR plots of acceleration recordings for lower floor of TEE building (left figure), ground floor of TEE building (middle figure) and first of TEE building (right figure).

on each level. The metallic columns of model is connected with the racks through screws. Two damage scenarios are applied in the specific case study. Initially there is no damage at all. All screws are absolute screwed. HVSR recordings of 1-hour time duration are applied in every level during day and night. The same HVSR recordings of 1 h are applied also for the second scenario (damaged scenario). In this case study artificial damage has been introduced in the model through the relax of specific screws in the right corner of the third rack. The results of the data analysis reveal that HVSR rises as the damaged introduced to the system (as stiffness of model is reduced). This is also correlated with higher structural vulnerability of the model. The accelerometer sensors have specific maze which is added on the total maze of the

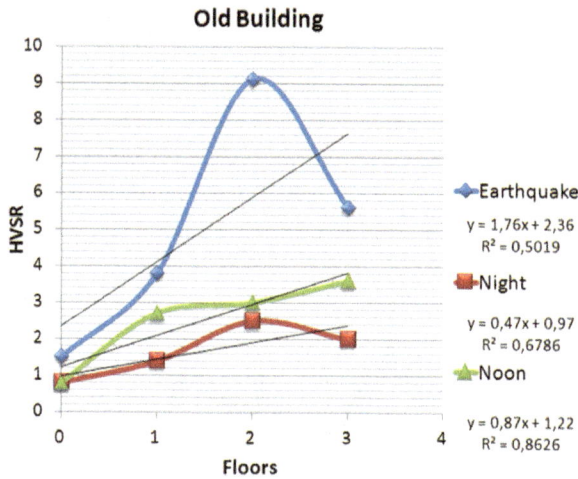

Figure 23. HVSR recordings from the 9 March 2013, 07:43 UMT time seismic event (blue line), the 12 March 2013 midnight recording (red line) and the 12 March 2013 noon recording of the old building of TEI in Chania.

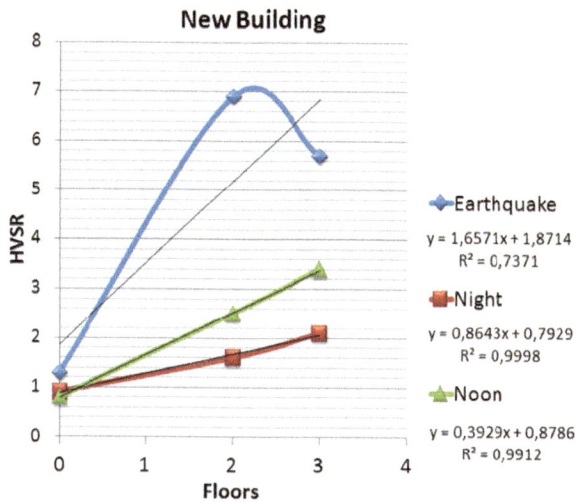

Figure 24. HVSR recordings from the 9 March 2013, 07:43 UMT time seismic event (blue line), the 12 March 2013 midnight recording (red line) and the 12 March 2013 noon recording of the new building of TEI in Chania

rack-based model. Also they are stable related the metallic model. On Fig. 25 is presented the photo of metallic Dexion. Figures 26 and 27 present the HVSR analysis of ambient noise acceleration recording of the metallic model, for the undamaged scenario. Figures 28 and 29 present the increase of the maximum amplitude of the HVSR value, for the damaged scenario. Also on Figs. 26, 27, 28 and 29 the HVSR plot graphs with label 1_LEV is the recording of first (lower) rack of the metallic model, and respectively label 2_LEV is the recording of the second rack, 3_LEV the recording of the third rack and 4_LEV the recording of the fourth (highest) rack.

Figure 25. Metallic Dexion model.

Resonance frequencies (at 6 and 8 Hz) are almost stable and very close to the value of 2 for all levels. The HVSR rise is very low. The range of the values is from 1.5 until 2.5.

Resonance frequencies have an HVSR range from 1.8 up to 3.5. Also there is frequency shift at all response frequencies in damaged scenario related the undamaged scenario. At damaged scenarios, ambient noise creates HVSR rate much higher than in undamage scenario. Also the increase of the HVSR index from level to level is higher in damage case than the undamaged.

7 Discussion

HVSR technique has been used in order to present the impact of seismic activity in two building (different age), and how it differs from floor to floor. In this work there is a prototype approach of HVSR method in SHM recordings. SHM data gathered from wired SHM system are analysed by HVSR method. These data present the effect of ambient noise on the ground floor, the second and the third of two university buildings and also present the effect of amplification site which could define vulnerability of each floor and the whole building. This work uses HVSR technique to compare ambient noise in both buildings (an old 19 years and a new 7 years), search the site amplification in ground floor, second and third and try to find out if these recordings extract interesting findings in terms of site amplification. Also is trying to find out possible differences and similarities in the structural response of both buildings under

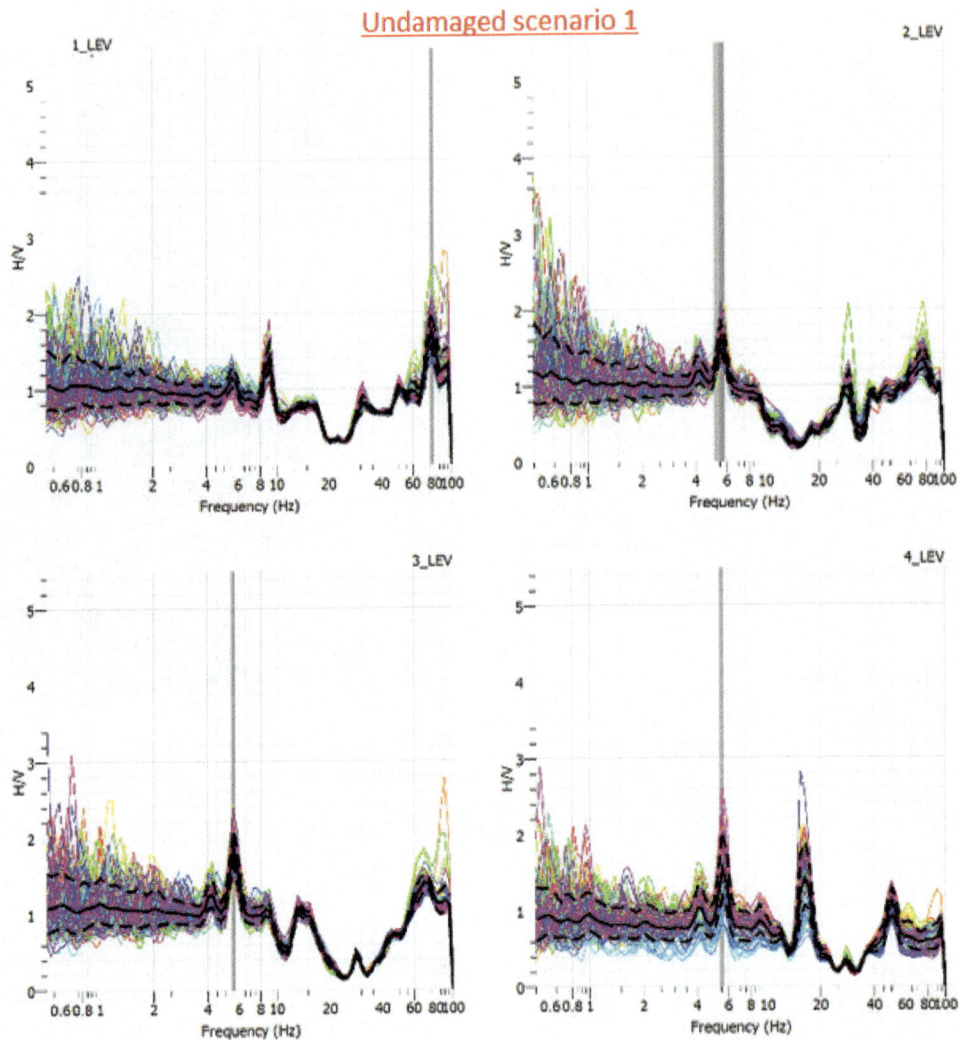

Figure 26. HVSR analysis of ambient noise recording, for undamaged scenario 1 of the metallic model.

seismic and ambient excitation. From the earthquake data it is observed that HVSR is much higher in every floor than the programmed time measurements and also that these values follow the pattern of the seismic acceleration of the building (the transfer function of the building) where the amplitude of seismic acceleration of the second floor is higher than the amplitude of the third floor.

8 Conclusions

Horizontal to vertical spectral ratio (HVSR) and receiver function (RF) methods, have been applied in microtremors and earthquake acceleration recordings, in order to study the resonances frequencies and their spectral amplitude, that exist in two concrete buildings, in a high seismogenic region. These frequencies are in the range 5.5–6.5 Hz. The site amplification on the area that case study buildings are located, is much lower (around 0.7 Hz). The HVSR rise as floor gets

higher. In this study the increase of HVSR is strongly related to the age of the buildings and the visible cracks in the beams. HVSR also indicates higher differential acceleration from floor to floor and such higher structural vulnerability. This work presents for first time an approach of HVSR by implementation of the method for structural health monitoring. More specifically it applies HVSR in each floor of buildings, finds out the different HVSR values and suggests a new index which compares and analyses the HVSR of the fundamental frequency in each floor of a building and how this value changes. Also it searches the possibility of correlation of this value with the vulnerability of a building and presents that as HVSR rises as floor gets higher the vulnerability of the building could rise for these floors. Analysis of building vulnerability provided by HVSR and/or RF method, is a very cost effective and fast method which use simple acceleration recordings and provides information for structural vulnerability. The data results of this study reveal that earthquake

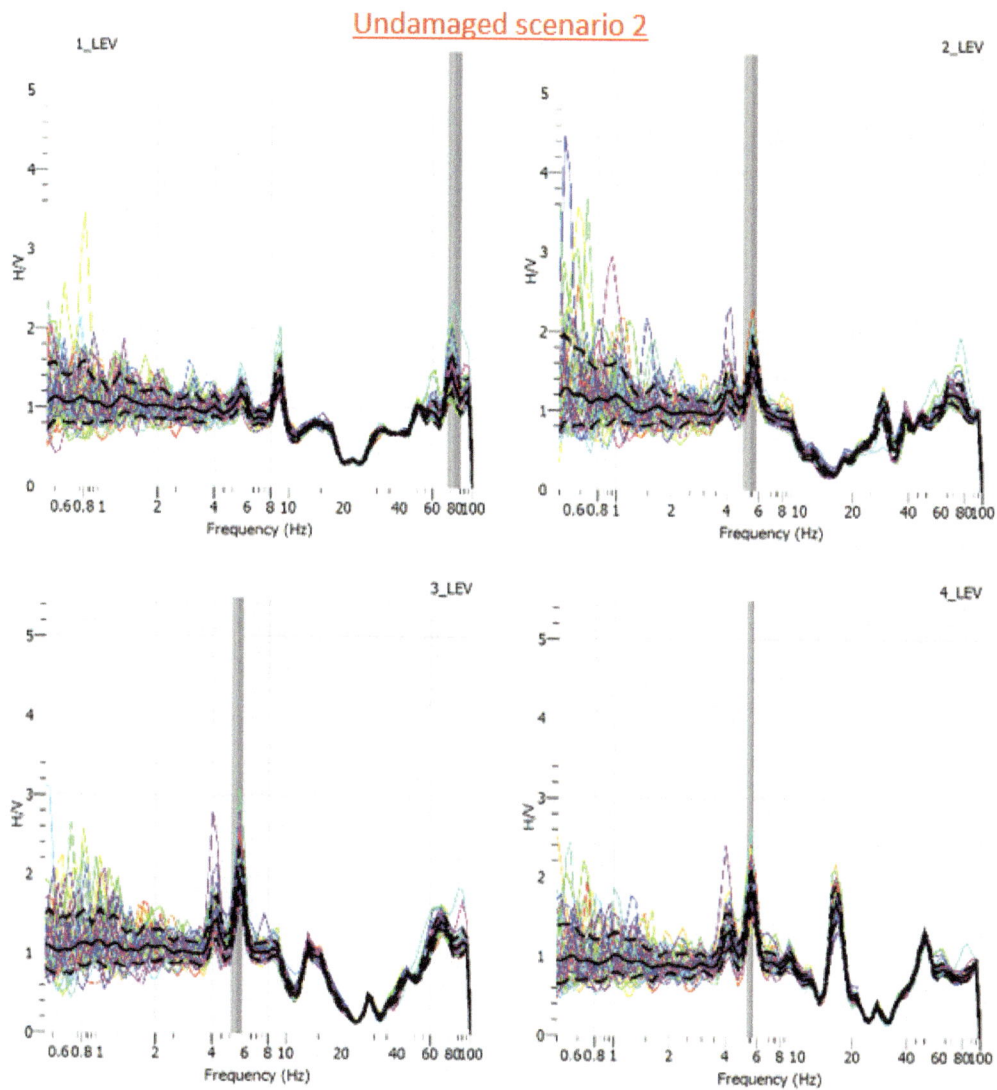

Figure 27. HVSR analysis of ambient noise recording, for undamaged scenario 2 of the metallic model.

excitation follow the same analogy of RF rise (as the floor rises) as the HVSR of ambient noise excitation. This could indicate the way that a structure could response under strong seismic excitation, by recording simple environmental noise (microtremors).

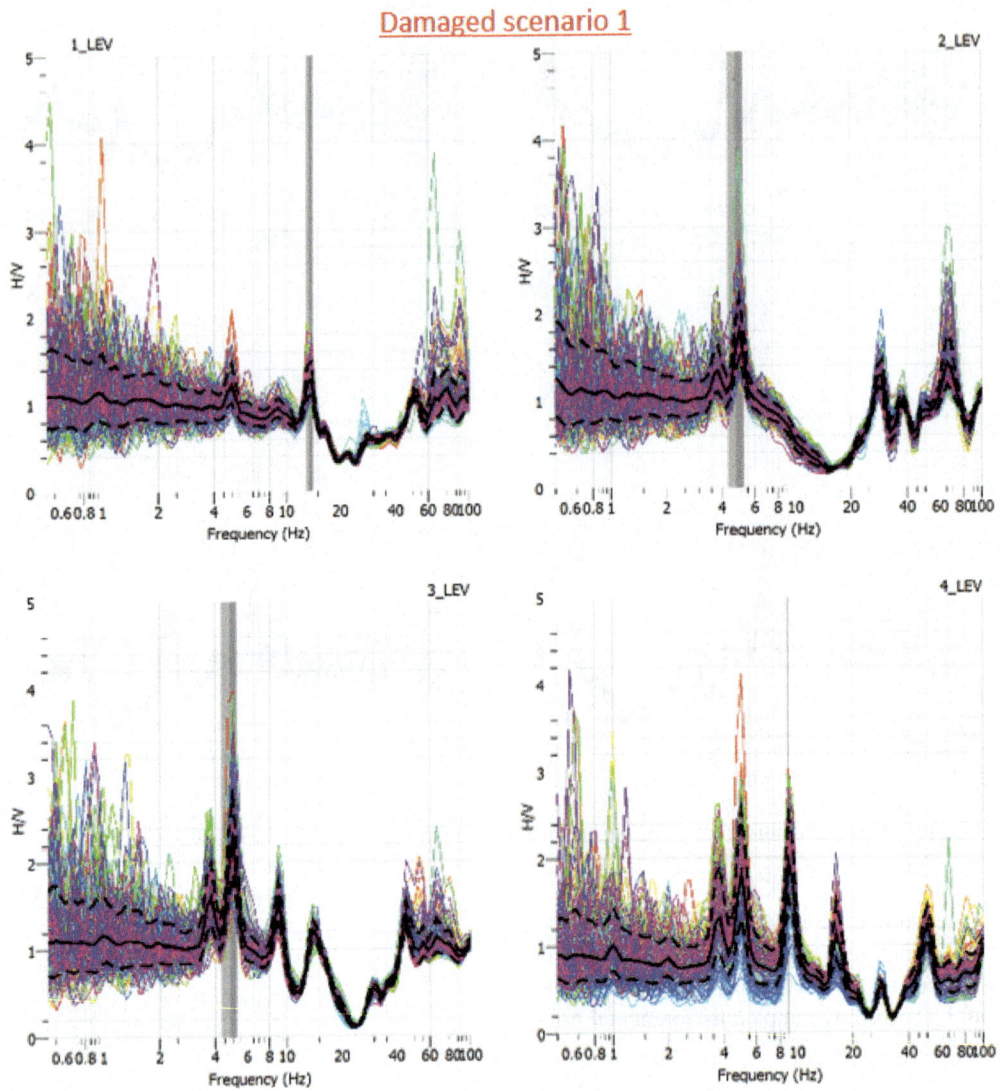

Figure 28. HVSR analysis of ambient noise recording for damaged scenario 1 of the metallic model.

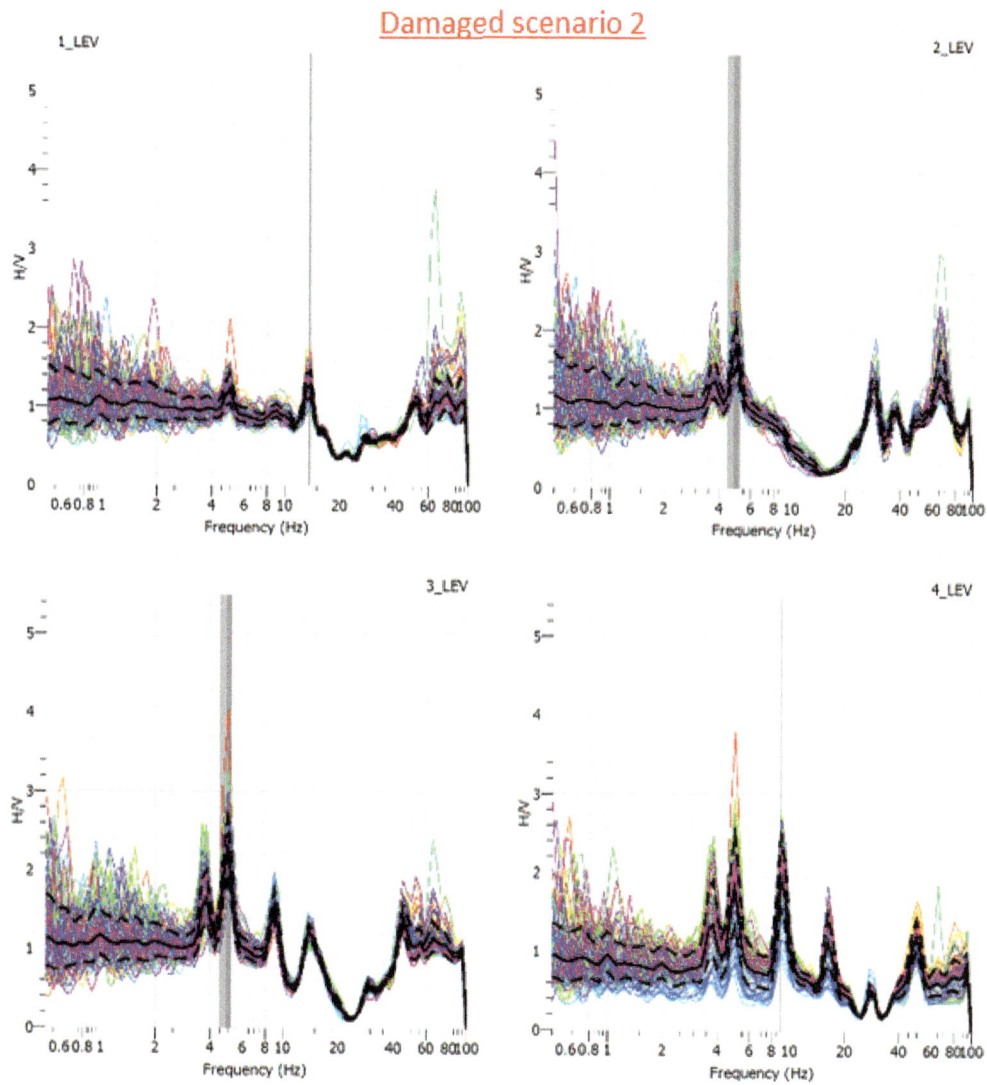

Figure 29. HVSR analysis of ambient noise recording for damaged scenario 2 of the metallic model.

Acknowledgements. This work was supported by State Scholarships Foundation of Greece (IKY).

Edited by: N.-T. Nguyen
Reviewed by: two anonymous referees

References

Carniel, R., Malisan, P., Barazza, F., and Grimaz, S.: Improvement of HVSR technique by wavelet analysis, Soil Dynam. Earthq. Eng., 28, 321–327, 2008.

Carniel, R., Barbui, L., and Malisan, P.: Improvement of HVSR technique by self-organizing map (SOM) analysis, Soil Dynam. Earthq. Eng., 29, 1097–1101, 2009.

Chávez-García, F. J. and Tejeda-Jácome, J.: Site response in Tecoman, Colima, Mexico – I: Comparison of results from different instruments and analysis techniques, Soil Dynam. Earthq. Eng., 30, 711–716, 2010.

Dimitriu, P., Kalogeras, I., and Theodulidis, N.: Evidence of nonlinear site response in horizontal-to-vertical spectral ratio from near-field earthquakes, Soil Dynam. Earthq. Eng., 18, 423–435, 1999.

Fäh, D., Kind, F., and Giardini, D.: A theoretical investigation of average H/V ratios, Geophys. J. Int., 145, 535–549, 2001.

Liu, L., Chen, Q. F., Wang, W., and Rohrbach, E.: Ambient noise as the new source for urban engineering seismology and earthquake engineering: A case study from Beijing metropolitan area, Earthquake Sci., 27, 89–100, 2014.

Lombardo, G. and Rigano, R.: Local seismic response in Catania (Italy): A test area in the northern part of the town, Eng. Geol., 94, 38–49, 2007.

Luo, G. C., Liu, L. B., Qi, C., Chen, Q. F., and Chen, Y. P.: Structural response analysis of a reinforced concrete building based on excitation of microtremors and passing subway trains, Chin. J. Geophys., 54, 2708–2715, 2011.

Mucciarelli, M., Contri, P., Monachesi, G., Calvano, G., and Gallipoli, M.: An empirical method to assess the seismic vulnerability of existing buildings using the HVSR technique, Pure Appl. Geophys., 158, 2635–2647, 2001.

Nakamura, Y.: Method for dynamic characteristics estimation of subsurface using microtremor on the ground surface, Quarterly Report of RTRI (Railway Technical Research Institute) (Japan), 30, 25–33, 1989.

Nogoshi, M. and Igarashi, T.: On the amplitude characteristics of microtremor (part 2), J. Seismol. Soc. Jpn., 24, 26–40, 1971.

Parolai, S., Richwalski, S. M., Milkereit, C., and Bormann, P.: Assessment of the stability of H/V spectral ratios from ambient noise and comparison with earthquake data in the Cologne area (Germany), Tectonophysics, 390, 57–73, 2004.

Panou, A. A., Theodulidis, N., Hatzidimitriou, P., Stylianidis, K., and Papazachos, C. B.: Ambient noise horizontal-to-vertical spectral ratio in site effects estimation and correlation with seismic damage distribution in urban environment: the case of the city of Thessaloniki (Northern Greece), Soil Dynam. Earthq. Eng., 25, 261–274, 2005.

Papadopoulos, I.: Experimental and theoretical study of site amplification by usage of microtremors and field geophysical measurements, PhD, Dept. of Geology, University of Thessaloniki, Thessaloniki, 2013.

Pentaris, F. P., Stonham, J., and Makris, J. P.: A review of the state-of-the-art of wireless SHM systems and an experimental set-up towards an improved design, in: IEEE EuroCon'13, Zagreb, 2013, 275–282, 2013.

Sokolov, V. Y., Loh, C.-H., and Jean, W.-Y.: Application of horizontal-to-vertical (H/V) Fourier spectral ratio for analysis of site effect on rock (NEHRP-class B) sites in Taiwan, Soil Dynam. Earthq. Eng., 27, 314–323, 2007.

Triwulan, W., Utama, D., Warnana, D., and Sungkono: Vulnerability index estimation for building and ground using microtremor, The second International Seminar on applied Technology, Science and Arts, 2010.

Yuncha, Z. A. and Luzon, F.: On the horizontal-to-vertical spectral ratio in sedimentary basins, Bull. Seismol. Soc. Am., 90, 1101–1106, 2000.

12

Thermal imaging as a modern form of pyrometry

U. Kienitz

Optris GmbH, Berlin, Germany

Correspondence to: U. Kienitz (ulrich.kienitz@optris.de)

Abstract. Pyrometers and thermography cameras used to be characterized by different specifications and technical definitions. After an analysis of the market situation and the physical basics, the following article describes common methods to determine optical and thermal key parameters. Based on this, aspects of future sensor developments and certain applications of infrared (IR) cameras are discussed.

1 Market and trends

Military and security applications will continue to be the main markets for infrared (IR) thermometry. The night combat capabilities provided by this technology have completely changed combatants' attacking strategies in conflicts like those in the Middle East. Another use for target acquisition devices which are immune to adverse weather and light conditions can be found in drone applications (FLIR Corporation, 2013). In the civilian sector, however, the current use of thermal imaging cameras in motor vehicles – with an installation rate below 1 % – is still far short of initial expectations (YOLE Developement, 2010). But the market for thermal imaging measurement systems has nevertheless greatly expanded. The sole driver of growth has been the price developments of these mainly hand-held IR cameras. More efficient production methods, particularly the wafer-level vacuum packing for the widely used thin-layer bolometers, have significantly reduced product costs. Hand-held units have traditionally been used for electrical and mechanical maintenance applications, as well as for detecting thermal leakages in buildings. Stationary IR cameras, when used for test and measurement and in automation applications, have benefited from the availability of inexpensive IR focal plane arrays.

2 Imaging devices are displacing pyrometers

Point-measuring radiation thermometers have been preferred over cameras for industrial measurement applications because of their ruggedness, ease of use and ease of integration. The advantages that cameras offer in locating thermal inhomogeneities have been tempered by their significantly higher costs. Ten years ago, hand-held pyrometers maintained a predominant market share for maintenance applications. But since then, the market has shifted such that more and more imaging devices are being used. This shift away from the use of pyrometers is also occurring for the temperature measurement systems used in stationary applications by machine and equipment builders. New systems often use cameras instead of pyrometers because the cameras offer better variability and programmability of measurement positions.

The technical properties of cameras and pyrometers differ significantly. The VDI 3511 standard describes basic parameters of IR thermometers (VDI/VDE 3511, 2004). Similar definitions for thermal cameras only exist currently in draft form (VDI/VDE 3511, 2010). The key measurement criteria for both types of devices are compared below. We are limiting our focus here on economically relevant measurement systems used to detect objects within the standard ambient temperature range. For reasons of physical radiation, such devices usually operate in the 8–14 μm spectrum. For their detectors, the IR cameras use uncooled thermal micro-bolometer arrays that range from 80×80 pixels to 1024×768 pixels, with pixel sizes ranging from 50×50 to 12×12 μm (Durand et al., 2011). Such focal plane arrays (FPAs) require a mechanical shutter for offset compensation. This is necessary due to the basic principle of

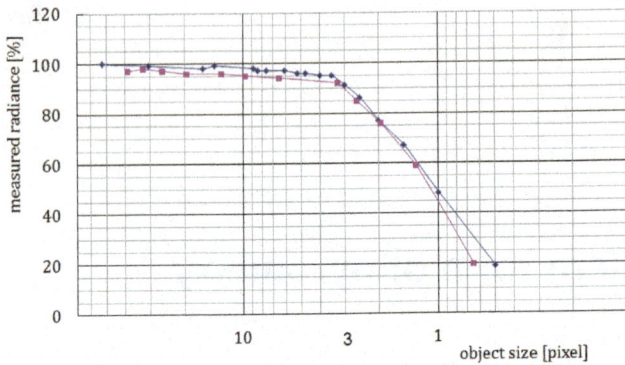

Figure 1. Geometric resolution of two commercially available IR cameras with an FOV of approximately 20°.

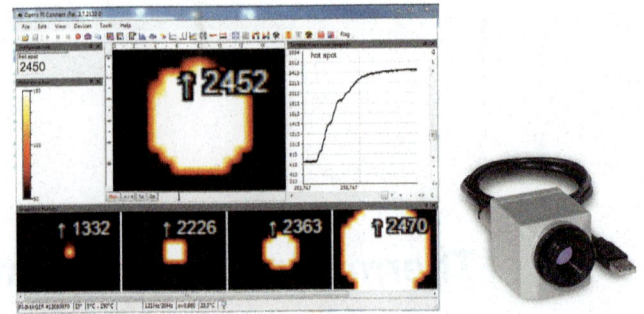

Figure 2. Radiation proportional values for the central pixel at illuminations of **(a)** 1×1, **(b)** 3×3 and **(c)** 5×5, and **(d)** a sufficient number of pixels (infrared camera PI160 from Optris; 23° lens; values from a 14 bit video AD converter).

bolometric resistance bridges and the non-negligible drift in the amplification and serialization electronics. Because of their constant light sensitivity, thermopiles with effective sensor diameters ranging from 0.4 to 2 mm are normally used for radiation thermometers (Datasheet Heimann Sensor GmbH, 2013; Datasheet Micro-Hybrid Electronic GmbH, 2013). Under similar performance parameters, pyro-electric sensors require chopper motors for light modulation. Based on a 63 % step response, the time constant for thermal sensors is in a range of 5–150 ms. Pyrometers with uncooled extended InGaAs (indium gallium arsenide) photodiodes are an exception; they have time constants well below 1 ms, but they only detect object temperatures starting at 50 °C. In addition, such photodiodes operate in the 1.6–2.3 μm range, which makes them sensitive to daylight.

3   Parameterization of IR cameras and radiation thermometers

3.1   Optical resolution

Based on the specific detectivity (D^*) of the core component of the IR measuring device – the thermal detector or the photodiode – its performance can be calculated when the type of optics used is also taken into consideration. Thus the optical resolution can be optimized at the expense of the thermal resolution across a wide range. The key parameter for a pyrometer is the distance ratio ($D_{/S\ \text{geo}}$):

$$D_{/S\ \text{geo}} = \frac{a_\text{m}}{d_\text{m}}. \tag{1}$$

The normal values for this ratio between measuring distance (a_m) and spot diameter (d_m) are 0.5–150 for devices measuring from −50 to 150 °C. For higher object temperatures beginning from 150 °C, pyrometers can be designed with distance ratios from 20 to 300, which are usually operating at shorter wavelengths. When determining the distance ratio ($D_{/S\ \text{meas}}$), the relevant factor at a specific measuring distance is the diameter of a black body at which the radiation

signal has decreased by 10 % compared to the signal of a sufficiently large black body. This definition takes into consideration the spherical and chromatic aberration of the optics, the effective detector area and the scattering effects of the optical channel. Taking into account solely the main parameters of the optics – their effective aperture respectively lens diameter (d_L) and their f-number (F) – according to Product brochure Optris GmbH (2013) a geometric distance ratio can be calculated using the sensor area (A_s):

$$D_{/S\ \text{geo}} = \frac{F d_L}{\sqrt{A_s}}. \tag{2}$$

For large sensor dimensions, compared to the wave length and coated IR glass lens systems, this geometric target distance corresponds in good approximation to the measured target distance. For imaging devices, single-pixel-based fields of view IFOV_geo are used instead of distance ratios. This value is calculated from the field of view (FOV) and the number of associated horizontal and vertical pixels ($p_\#$):

$$\text{IFOV}_\text{geo} = \frac{\text{FOV}}{p_\#}. \tag{3}$$

Values ranging from 0.3 to 15 mrad are typical. The following formula is used to calculate the distance ratio from this parameter:

$$D_{/S\ \text{geo}} = \frac{1}{\text{IFOV}_\text{geo}}. \tag{4}$$

Figures 1 and 2 show the dependence of the displayed radiance on the object dimensions shown as a multiple of the pixel size for commercially available IR cameras. Since the dimensions of the individual pixels are already similar to the wavelength being detected, the resolution of the optical system is limited.

So the object size must be at minimum three pixels, corresponding to a three-time IFOV_geo to specify a distance ratio ($D_{/S\ \text{meas}}$) with a sufficient 90 % radiation signal. This

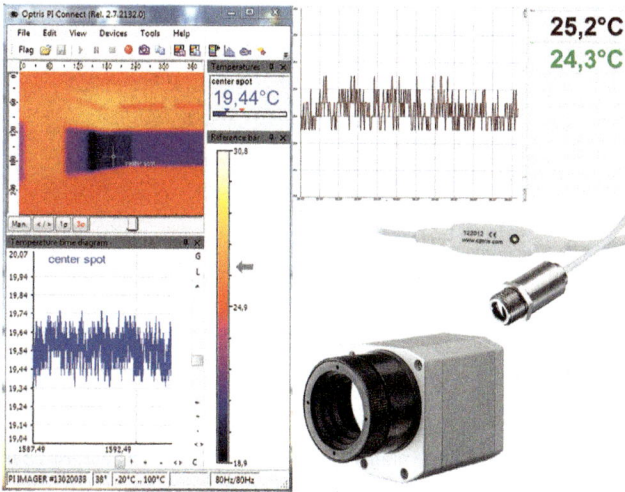

Figure 3. Noise measurements on an Optris PI400 camera with 80 Hz image frame rate; ΔT_{p-p} is 0.4 K and thus 67 mK NETD. The same noise level is obtained when using a CSmicroLT pyrometer with a response time (t_{95}) of 30 ms.

commonly used estimation allows an effective comparison between the performance parameters of cameras and pyrometers:

$$D_{/S\,\text{meas}} = \frac{1}{3\,\text{IFOV}_{\text{geo}}}. \qquad (5)$$

3.2 Thermal resolution

Caused by the noise of the IR detector, its preamplification and serialization electronics, IR measuring devices display temperature fluctuations that are normally distributed. The deviation of temperatures around a mean value can be specified as the noise-equivalent temperature difference (NETD). Although rather low values between 10 and 150 mK are listed in the data sheets, according to HAMEG Instruments GmbH (2013) the observed maximum and minimum differences resulting from the noise distribution are

$$\Delta T_{p-p} = 6\,\text{NETD}. \qquad (6)$$

Various noise reduction mechanisms – such as the averaging of image or measurement sequences and measurement time constants that are dependent on signal changes – can be used to stabilize the displayed temperatures. This noise signal still differs significantly from the reproducibility specified in data sheets, which includes additional systematic measurement error resulting from varying environmental conditions. Typically, the reproducibility is double the observed minimum–maximum noise value. With regard to measurement accuracy (which is linked to calibration standards), a further doubling of the error range can be expected: using the radiation thermometer in Fig. 3 for an object at 23 °C at 67 mK NETD, a reproducibility of ± 0.5 K and an accuracy

of ± 1 K can be attained. Often, a temperature coefficient for the measurement error is specified which describes the accuracy of the thermometer within the entire operating temperature range. The measurement accuracy of cameras is strongly influenced by varying ambient temperatures. A signal drift occurs during the time interval between two offset compensations. This drift can be described with a reproducibility of ± 1 K and measurement accuracy of ± 2 K.

3.3 Time behavior

The measurement time constant for pyrometers is generally described as the response time (t_{95}) that a radiance proportional signal requires after a signal jump to reach 95 % of its final value (S_{∞}; Piotrowski and Rogalski, 2013). Thus for the corresponding bandwidth (Δf) of the measurement channel,

$$\Delta f = \frac{3}{t_{95}}. \qquad (7)$$

The acquisition time (t_{ex}) is a parameter that is independent of any time lag between the input and output signals. It is defined as the minimum amount of time that an object must remain in the field of view to be detected at a defined signal amplitude (of x percent). For imaging devices, the imaging frame rate (f_B) is an indicator of the time behavior. Due to the sampling theorem the maximum transferable signal bandwidth is limited to half of this value.

However, the acquisition time for short-term events is surprisingly low. Figure 5 shows a hot object, which has been exposed by the focal-plane shutter of a camera for 1 ms. The signal amplitude has decreased to about 8 % of the original value. However, it can be corrected by using an appropriate signal correction method (based on known reference temperatures in the image). After the exposure, three additional images are taken with the same relative image content and decreasing amplitude. In order to estimate the microbolometer's time constant, the assumption can be made that

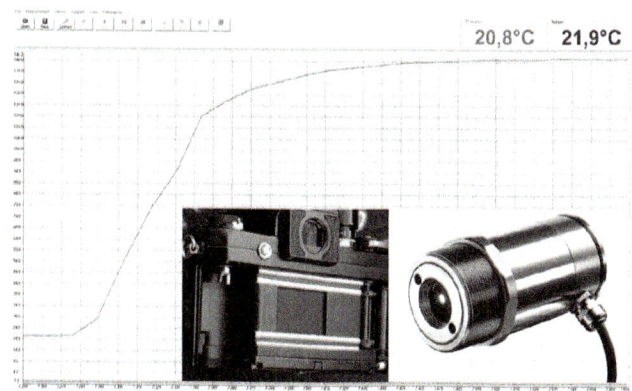

Figure 4. Measuring the response time for a CTlaser LT radiation thermometer using the focal-plane shutter of a traditional camera, according to VDI/VDE standard 3511, Sect. 4.3. (One abscissae interval represents 10 ms; the ordinate is split in 5 K steps).

a)

b)

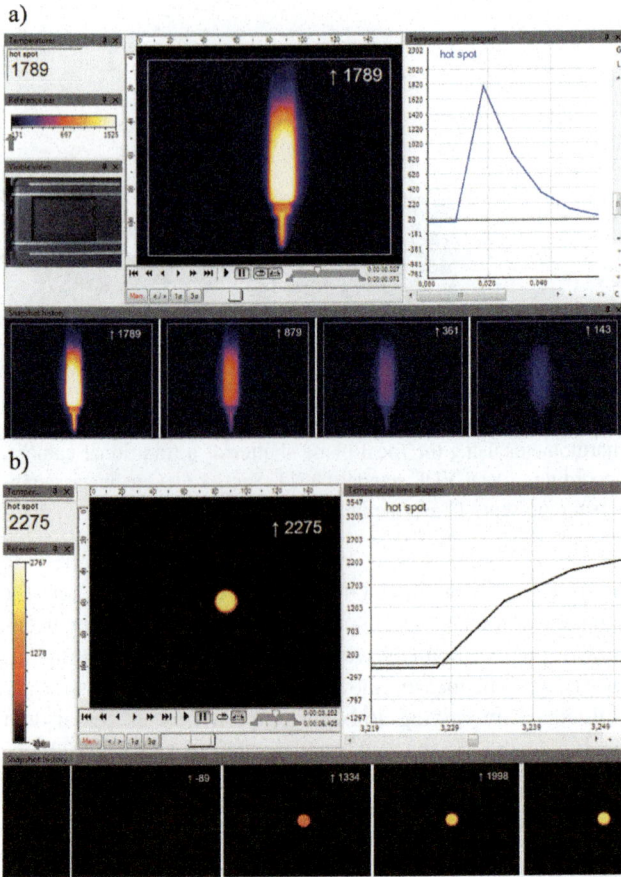

Figure 5. (a) Snapshot representation of a moving soldering iron after a 1 ms exposure in front of a 120 Hz frame rate PI160 thermal imaging camera. **(b)** Time behavior after an instantaneous opening of a photographic focal-plane shutter in front of a black body at 147 °C. Displayed numbers are radiation proportional values similar to signal voltages measured by the video AD converter of the imager.

it exhibits a simple low-pass behavior. The sensor time constant can then be shown as

$$t_{63} = \frac{1}{f_B} \ln \left(\frac{S_\infty - S_1}{S_\infty - S_2} \right). \tag{8}$$

Here, S_1 and S_2 denote the radiance proportional values from two consecutive images. S_∞ is the final value which results from the decreasing or increasing signal behavior (as shown in Fig. 5). The calculated bolometer time constant of about 8 ms corresponds fairly well with the data from Durand et al. (2011). The following detectivity analysis assumes that the imaging devices have a signal bandwidth of

$$\Delta f = \frac{1}{t_{63}}. \tag{9}$$

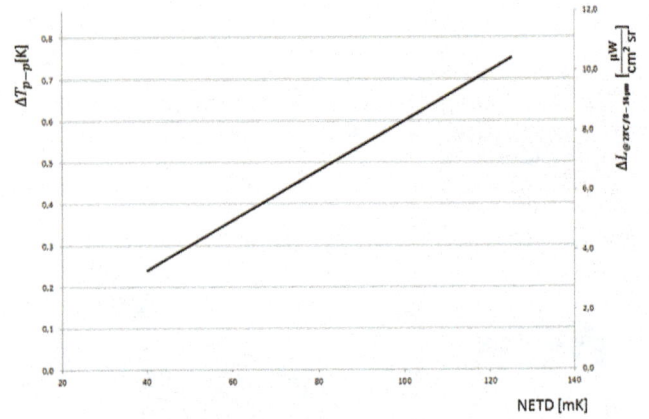

Figure 6. Noise-equivalent radiance difference and peak-to-peak noise as a function of the NETD at 23 °C in the 8–14 µm spectrum.

3.4 Detectivity

In order to compare the performance of pyrometers with infrared cameras, it is necessary to analyze the specific detectivity (D^*) as the main sensor parameter as well as the characterization of the optics. The detector and optics are the two most important cost factors.

The detectivity is a measure of the noise-equivalent beam flux (NEP), normalized with respect to area and bandwidth:

$$D^* = \frac{\sqrt{\Delta f A_s}}{\text{NEP}}. \tag{10}$$

A target NETD for the IR measuring device is assumed for the following analyses. According to Planck's radiation law, a surface radiates with a radiance L which is dependent on the spectrum range in use and the surface temperature (Piotrowski and Rogalski, 2013). A radiance difference (ΔL) is defined which corresponds to the temperature difference NETD at an ambient temperature of 23 °C (296 K). Thus

$$\Delta L = \frac{c_1}{\Omega_0} \left(\int_{8\,\mu m}^{14\,\mu m} \lambda^{-5} \left(e^{\frac{c_2}{\lambda (296\,K + NETD)}} - 1 \right)^{-1} d\lambda \right.$$
$$\left. - \int_{8\,\mu m}^{14\,\mu m} \lambda^{-5} \left(e^{\frac{c_2}{\lambda (296\,K)}} - 1 \right)^{-1} d\lambda \right), \tag{11}$$

with $c_1 = 1.191 \times 10^{-16}$ W m^2 and $c_2 = 0.01439$ m K. Figure 6 demonstrates that Eqs. (5) and (10) can be used to calculate the noise-equivalent radiance difference (ΔL) from a predefined NETD in addition to a max-to-min temperature fluctuation. The NEP can be calculated from the ΔL via the transmittance (τ) of the optical system, the sensor surface (A_s) and the solid angle that the optics have relative to the sensor. According to Bernhard (2004), we can see that the detectivity is determined by the quality of the optics (expressed by the f-number (F), transmittance and the optical diameter (d_L)) as well as the temporal, geometric and thermal resolutions:

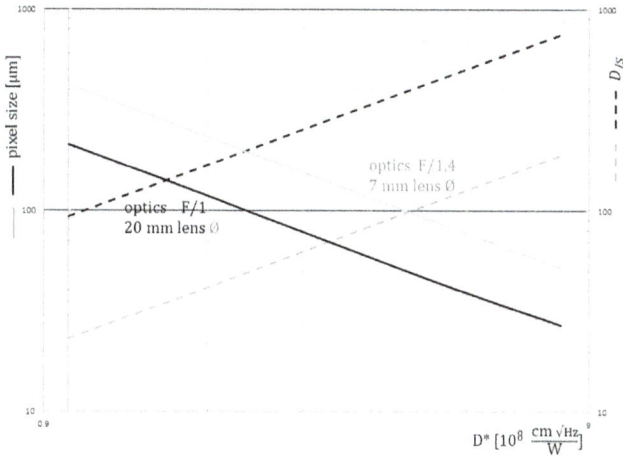

Figure 7. Sensor dimensions and achievable distance ratios as a function of the detectivity level of the thermal sensor.

$$D^* = \frac{4F}{\pi \tau d_L} \sqrt{\Delta f} \; \frac{D_{/S \, \text{geo}}}{\Delta L}. \qquad (12)$$

Table 1 lists the interdependent performance parameters for mass-produced IR measuring devices. The resulting detectivity reveals substantial differences between the thermopiles used in the pyrometers and the cameras' microbolometers. Since the f-number of the optics can not be chosen arbitrarily large (due to power limitations) and since the lens diameter is limited by cost and handling restraints, greater distance ratios can only be achieved with a smaller sensor geometry. Thus more powerful sensors with improved detectivity are inevitably designed with smaller sensor dimensions. Figure 7 clarifies this relationship: for a NETD of 70 mK, different specific detectivities and various lens performance classes place specific demands on the sensor dimensions associated with the achievable distance ratios. Developers of sensors should always take these relationships into account in order to target, in addition to an increased sensitivity and reduced noise, a geometric miniaturization.

4 Application examples for industrial cameras

Such cameras are used in industrial applications to detect defects (such as hot spots in solar facilities) and also for inspecting the uniformity of heating and cooling processes (when using plastic molds, for example). In research and development, the analysis of thermal designs is a key area, as is solving the challenge of dissipating heat away from miniature electronic components and high-power LEDs.

Similar to the trend of cameras replacing pyrometers, the following specific application indicates that imagers may also influence the market of line scanning devices. Often the optical access to the object may be limited, as is the case

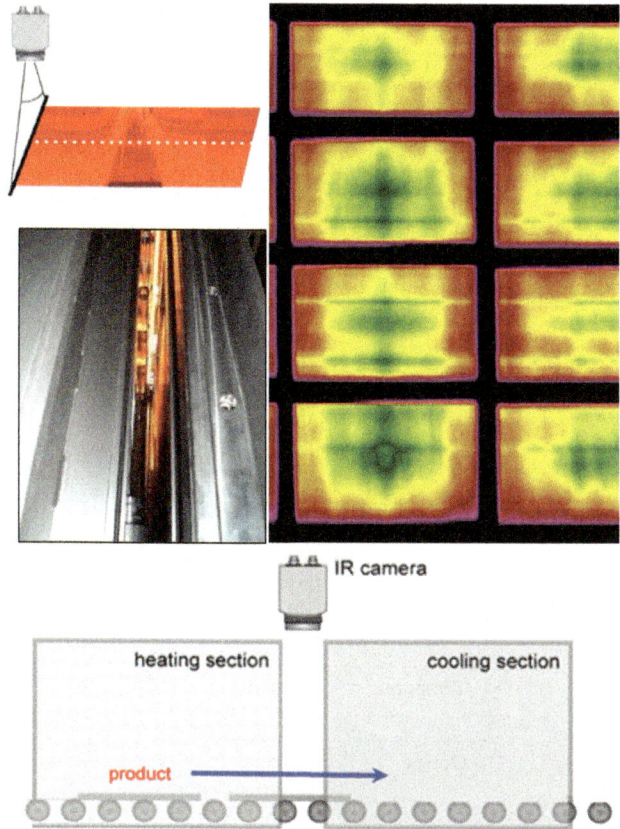

Figure 8. IR cameras used as line scanners allow the object to be sighted through extremely narrow viewing slits. Large-scale measurement images can thus be reconstructed by concatenating the lines of consecutive images.

at the exit of tunnel furnaces. Here the high optical resolution permits the object to be sighted through narrow viewing slits. It is sufficient then to measure the object's temperature in only a few lines per image. Due to the high image frequency (and thus high line frequency) of 120 Hz, it is possible to reconstruct a variably long thermal image of the object in the furnace even when dealing with large material feed rates. The software permits variable image construction, in particular diagonally arranged measurement lines, in which the visual field is 25 % larger relative to the horizontal view. Corresponding digital algorithms correct any barrel-shaped distortions in the optical system. Figure 8 shows an example of the use of thermal imaging cameras for curing glass surfaces. Only a very narrow gap is used for measuring the glass sheets as the 580 °C heated plates move by 2 m s^{-1} between the heating and cooling sections of the curing facility.

The sheets are measured before they are cooled to inspect their homogeneity and thus evaluate the degree of thermal stress. The thermal images then created show the heat distribution in the furnace and allow the individual heating sections to be controlled more precisely.

Table 1. A comparison of basic performance parameters of radiation thermometers and IR cameras (at 23 °C object temperature, 8–14 μm range).

Parameter		Radiation thermometer		IR camera
		Optris CTLT	Optris LS	Optris PI400
Thermal	NETD [mK]	67	40	67
	ΔT_{p-p} [K]	0.40	0.24	0.40
	Reproducibility [°C]	±0.5	±0.5	±1
	Accuracy [°C]	±1	±0.75	±2
Geometrical	D/S_{geo}	15	75	560
	D/S_{meas}	15	75	187
	$IFOV_{geo}$ [mrad]	66.6	13.3	1.8
	FOV [°]	3.8	0.8	39
	Pixel $p_{\#}$	1	1	382
Temporal	t_{95} [ms]	30	120	25
	$\Delta f / f_B$ [Hz]	100	25	120
Optics	f-number F	1.50	1.60	0.80
	Aperture d_L [cm]	0.8	2.1	1.8
	Transmission τ	0.7	0.6	0.75
Detector	Sensor area A_s [μm²]	Ø 750	Ø 450	25 × 25
	Detectivity [$\frac{cm\sqrt{Hz}}{W}$]	1.1×10^8	2.0×10^8	8.5×10^8

5   The outlook

The calculation methods presented here accurately describe measurements obtained from series products. Smaller sensors with higher sensitivity will continue to influence the key parameters for one-dimensional and two-dimensional non contact radiation thermometers. The significantly improved performance of bolometers compared to thermopiles will open up new application fields in which small measuring points will be measured with variable positioning and small thermal contrast over large relative distances. Ultimately it will be the cost of such devices that plays the greatest role in widening or limiting their use.

Brief biography

Dr. Ulrich Kienitz is the General Manager of Optris GmbH, one of Germany's leading companies in the field of non-contact temperature measurement technology. Previously he managed the European division for a major American IR instrumentation company for over 10 years. As the holder of numerous patents, he has significantly influenced the design of many of the major infrared thermometry devices on the market.

Edited by: G. S. Aluri
Reviewed by: two anonymous referees

References

Bernhard, F.: Technische Temperaturmessung, Physikalische und meßtechnische Grundlagen, Sensoren und Meßverfahren, Meßfehler und Kalibrierung, Berlin, Heidelberg: Springer-Verlag, 2004.

Datasheet Heimann Sensor GmbH: Miniature Thermopile Sensors for Remote Temperature Measurement and Gas Analysis, available at: http://www.heimannsensor.com/Datasheet-2b-TO18_rev2.pdf, last access: 19 November 2013.

Datasheet Micro-Hybrid Electronic GmbH: One Channel Thermopile Detector TS1x80B-A-D0.48, available at: http://www.micro-hybrid.de/fileadmin/user/IR-systems-documents/Datenblaetter/Thermopiles/MH_TS1x80B-A-D0.48.pdf last access: 19 November 2013.

Durand, A., Tissot, J. L., Robert, P., Cortial, S., Roman, C., Vilain, M., and Legras, O.: VGA 17 µm development for compact, low power systems, available at: http://www.ulis-ir.com/uploads/Documents/8012-43_ULIS.pdf (last access: 8 November 2013), 2011.

FLIR Corporation – INVESTOR PRESENTATION, March 2013 forward looking, available at: http://www.flir.com/investor/, last access: 15 March 2013.

HAMEG Instruments GmbH: Professional Article, What is noise?, available at: http://www.hameg.com/articles.0.html?&no_cache=1&L=1&tx_mdownloads_pi1[mode]=download&tx_hmdownloads_pi1[uid]=3046, last access: 19 November 2013.

Piotrowski, J. and Rogalski, A.: High-Operating-Temperature Infrared Photodetectors, available at: http://spie.org/samples/PM169.pdf, last access: 19 November 2013.

Product brochure Optris GmbH: Infrared cameras, The most portable infrared online camera, Version D2013-06-A, available at: http://www.optris.com/downloads-infrared-cameras?file=tl_files/pdf/Downloads/Infrared20Cameras/PI_Brochure.pdf, last access: 19 November 2013.

VDI/VDE 3511, Sheet 4.3: Radiation Thermometry/Calibration of Radiation Thermometers, Edn. Dec., 2004.

VDI/VDE 3511, Sheet 4.4: Anwendung der Thermografie zur Diagnose in der Instandhaltung, Edn. June, 2010.

YOLE Developement: Uncooled Infrared Cameras and Detectors; Report, Lyon, 2010.

13

Data fusion of surface normals and point coordinates for deflectometric measurements

B. Komander[1], D. Lorenz[1], M. Fischer[2], M. Petz[2], and R. Tutsch[2]

[1]TU Braunschweig, Institut für Analysis und Algebra, Braunschweig, Germany
[2]TU Braunschweig, Institut für Produktionsmesstechnik, Braunschweig, Germany

Correspondence to: B. Komander (b.komander@tu-braunschweig.de)

Abstract. Measuring specular surfaces can be realized by means of deflectometric measurement systems with at least two reference planes as proposed by Petz and Tutsch (2004). The results are the point coordinates and the normal direction of each valid measurement point. The typical evaluation strategy for continuous surfaces involves an integration or regularization of the measured normals. This method yields smooth results of the surface with deviations in the nanometer range but it is sensitive to systematic deviations. The measured point coordinates are robust against systematic deviations but the noise level is in the order of micrometers. As an alternative evaluation strategy a data-fusion process that combines both the normal direction and the point coordinates has been developed. A linear fitting technique is proposed to increase the accuracy of the point coordinate measurements by forming an objective functional as the mean squared misfit of the gradients with respect to the point coordinates on the one hand and to the normals on the other hand. Moreover, a constraint on the maximal change of the coordinate measurements is added to the optimization problem. To minimize to objective under the constraint a projected gradient method is used. The results show that the proposed method is able to adjust the point coordinate measurement to the measured normals and hence decrease the spatial noise level by more than an order of magnitude.

1 Introduction and measurement principle

The combination of close range photogrammetry and structured illumination is well established in the field of three-dimensional measurements of diffusely reflecting surfaces. The so-called fringe projection technique is based on the triangulation principle and makes use of at least one electronic camera and one projector. By means of structured illumination, which especially involves phase shifting techniques, the surface under test is optically coded. Based on this spatial coding, the camera images can be evaluated in a way that delivers a three-dimensional object point for each image pixel.

This fringe projection technique depends on the optical imaging of the surface under test onto the image sensor of the camera. This requires a minimum amount of diffusely scattered light from the surface. For reflecting surfaces the amount of diffusely scattered light is very low – in the ideal case there should be no scattered light at all. Reflecting surfaces therefore are not directly visible, but their presence can affect the propagation of light in a characteristic way (Tutsch et al , 2011).

It is a common experience that an image generated by reflection at a curved mirror shows characteristic deformations. These optical deformations of an object's mirror image carry information about the geometry of the reflective surface. If the imaged object is a known regular pattern, observed deformations of the pattern can be taken as an indication of form deviations or defects of the surface under test. This technique, known as deflectometry, is well suited for the qualitative inspection of reflective surfaces. An absolute shape measurement, however, is not possible directly. As can be seen in Fig. 1, in general there are various combinations of position and orientation of a mirror element that lead to the same observation.

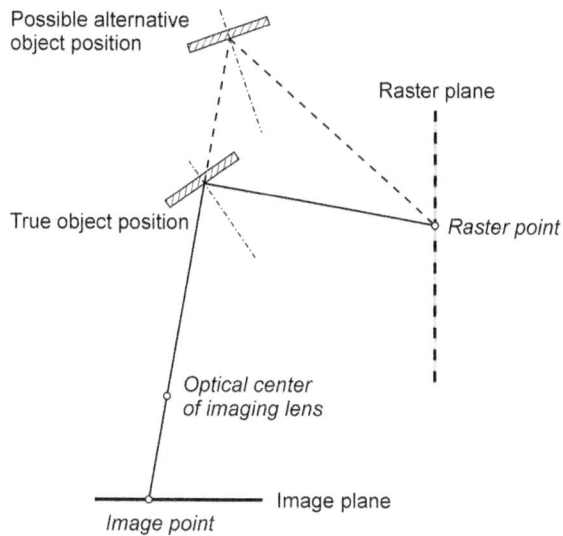

Figure 1. Ambiguity of determination of position and orientation of a mirror element in deflectometry (Tutsch et al., 2011).

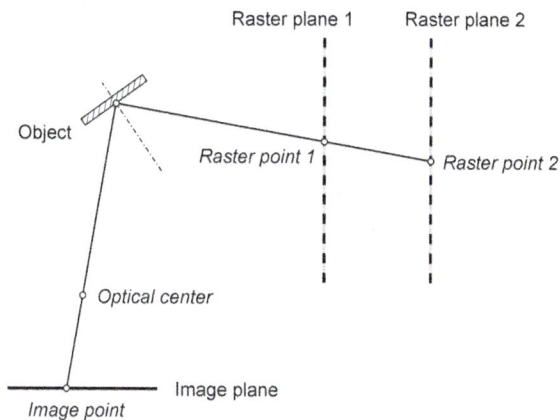

Figure 3. Setup for deflectometric measurements according to the enhanced approach shown in Fig. 2 (Petz, 2006).

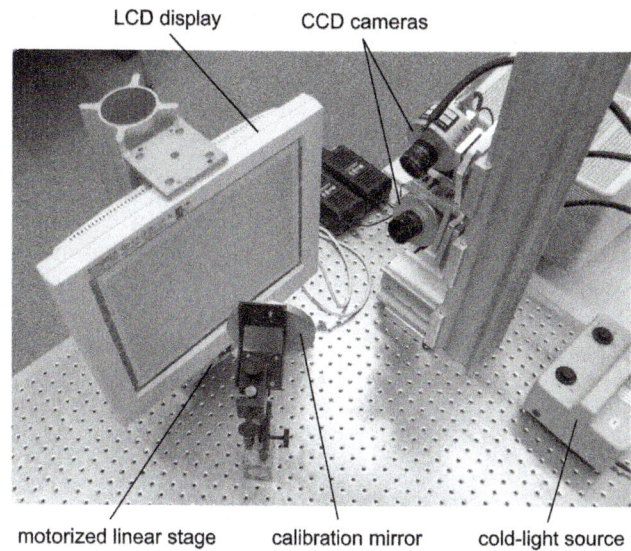

Figure 2. Principle of an enhanced deflectometric approach using two reference pattern positions (Tutsch et al., 2011).

from the spatially coded reference patterns delivers an unambiguous ray. The intersection of these two rays defines the location of the object point. In addition to the three-dimensional position of the object point also the surface normal in that point can be determined, as for the reflective surface the law of reflection must be valid.

A technical realization of the approach illustrated in Fig. 2 can be seen in Fig. 3 (Petz, 2006). As reference structure a common LCD-TFT display is used. The absolute two-dimensional spatial coding of the display area is accomplished by a combination of phase shifting techniques and a heterodyne approach for deconvolution. The display is mounted onto a motorized linear stage and can thus be moved to various positions during the measurement process. The object under test will be mounted in the position where the figure shows a calibration mirror, so that the camera observes a mirror image of the LCD display.

As mentioned above, the enhanced deflectometric approach directly delivers three-dimensional coordinates of the observed surface points. Thus there is no need for surface reconstruction by means of regularization or integration of the surface slope. However, a closer investigation of the measurement data reveals that the direct measurement of coordinates is more sensitive to noise than the measurement of the surface normals. In earlier evaluations Petz (2006) has shown that the noise of the measured surface is approximately 3 orders of magnitude higher for the direct coordinate measurement than it is for the reconstruction based on the integration of the surface slopes.

However, the surface reconstruction based on integration is susceptible to local disturbances and systematic errors. Therefore an advantageous approach for the surface measurement in deflectometry should consider both, the absolute

This ambiguity can only be resolved by additional information. For instance assumptions can be made concerning the properties of the mirror surface. Then the surface can be reconstructed by means of regularization (Werling et al., 2009). However, the parameters of this reconstruction process have to be determined by other means, such as approximation or additional measurements for at least one point on the surface (Li et al., 2012).

Petz and Ritter (2011), Petz and Tutsch (2004), and Petz (2006) therefore proposed an enhanced deflectometric approach, shown in Fig. 2, that solves the ambiguity discussed for the basic approach. The enhanced approach uses at least two different positions of the reference pattern, which can easily be realized by moving the reference structure with a linear stage. With this approach not only the camera defines a unique ray for each image point, but also the information coordinates determined by triangulation and the observed surface normals. A correspondent approach is the subject of the present paper.

2   Notation and data type

In this paper the Euclidean norm $\| \cdot \|$ on $\mathbb{R}^3$ will be used. Vectors v will be written in italic, bold letters and matrices $\mathbf{M}$ in bold, capital letters. Scalars c will be written in italic letters. A matrix $\mathbf{A}$, when of dimension $m \times n \times 3$, is of form $\mathbf{A} = (\mathbf{A}^x, \mathbf{A}^y, \mathbf{A}^z)$ with $\mathbf{A}^x$, $\mathbf{A}^y$, $\mathbf{A}^z$ matrices of dimension $m \times n$. The vector in the ith row and the jth column of $\mathbf{A}$ is denoted with the lowercase, italic, bold letter $v_{i,j} = (v_{i,j}^x, v_{i,j}^y, v_{i,j}^z)^T \in \mathbb{R}^3$ with its x, y and z coordinates. If any vector w or matrix $\mathbf{A}$ is corresponding to $m_{i,j}$ it will be written as $w^{i,j}$ or $\mathbf{A}^{i,j}$.

Furthermore, $\mathrm{vec}(\mathbf{A})$ is the vector which contains the entries of $\mathbf{A} \in \mathbb{R}^{m \times n}$ column wise, i.e., $\mathrm{vec}(\mathbf{A}) = \left(a_{1,1}, a_{1,2}, \ldots, a_{m,1}, a_{m,2}, \ldots, a_{m,n}\right)^T$.

The type of data set being worked with is the result of a deflectometric measurement process as described earlier in Sect. 1. After postprocessing the data set contains a matrix $\mathbf{P} = (\mathbf{P}^x, \mathbf{P}^y, \mathbf{P}^z)$ of measured point coordinates together with its information of neighborhood and a matrix $\mathbf{N} = (\mathbf{N}^x, \mathbf{N}^y, \mathbf{N}^z)$ of measured normal vectors. The normal vectors $n^{i,j}$ are corresponding to the point coordinates $p^{i,j}$.

If some measurement point did not yield a valid measurement, the corresponding entries in $\mathbf{P}$ and $\mathbf{N}$ will be undefined. Hence, each point coordinate has at most eight neighbors, i.e., the eight neighborhood. Point coordinates can have fewer neighbors because of invalid measurements and also at the boundary of the measured object.

3   Objective functional for data fusion

As mentioned in Sect. 1 the accuracy of the normal vectors is approximately 3 orders of magnitude higher than the accuracy of the measured point coordinates. The purpose of this section is to combine both measured data sets with the objective to get consistent data.

To that end an objective functional will be defined that will fuse the data in a process of optimization. According to the measurement process, it is known that the x and the y coordinates of the point coordinates are more trustworthy compared to the z coordinates. Hence, the objective is to adjust the z coordinate with respect to the orientation of the measured surface.

The first idea is to take a look at the normal vectors. For every measured point coordinate a calculated normal can be defined by the cross product with two neighboring point coordinates. Then the normals can be compared by minimizing the mean squared misfit of the measured and the calculated normals. However, this objective functional leads to an optimization problem which is nonlinear with respect to the

z coordinate. This is because the mapping from the point coordinates to the normals is highly nonlinear. This kind of optimization problem is surely solvable, but costly in realization.

By rethinking about information of surface orientation beside normal vectors, gradient vectors come to mind. Gradients carry the same information about the surface structure as normals. Hence, gradients can be calculated separately with respect to each measured normal and to the z coordinates of each point coordinate.

Thus, an objective functional can be formed as the mean squared misfit of two kinds of calculated gradients. On the one hand, gradients calculated with respect to the normal vectors and, on the other hand, to the point coordinates. With this approach a quadratic objective functional will be obtained.

First the gradient vector g corresponding to the normal vectors is formed by rearranging the vectors. The first mn rows of g are the partial derivatives of the z coordinate with respect to the x coordinate and the second mn rows are the partial derivatives with respect to the y coordinate. Thus, g is given by

$$g = \nabla \mathbf{P}^z = \begin{pmatrix} -\mathbf{N}^x / \mathbf{N}^z \\ -\mathbf{N}^y / \mathbf{N}^z \end{pmatrix},$$

and is of dimension $2mn$.

Next the gradient with respect to the point coordinates will be calculated. In order to do that, all neighbors in the eight neighborhood of one point coordinate $p_{i,j}$, for all $i = 1, 2, \ldots, m$ and $j = 1, 2, \ldots, n$, will be located, and build the vectors $v_1, \ldots, v_8$ from $p_{i,j}$ to its neighbors. The vectors $v_1, \ldots, v_8$ have form

$$v_1 = \begin{pmatrix} p_{i-1,j-1}^x - p_{i,j}^x \\ p_{i-1,j-1}^y - p_{i,j}^y \end{pmatrix},$$

$$\vdots$$

$$v_8 = \begin{pmatrix} p_{i+1,j+1}^x - p_{i,j}^x \\ p_{i+1,j+1}^y - p_{i,j}^y \end{pmatrix}.$$

The point coordinate $p_{i,j}$ with its possible neighbors and the vectors v_k, $k = 1, \ldots, 8$, are shown in Fig. 4. Note that the grid defined by the point coordinates is usually not regular.

On the one hand the directional derivatives of $p_{i,j}$ in direction to its neighbors can be calculated with $v_1, \ldots, v_8$ as follows:

$$\partial_{v_k} p_{i,j}^z = v_k^T \cdot \nabla p_{i,j}^z \qquad \text{for } k = 1, \ldots, 8. \tag{1}$$

All directional vectors v_k^T can be written line by line in a matrix $\mathbf{V}^{i,j}$, which is then at most of dimension 8×2. All neighbors of a point coordinate are considered into the calculation to obtain a more robust solution. In order to be able to calculate the matrix $\mathbf{V}^{i,j}$ at all, at least two neighbors are

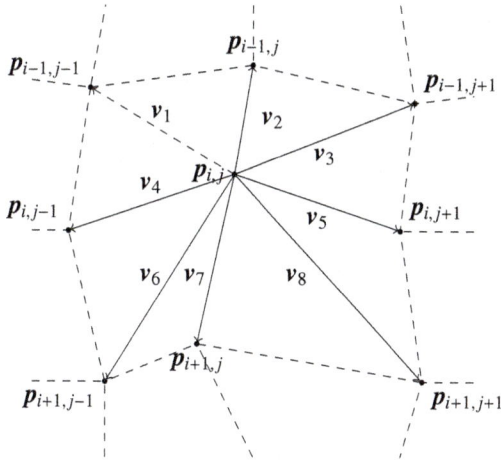

Figure 4. Neighborhood of $p_{i,j}$ with vectors $v_1, \ldots, v_8$ pointing to all direct neighbors of $p_{i,j}$.

needed. It can happen that a point coordinate has fewer than eight neighbors, e.g., at boundary points, so that $\mathbf{V}^{i,j}$ has a smaller number of rows.

On the other hand, the directional derivatives in direction of the neighboring points can be calculated by taking the differences of the z coordinates of the neighbors of $p_{i,j}$ and $p_{i,j}$ itself:

$$\partial_{v_1} p_{i,j}^z = p_{i-1,j-1}^z - p_{i,j}^z,$$

$$\vdots$$

$$\partial_{v_8} p_{i,j}^z = p_{i+1,j+1}^z - p_{i,j}^z. \tag{2}$$

Putting both Eqs. (1) and (2) for the directional derivatives together, the result is a linear system for $\nabla p_{i,j}^z$.

$$\mathbf{V}^{i,j} \cdot \nabla p_{i,j}^z$$

$$= \begin{pmatrix} 1 & 0 & 0 & 0 & -1 & 0 & 0 & 0 & 0 \\ 0 & 1 & 0 & 0 & -1 & 0 & 0 & 0 & 0 \\ 0 & 0 & 1 & 0 & -1 & 0 & 0 & 0 & 0 \\ 0 & 0 & 0 & 1 & -1 & 0 & 0 & 0 & 0 \\ 0 & 0 & 0 & 0 & -1 & 1 & 0 & 0 & 0 \\ 0 & 0 & 0 & 0 & -1 & 0 & 1 & 0 & 0 \\ 0 & 0 & 0 & 0 & -1 & 0 & 0 & 1 & 0 \\ 0 & 0 & 0 & 0 & -1 & 0 & 0 & 0 & 1 \end{pmatrix} \cdot \begin{pmatrix} p_{i-1,j-1}^z \\ p_{i-1,j}^z \\ p_{i-1,j+1}^z \\ p_{i,j-1}^z \\ p_{i,j}^z \\ p_{i,j+1}^z \\ p_{i+1,j-1}^z \\ p_{i+1,j}^z \\ p_{i+1,j+1}^z \end{pmatrix}$$

$$= \mathbf{C} \cdot \qquad\qquad z^{i,j}$$

The least-squares solution of this overdetermined system for the partial derivatives $\nabla p_{i,j}^z$ is given by

$$\nabla p_{i,j}^z = \begin{pmatrix} \nabla_x p_{i,j}^z \\ \nabla_y p_{i,j}^z \end{pmatrix} = \left((\mathbf{V}^{i,j})^+ \cdot \mathbf{C} \right) \cdot z^{i,j},$$

with $(\mathbf{V}^{i,j})^+$ the pseudo inverse of $\mathbf{V}^{i,j}$. The matrix $\mathbf{D}^{i,j} = (\mathbf{V}^{i,j})^+ \cdot \mathbf{C}$ is the calculated difference matrix of the point coordinate $p_{i,j}$.

Since the objective is not to calculate the partial derivatives separately but to calculate all in one, the entries of each difference matrix $\mathbf{D}^{i,j}$ can be written in the proper row and column of a matrix $\mathbf{D}$, which contains all differences for the partial derivatives with respect to x in its first mn rows, and all differences for the partial derivatives with respect to y in the second mn rows. So $\mathbf{D}$ is of dimension $2mn \times mn$. Note that since one point coordinate has at most eight direct neighbors, $\mathbf{D}$ will be a sparse matrix with at most nine nonzero entries per row.

In that way, an objective functional is given as the misfit of the gradients corresponding to the point coordinates and to the normals,

$$F(\mathbf{P}^z) = \frac{1}{2} \left\| g - \mathbf{D} \cdot \text{vec}(\mathbf{P}^z) \right\|^2,$$

which will be minimized with respect to $\mathbf{P}^z$. The objective functional F is quadratic in the vectorized z coordinates of the points.

Note that the gradient vector g and the difference matrix $\mathbf{D}$ have only to be calculated once at the beginning of the process.

4 Linear fitting with steepest descent method

The objective functional was formed in Sect. 3 and it is quadratic in $\mathbf{P}^z$. In this section the focus goes to the measured data and the optimization process itself.

An optimal $\mathbf{P}^z$ corresponding to the x and y coordinates can easily be found by shifting the z coordinates. Since the point coordinates are given from a deflectometric measurement process, the z coordinates cannot be shifted without considering the original position of the measured point coordinates and the occurring measurement uncertainties. Hence, the points can only be shifted in a given maximal range.

So the estimated accuracy of the coordinate measurement is used to add a constraint to the optimization problem on the maximal change of the measured points. We assume that the point coordinates are disturbed by noise that follows a Gaussian distribution in the z direction with known variance σ. Hence, the distribution of the sum of all squared distances in the z direction follows from a chi-squared distribution with the number of degrees of freedom equal to the number N of valid measurement points. We deduce that the expected value of this sum is equal to $N\sigma^2$. We use this value as tolerance $\delta = N\sigma^2$ and define the set in which the point coordinates are allowed to be shifted as $B = \{ \| \mathbf{P}^z - \widetilde{\mathbf{P}}^z \|^2 \le \delta \}$, the sphere with radius δ around the measured z coordinate $\mathbf{P}^z$. The shifted z coordinate is given by $\widetilde{\mathbf{P}}^z$.

Therefore, with objective functional $F(\mathbf{P}^z) = \frac{1}{2} \| g - \mathbf{D} \cdot \text{vec}(\mathbf{P}^z)\|^2$, the optimization problem becomes

$$\min_{\mathbf{P}^z} F(\mathbf{P}^z) \quad \text{s.t} \quad \left\| \mathbf{P}^z - \widetilde{\mathbf{P}}^z \right\|^2 \leq \delta.$$

In order to minimize the objective functional under the constraint, a projected gradient method will be used. Therefore, the search direction for the optimum is given by the negative gradient of F, which is

$$-\nabla F(\mathbf{P}^z) = -\mathbf{D}^T \left(\mathbf{D} \cdot \text{vec}(\mathbf{P}^z) - g \right),$$

and, hence, the update for each optimization step k with respect to the optimization constraint is

$$\mathbf{P}_{k+1}^z = \text{Proj}_B \left(\mathbf{P}_k^z - \tau_k \cdot \nabla F(\mathbf{P}_k^z) \right),$$

with τ_k the step size for the iteration k.

The projection of the point coordinates $\mathbf{P}_k^z$ onto the set B is given by

$$\text{Proj}_B \left(\mathbf{P}_k^z \right) =$$
$$\begin{cases} \mathbf{P}_k^z & , \|\mathbf{P}^z - \mathbf{P}_k^z\|^2 \leq \delta \\ \mathbf{P}_k^z - \left(1 - \frac{\sqrt{\delta}}{\|\mathbf{P}_k^z - \mathbf{P}^z\|} \right) \left(\mathbf{P}^z - \mathbf{P}_k^z \right) & , \|\mathbf{P}^z - \mathbf{P}_k^z\|^2 > \delta. \end{cases}$$

Now, a sensible choice of the step size τ_k is to be made. By observing the objective functional F, it can be seen that a convex quadratic problem is to be solved.

The valid step size is

$$\tau = \frac{\| \nabla F(\mathbf{P}^z)\|^2}{\| \mathbf{D} \cdot \nabla F(\mathbf{P}^z)\|^2}.$$

The step size with respect to the gradient of F at $\mathbf{P}_k^z$ has to be calculated for each step k.

Putting all results together, the data-fusion algorithm is as follows.

Algorithm Data-Fusion

Input: Point coordinates $\mathbf{P}$, normals $\mathbf{N}$
Output: Fitted z-coordinate $\widetilde{\mathbf{P}}^{\mathbf{z}}$
    Calculate g, $\mathbf{D}$, ∇F
    Let $\epsilon > 0$, $k = 0$, $\widetilde{\mathbf{P}}_0^{\mathbf{z}} = \mathbf{P}^z$
    while $\left\| \widetilde{\mathbf{P}}_{k+1}^{\mathbf{z}} - \widetilde{\mathbf{P}}_k^{\mathbf{z}} \right\| > \epsilon$ **do**
        Calculate $\tau_k = \frac{\|\nabla F(\widetilde{\mathbf{P}}_k^{\mathbf{z}})\|^2}{\|\mathbf{D} \cdot \nabla F(\widetilde{\mathbf{P}}_k^{\mathbf{z}})\|^2}$
        $\widetilde{\mathbf{P}}_{k+1}^{\mathbf{z}} = \text{Proj}_B \left(\widetilde{\mathbf{P}}_k^{\mathbf{z}} - \tau^E \nabla F(\widetilde{\mathbf{P}}_k^{\mathbf{z}}) \right)$
        Calculate $\nabla F(\widetilde{\mathbf{P}}_{k+1}^{\mathbf{z}})$
        $k \leftarrow k + 1$
    end while

---

5 Experimental results and conclusions

To verify the feasibility of the algorithm, a MATLAB implementation of the algorithm has been created. Two sets of measurement results were used. On the one hand a plane mirror and on the other hand a spherical mirror. In the first case all normal directions of the point coordinates are almost the same, since the measured object is plane. But, a calculation of normals with respect to the measured point coordinates leads to deviations of up to 20° at some points. With the second data set similar results in deviations are observed. Since problems can occur during data fusion, further examinations of the data type itself will be done.

5.1 Implementation

In practice, a point coordinate at the boundary of the measured object will have fewer neighbors than eight. Also, point coordinates within the object can have fewer neighbors, e.g., if the measured object has a hole within, has scratches or if some coordinate points could not be measured.

Thus, when the matrix $\mathbf{V}^{i,j}$ is formed, there can be entries that are not defined and so $\mathbf{V}^{i,j}$ (cf. Sect. 3) will have a smaller number of rows. Thus, the matrices $\mathbf{C}^{i,j}$ and $\mathbf{D}^{i,j}$ and the vector $z^{i,j}$ will also be smaller during the calculation of the gradient of $p_{i,j}^z$. In this case, all rows and columns corresponding to the undefined point coordinates have to be deleted.

After calculating $\mathbf{D}^{i,j}$, the deleted rows and columns of the matrix will be reinserted by writing zeros in the proper row and column, thereby acquiring its original size of 2×8. For insertion, zeros were chosen, because these entries should not and will not be changed during the calculation process. After that the entries of $\mathbf{D}^{i,j}$ can be written within the matrix $\mathbf{D}$.

The matrix $\mathbf{D}$ is a sparse matrix, since one point coordinate has at most eight direct neighbors. This matrix is of dimension $2m \times n$, the first m rows are corresponding to the partial derivatives with respect to x, the second to the partial derivatives with respect to y. The principal diagonal of the first m rows and of the second m rows of $\mathbf{D}$ has entries corresponding to the point coordinate $p_{i,j}^z$ and so on. So the entries of each matrix $\mathbf{D}^{i,j}$ can be saved in vectors, which can be written on the diagonals of $\mathbf{D}$.

5.2 Results

Different data sets were used for the tests. Here, results of the data-fusion process will be shown in direct comparison of the data before and after the calculation. The comparison will be shown by two different measured objects. These objects are a plane reference mirror and a spherical reference mirror.

In Fig. 5 the logarithm of the value of the objective functional for both measured objects is shown. In both semi-logarithmic plots it can be seen that the value is falling even by 1000 iteration steps.

As an a posteriori validation of the assumption that the noise in the z coordinate of the coordinate measurement was indeed Gaussian distributed, we show histograms of the residuum of the originally measured z coordinate with the

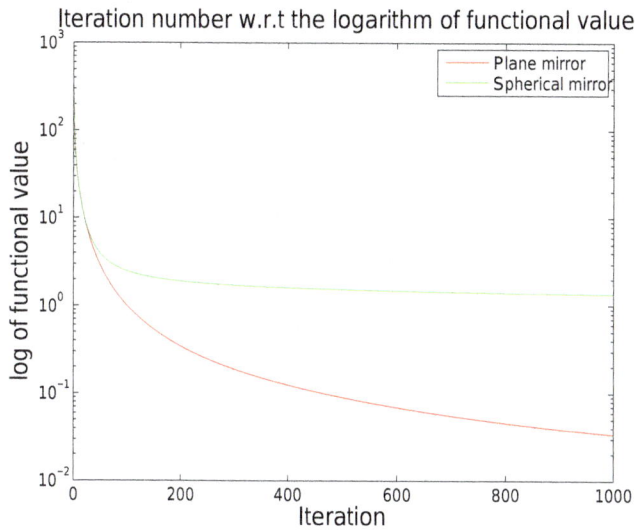

Figure 5. Logarithm of functional value during iteration for a measured plane reference mirror (lower curve) and for the spherical reference mirror (upper curve).

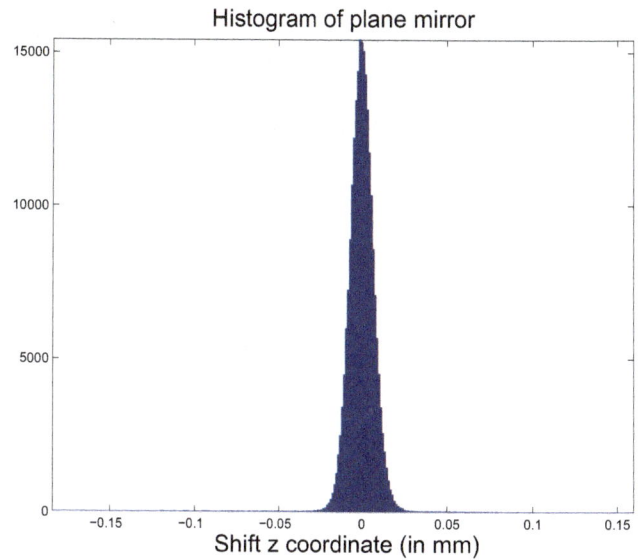

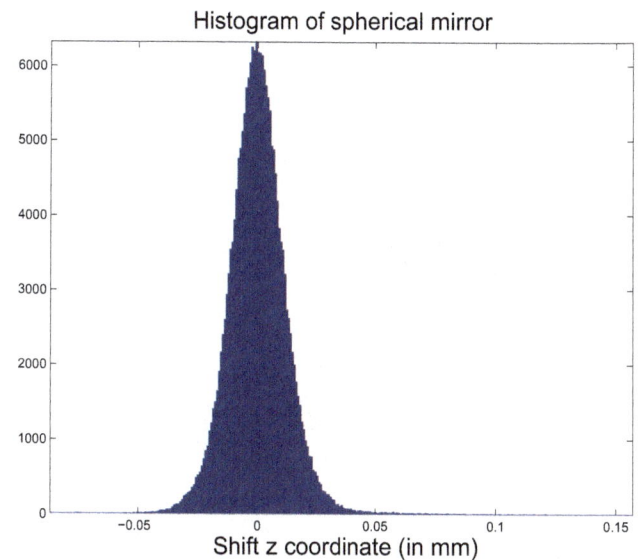

Figure 6. Top panel: residuum of the measured z coordinate and the z coordinate after data fusion in the plane reference mirror case. Bottom panel: residuum of the measured z coordinate and the z coordinate after data fusion in the spherical reference mirror case.

new z coordinate after data fusion in Fig. 6. It turns out that the histograms are approximately Gaussian and, since our data-fusion model in fact produces consistent data of point coordinates and normals, this verifies a posteriori that the assumption of a Gaussian distribution was valid.

In Fig. 7 the gradients of the measured plane and measured spherical reference mirror were plotted. The two plots on the left-hand side show the gradients of the plane mirror before and after the data-fusion process. The two plots on the right-hand side show the same for the spherical mirror. In every plot of this figure the gradients with respect to the measured normals are given in red and the calculated gradients with respect to the point coordinates are given in green. As one can see, the measured gradients all are pointing almost in the same direction, which is reasonable because the two measured objects did not have any visible jumps on the surface. The calculated gradients before data fusion otherwise differ considerably from the calculated ones. After data fusion the gradients with respect to the fitted point coordinates were calculated. On each of the right-hand side plots it can be seen that the new calculated gradients lie on the measured gradients. Hence, the data-fusion process leads to same information of the orientation of the measured surface.

In Fig. 8, the point coordinates before and after data fusion are shown. The surface is colored according to the deviation of the angle between the measured normal and the normal calculated with respect to the point coordinates before and after the optimization process. The color bar is given in degrees.

Before data fusion, the error between the normal vectors on the surface defined by the point coordinates and the measured normal vectors was up to $20°$ at some coordinates.

After optimization the maximal error was below $1°$ in all coordinates.

In Fig. 9 the flatness deviation of a cross section through the plane mirror is shown. The original data obtained from triangulation shows a noise amplitude of about $20\,\mu m$. The data-fusion process reduces the noise level by approximately an order of magnitude. For the spherical mirror very similar results are obtained.

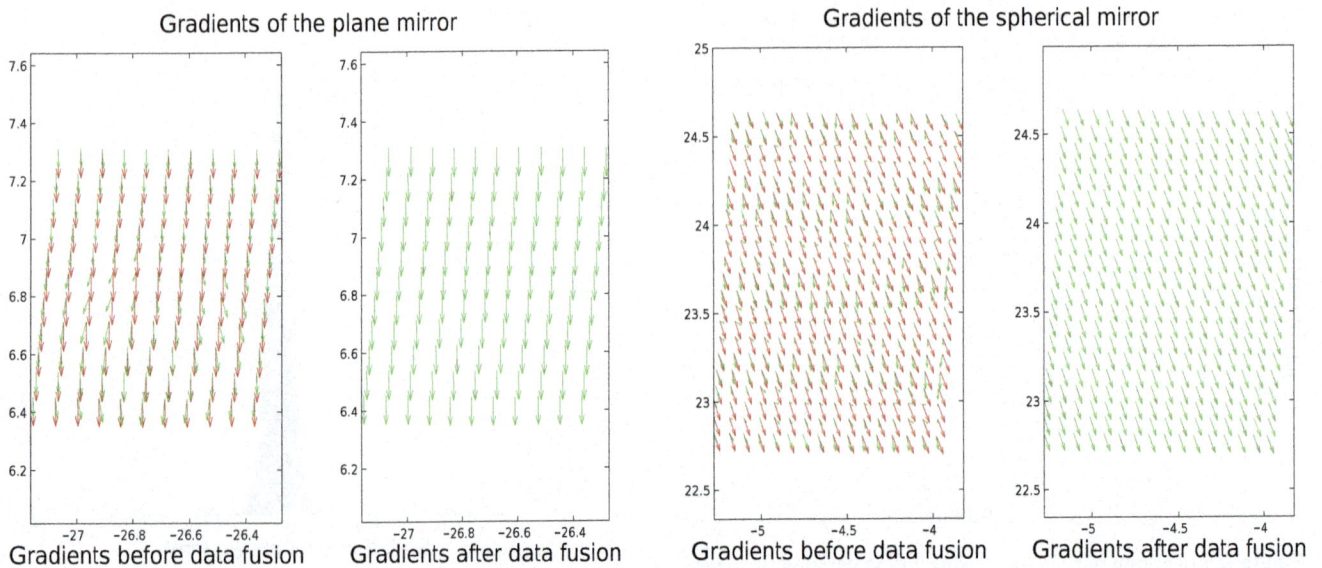

Figure 7. Clipped area of plane and spherical mirror gradients before and after the data-fusion process with 1000 iterations. Measured gradients in red, calculated gradients from the point coordinates in green.

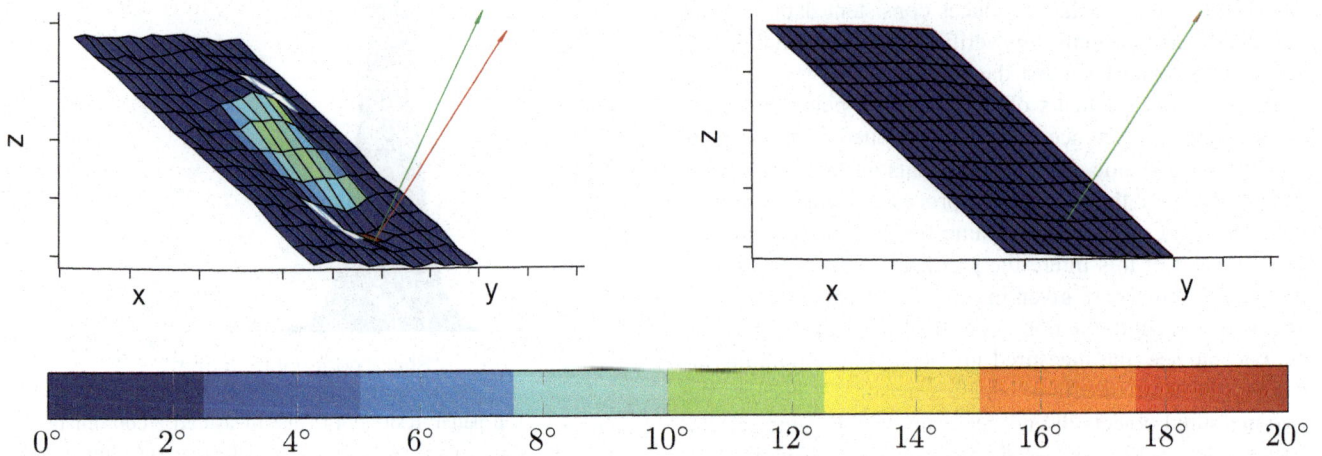

Figure 8. Left: measured point coordinates. Right: point coordinates after optimization. The red vector is a measured normal, the green a calculated normal. The color encodes the error between calculated normals and measured normals in degrees.

5.3   Conclusions

In this paper a novel evaluation approach for a deflectometric measurement has been developed. The results of such a measurement, when carried out with multiple reference planes and active triangulation of the incident and reflected light rays, are the three-dimensional point coordinates and normal directions for each valid pixel of the camera. Due to the local and absolute measurement, the point coordinates are very robust against systematic deviations but with stochastic deviations in the order of 10 μm. However, the commonly applied integration or regularization of the normal directions yields a smooth surface with stochastic deviations in the nanometer range but is sensitive to global form deviations due to systematic influences. The data-fusion method proposed in this work combines both data sets by adjusting the point coordinates to the measured normal directions such that the data is consistent up to the measured tolerances. This adjustment is done in a uniform way such that no systematic deviations are expected to occur.

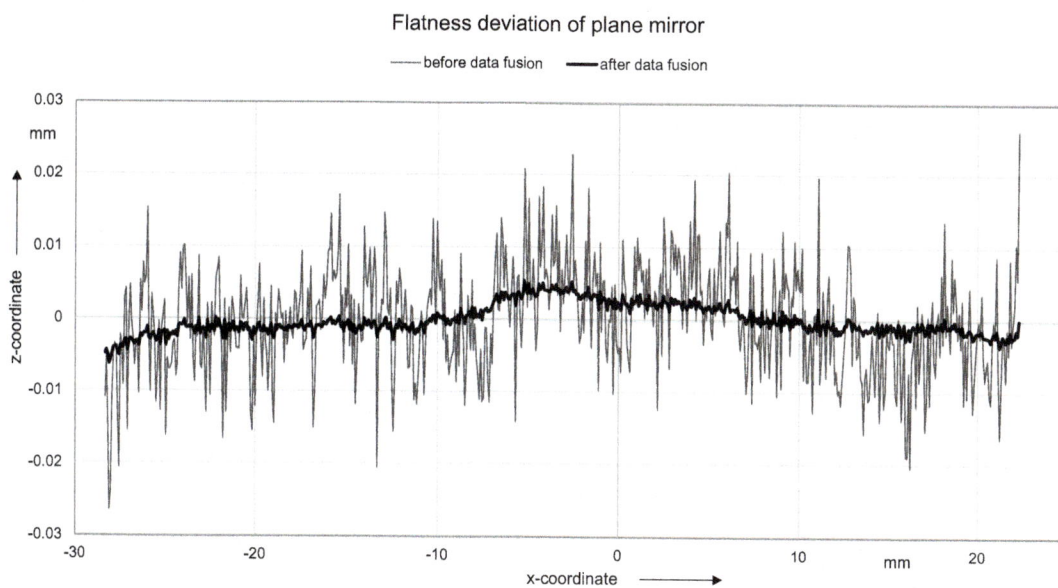

Figure 9. Flatness deviation of one row of the plane mirror before and after data fusion.

The method has been tested with real data sets for a planar and a spherical mirror. It has been shown that the spatial noise level of the resulting point cloud could be reduced by more than an order of magnitude without sacrificing any of the major advantages of a local and absolute measurement, such as the ability to measure nonsmooth surfaces.

Appendix A

A list of all the important variables used.

$B = \left\{ \left\| \mathbf{P}^z - \widetilde{\mathbf{P}}^z \right\|^2 \leq \delta \right\}$	The sphere with radius δ around the measured z coordinate $\mathbf{P}^z$.
$\mathbf{C}$	Matrix.
$\mathbf{D}^{i,j} = (\mathbf{V}^{i,j})^+ \cdot \mathbf{C}$	Difference matrixcorresponding to the point coordinate $\boldsymbol{p}_{i,j}$.
$\mathbf{D}$	Difference matrix summarized from alldifference matrices $\mathbf{D}^{i,j}$.
δ	Tolerance value for shiftingdistance of the point coordinates.
$F\left(\mathbf{P}^z\right)$	Objective functionalwith respect to the z coordinates of the measured points.
g	Gradient with respect to the z coordinates of the points.
$\mathbf{N}$	Measured normal vectors.
$\mathbf{N}^x, \mathbf{N}^y, \mathbf{N}^z$	The x, y, z coordinates of the measurednormal vectors.
$\mathbf{P}$	Measured point coordinates.
$\mathbf{P}^x, \mathbf{P}^y, \mathbf{P}^z$	The x, y, z coordinates of the measured point coordinates.
$\text{Proj}_B(X)$	The projection of the set X of point coordinates into the sphere B.
v_k	The vectors which point from a pointcoordinate $\boldsymbol{p}_{i,j}$ to its neighbors.
$\mathbf{V}^{i,j}$	Matrix containing all vectors v_k.
$\left(\mathbf{V}^{i,j}\right)^+$	Pseudo inverse of matrix $\mathbf{V}^{i,j}$.

Edited by: U. Schmid
Reviewed by: two anonymous referees

References

Li, W., Sandner, M., Gesierich, A., and Burke, J.: Absolute optical surface measurement with deflectometry, in: Proc. SPIE 8494, Interferometry XVI: Applications, 84940G, 13 September, 2012.

Petz, M.: Rasterreflexions-Photogrammetrie – Ein neues Verfahren zur geometrischen Messung spiegelnder Oberflächen. Dissertation, Technische Universität Braunschweig, 2006, Schriftenreihe des Instituts für Produktionsmesstechnik, Band 1, Aachen, Shaker, ISBN 3-8322-4944-3, 2006.

Petz, M. and Ritter, R.: Reflection grating method for 3D measurement of reflecting surfaces, in: Proc. SPIE Vol. 4399 (2001) – Optical Measurement for Industrial Inspection II: Applications in Production Engineering, edited by: Höfling, R., Jüptner, W., and Kujawinska, M., 35–41, 2011.

Petz, M. and Tutsch, R.: Rasterreflexions-Photogrammetrie zur Messung spiegelnder Oberflächen, tm – Technisches Messen, Heft 71, 389–397, 2004.

Tutsch, R., Petz, M., and Fischer, M.: Optical three-dimensional metrology with structured illumination, Opt. Eng., 50, 101507–101507-10, 2011.

Werling, S., Mai, M., Heizmann, M., and Beyerer, J.: Inspection of specular and partially specular surfaces, Metrol. Meas. Syst. Vol. XVI (2009), 3, 415–431, 2009.

14

Selective detection of hazardous VOCs for indoor air quality applications using a virtual gas sensor array

M. Leidinger[1], T. Sauerwald[1], W. Reimringer[2], G. Ventura[3], and A. Schütze[1]

[1]Saarland University, Lab for Measurement Technology, Saarbrücken, Germany
[2]3S GmbH, Saarbrücken, Germany
[3]IDMEC – Institute of Mechanical Engineering, Porto, Portugal

Correspondence to: M. Leidinger (m.leidinger@lmt.uni-saarland.de)

Abstract. An approach for detecting hazardous volatile organic compounds (VOCs) in ppb and sub-ppb concentrations is presented. Using three types of metal oxide semiconductor (MOS) gas sensors in temperature cycled operation, formaldehyde, benzene and naphthalene in trace concentrations, reflecting threshold limit values as proposed by the WHO and European national health institutions, are successfully identified against a varying ethanol background of up to 2 ppm. For signal processing, linear discriminant analysis is applied to single sensor data and sensor fusion data.

Integrated field test sensor systems for monitoring of indoor air quality (IAQ) using the same types of gas sensors were characterized using the same gas measurement setup and data processing. Performance of the systems is reduced due to gas emissions from the hardware components. These contaminations have been investigated using analytical methods. Despite the reduced sensitivity, concentrations of the target VOCs in the ppb range (100 ppb of formaldehyde; 5 ppb of benzene; 20 ppb of naphthalene) are still clearly detectable with the systems, especially when using the sensor fusion method for combining data of the different MOS sensor types.

1 Introduction

The quality of indoor air (IAQ) is determined by the contamination of the air with various chemical compounds, such as carbon dioxide (CO_2), carbon monoxide (CO), nitrogen dioxide (NO_2) and volatile organic compounds (VOCs). Several investigations have been performed to determine the occurrence of these substances in indoor air, e.g., by Bernstein et al. (2008) or in European projects like the Airmex study (Geiss et al., 2011) and the INDEX project (Koistinen et al., 2008).

Negative health effects of exposure to these substances, even at low concentrations, mainly including the respiratory system and skin irritations, have been observed (Jones, 1999). Additionally, some VOCs (e.g., benzene) are carcinogenic, while others (e.g., formaldehyde) are suspected to be carcinogenic (Gou et al., 2004).

Hazardous VOCs pose a special problem. Despite that threshold limits for single substances are recommended for indoor air, e.g., by the WHO (World Health Organization, 2010), there is currently no online measurement technology commercially available to identify and quantify different volatile organic substances reliably and at reasonable cost. Monitoring total VOC (TVOC) concentrations is state of the art (Umweltbundesamt, 2007), but this parameter is not significant in terms of health effects since it also includes benign substances and cannot be attributed to symptoms like the sick building syndrome (Burge, 2004; Brinke et al., 1998). Selective VOC detection and quantification is today based on gas sampling and analytical techniques, especially gas chromatography coupled with mass spectrometry (GC-MS; Wu et al., 2004). The resulting high cost for individual measurements prevents ubiquitous VOC monitoring in IAQ applications today.

A possible application for selective VOC monitoring is demand-controlled ventilation in smart buildings. VOC levels can be used as an additional parameter for controlling indoor ventilation in addition to other indicators like

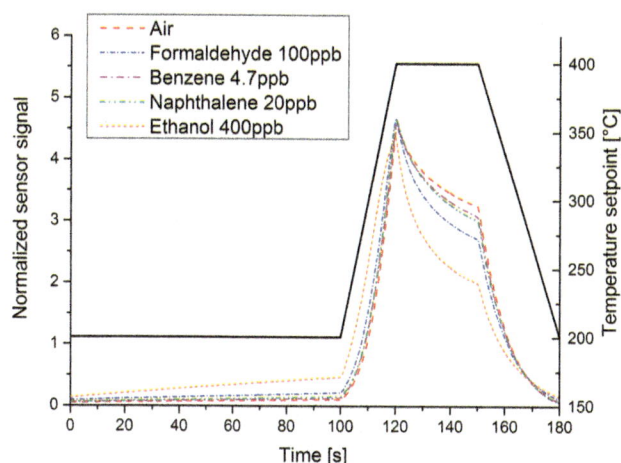

Figure 1. Temperature cycle (solid line) and normalized temperature cycle sensor signals (UST GGS 1330) in the presence of different gases.

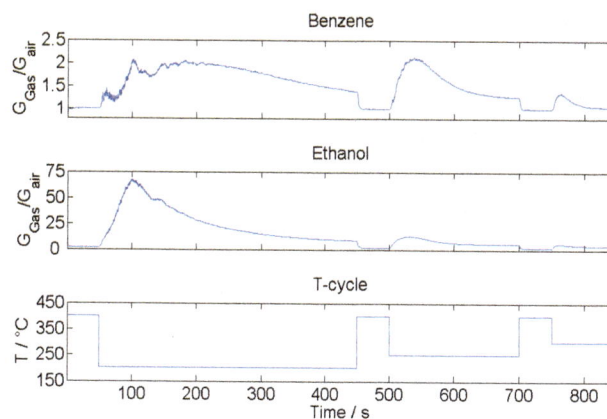

Figure 2. Sensor responses to 25 ppb of benzene and 500 ppb of ethanol during TCO optimization cycle at 12.5 % relative humidity (Fricke et al., 2014).

temperature and CO_2 levels. Then, selective measurement of single VOCs is necessary since ventilation should be increased only if thresholds of hazardous VOCs are exceeded.

From the wide range of VOCs, three compounds were selected for further investigations on selective detection: formaldehyde, benzene and naphthalene, which are three of the first priority harmful VOCs (Koistinen et al., 2008; World Health Organization, 2010). The selected target concentrations of these gases are 10 ppb for formaldehyde, 0.5 ppb for benzene and 2 ppb for naphthalene, based on international and European national regulations (e.g., World Health Organization, 2010; French decree no. 2011-1727, 2011; Sagunski and Heger, 2004). For benzene, the World Health Organization even states that there is no safe level due to its high carcinogenicity (World Health Organization, 2010). Thus, not only a high selectivity is required for identifying these gases but also a very high sensitivity in order to detect ppb levels of these specific VOCs.

One type of sensors which can detect VOCs in this concentration range is a metal oxide semiconductor (MOS) gas sensor (Schüler et al., 2013). MOS sensors in temperature cycled operation (TCO) are used here to measure the selected VOCs against a high background of interfering gas, similar to Reimann and Schütze. (2012). These sensors were also integrated in low-cost sensor systems designed for field testing and as a basis for future commercial online VOC monitoring devices.

2 TCO optimization

Semiconductor gas sensors are very sensitive sensors, but usually they are broadband sensors and show little selectivity to specific gases. One method to improve selectivity, sensitivity and also stability is temperature cycled operation

(Lee and Reedy, 1999; Gramm and Schütze, 2003; Schüler et al., 2013). By modulating the operating temperature of the MOS sensing layer, different states of the sensor material itself (i.e., surface coverage with oxygen) and its interaction with gas molecules are activated, and thus different sensing characteristics are obtained. Figure 1 shows normalized sensor signals of the same temperature cycle when different gases are applied to a MOS gas sensor. The differences of the recorded signal shapes (i.e., slopes, average values in different sections) are obvious; these features are characteristic of specific gases.

Three types of ceramic substrate MOS gas sensors were evaluated for detection of the target VOCs: GGS 1330, GGS 2330 (both SnO_2 based) and GGS 5330 (WO_3 based) by UST Umweltsensortechnik GmbH (Geschwenda, Germany).

A method for optimizing the TCO cycle was evaluated. In order to find the most sensitive and most selective temperature transitions, the relaxation behavior from a high temperature to different lower temperatures was investigated. Specifically, temperature changes from 400 to 200 °C, 250 °C and 300 °C were performed with a GGS 1330 SnO_2 sensor with benzene and ethanol as test gases. The results are shown in Fig. 2.

The sensor response was calculated by dividing the sensor signal (conductivity of the sensitive layer) of a cycle in gas by the sensor signal of a cycle in pure air for each point of the cycle. The response has distinct peaks several seconds after the temperature steps from the high temperature to the lower temperatures – e.g., for ethanol 50 s after changing the sensor temperature from 400 to 200 °C. The sensor response after cooldown from 400 to 200 °C reaches approx. 67 for ethanol and then drops to approx. 9 at the steady state. Thus, the sensitivity is significantly increased in TCO mode due to non-equilibrium state of the sensor surface after temperature changes (Sauerwald et al., 2014). For benzene, the sensor

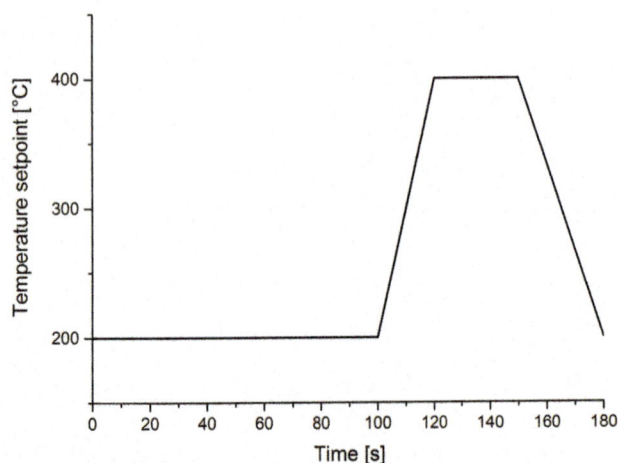

Figure 3. 180 s temperature cycle for the GGS 1330 and GGS 2330 SnO$_2$-based sensors.

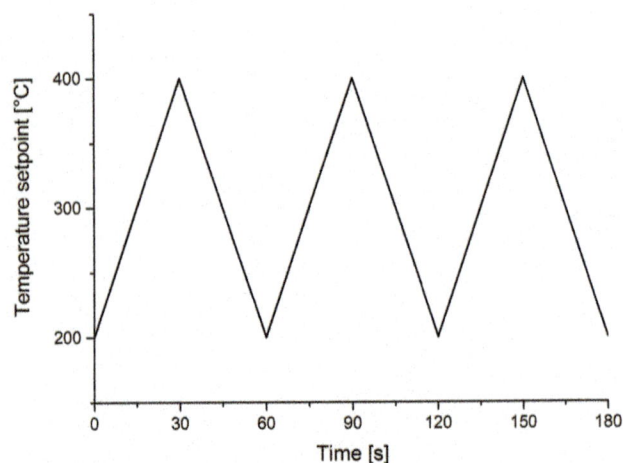

Figure 4. 60 s temperature cycle for the GGS 5330 WO$_3$-based sensor; three cycles are run in order to synchronize the signals with the 180 s cycle for the SnO$_2$ sensors.

response rises to 2.1 at the peak 36 s after the temperature transition from 400 to 250 °C compared to 1.3 at the steady state, corresponding to an almost 4-fold increase in sensitivity.

Based on these results, the temperature steps and the lengths of these steps were defined. For the SnO$_2$ sensors (GGS 1330 and GGS 2330) a two-step temperature cycle with ramp transitions was chosen (see Fig. 3). The ramps were implemented in order to achieve a defined heating up and cooling down of the sensitive layer independent of ambient temperature and humidity. The durations of the ramps are the result of the heating and cooling characteristics of the sensors. Due to the size of the ceramic substrates, heating up the sensors takes up to 20 s and cooling down even longer, up to 30 s. These are the values chosen for the respective ramps.

The length of the low temperature step is 100 s to cover all the response peaks of the previous optimization measurement. The total duration of the temperature cycle is 180 s, which is sufficiently short for the target application in IAQ monitoring.

The WO$_3$-based sensor (GGS 5330) did not show any delayed response maxima, but only a temperature-dependent response. Thus, a simple ramp up and down between 400 and 200 °C was selected covering the range of maximum sensitivity to the target gases (Fig. 4). The duration of a cycle is 60 s; to synchronize all sensors, three cycles of the WO$_3$-based sensor are run during one cycle of the two SnO$_2$-based sensors.

3   Sensor characterization measurements

The three target VOCs were applied in two concentrations each: one at the respective threshold limit value and one at the 10-fold value. Additionally, the measurements were performed with two concentrations of ethanol as a background

interference gas and two values for the relative humidity (RH). Table 1 gives an overview for all concentration and humidity values.

The measurements were conducted with a gas mixing system which was designed and set up specifically for trace gas generation with wide concentration ranges by Helwig et al. (2014). The VOCs were diluted into a carrier gas stream of synthetic air (purity 5.0) either from a gas cylinder or from a permeation furnace. Total gas flow was 200 mL min^{-1}; the three sensors were set up in a stainless steel sensor chamber. Each of the 36 VOC gas configurations was applied for 30 min; between the VOC exposures the sensors were flushed with background (humid air plus ethanol) for 30 min to allow their return to the baseline and prevent carryover. The complete data set contained 940 temperature cycles for the SnO$_2$-based sensors and 2820 cycles for the WO$_3$-based sensor. Not all of the cycles were used for signal processing; for the "background" groups without the target VOCs, six sections with a length of approx. 15 SnO$_2$ cycles each were selected, one after each change of the background conditions (humidity, ethanol).

4   Signal evaluation and data processing

4.1   Sensor characterization

As a first analysis of the data, quasistatic sensor signals are examined. These are generated by choosing one point of the temperature cycle and extracting the signal value at this point in the cycle for every cycle of the measurement. These values are then plotted over the respective cycle number, which generates a plot of the sensor signal of a specific point of the cycle over time. An example is given in Fig. 5.

The sensor reactions to all target gases and especially to the ethanol background are clearly visible. This method is

Table 1. Test gas setup

Gas	Conc./ppb ($\mu g\,m^{-3}$)	RH/%	EtOH background/ppm ($mg\,m^{-3}$)
Synthetic air		40; 60	0; 0.4; 2 (0; 0.21; 1.06)
Formaldehyde	10; 100 (12.3; 123)	40; 60	0; 0.4; 2 (0; 0.21; 1.06)
Benzene	0.5; 4.7 (1.6; 15)	40; 60	0; 0.4; 2 (0; 0.21; 1.06)
Naphthalene	2; 20 (10.5; 105)	40; 60	0; 0.4; 2 (0; 0.21; 1.06)

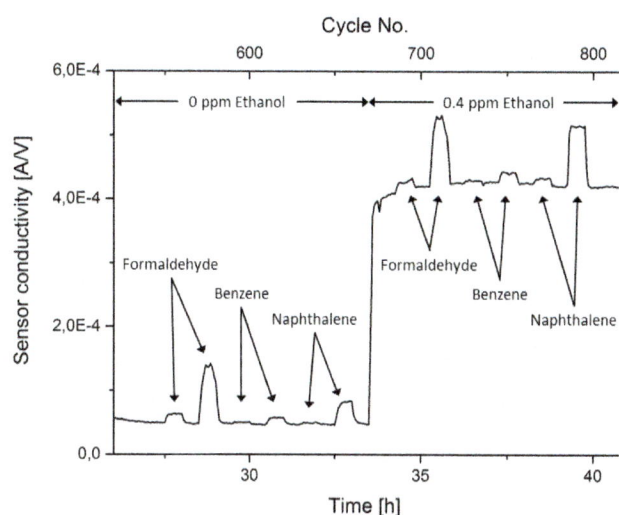

Figure 5. Section of the quasistatic sensor signal, UST GGS 1330, 60 %RH; the selected point of the cycle is the end of the low temperature step at 99 s (see Figs. 3/6).

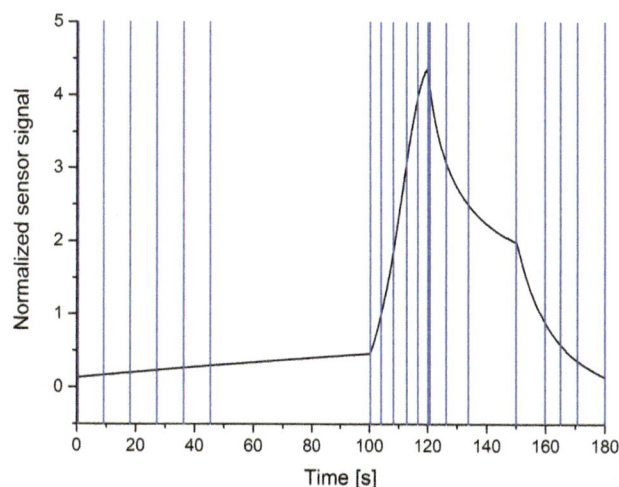

Figure 6. Selected feature ranges of the GGS 1330 and GGS 2330 sensor.

helpful for checking the nominal performance of the gas mixing system and to check the general response of the sensors to the gases. It is independent of the pattern recognition data analysis.

For further signal processing, the method of linear discriminant analysis (LDA) is applied (Gutierrez-Osuna, 2002). This pattern recognition technique can be used to separate different classes of input data while grouping data sets of the same type. In this case, it is used to assign the temperature cycle sensor signals to the different target gases. Thus, in the resulting plots, the algorithm should arrange all cycles of each target gas and background into one compact group while separating the groups of the different target VOCs and the background without VOCs from each other.

The approach used here is basically the same as presented by Bur et al. (2014). Input data sets for the LDA algorithm ("training", i.e., determination of LDA coefficients for the projection, and evaluation) are generated by extracting a set of features from each temperature cycle sensor signal. The temperature cycle is divided into several sections; 20 sections were chosen for the 180 s cycle for the GGS 1330/2330 sensors (see Fig. 6). From each section, features are calculated, in this case the mean value of the sensor signal and the slope of a linear fit. These features were chosen with regard to later

implementation of the LDA calculations on the field test system microcontroller since they are easy to calculate. For the GGS 5330 sensor, the 60 s cycle was divided into 14 sections. This generates a data set of 40 (28) values for each sensor for each temperature cycle, which is used as input for the LDA.

As mentioned above, in the presented measurement the aim is identification of the target VOCs. The extracted data sets were therefore assigned to four groups, one group for each target VOC and one for the background gas without any of the three targets. Each of the three VOC groups thus contains the cycles that ran during the application of one VOC with both VOC concentrations, both gas humidities and all ethanol backgrounds, i.e., a total of 12 different conditions. The "background" group contains sections of synthetic air with both humidities and all ethanol background concentrations.

The result of the LDA calculation for the GGS 1330 sensor is shown in Fig. 7. Separation of the four groups is quite successful, but there is still some overlap. As a validation of the results, leave-one-out cross-validation (LOOCV) is performed (Gutierrez-Osuna, 2002). This method checks how many feature vectors are classified correctly if the LDA is trained by all other vectors. For the GGS 1330 sensor, 98.9 % of the 435 used data sets are classified correctly if the method of k nearest-neighbors classification (kNN, $k = 5$) is applied. So despite the overlap of the groups, nearly all TCO feature sets are assigned to the correct gas.

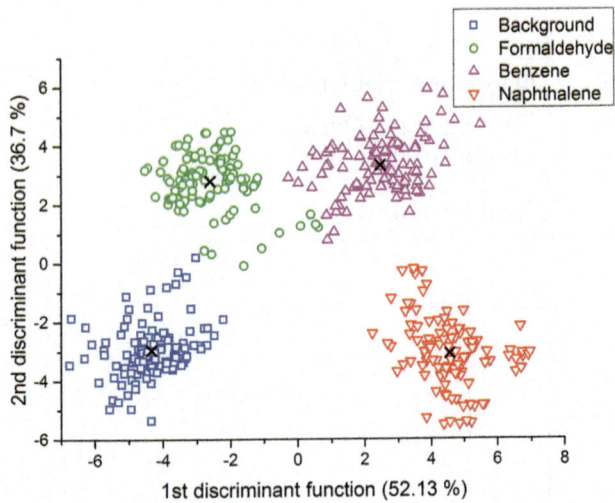

Figure 7. LDA plot of the UST GGS 1330 sensor.

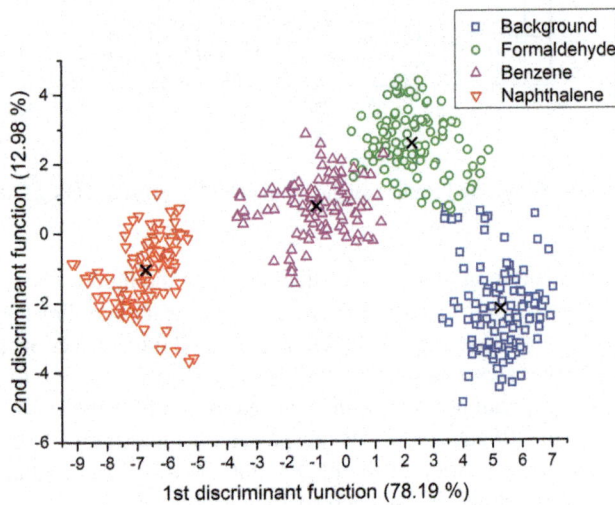

Figure 9. LDA plot of the UST GGS 5330 sensor.

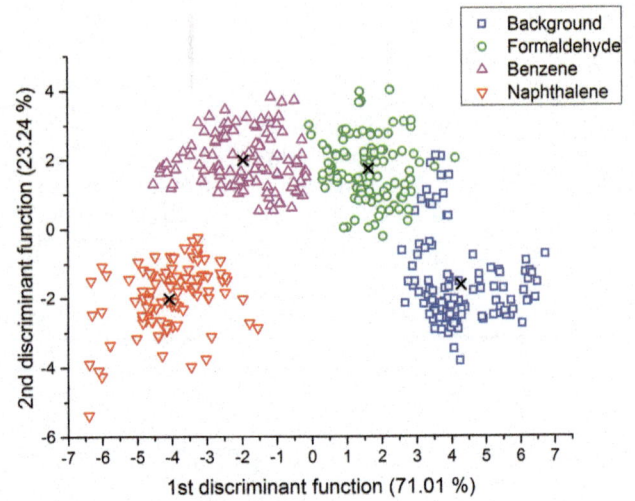

Figure 8. LDA plot of the UST GGS 2330 sensor.

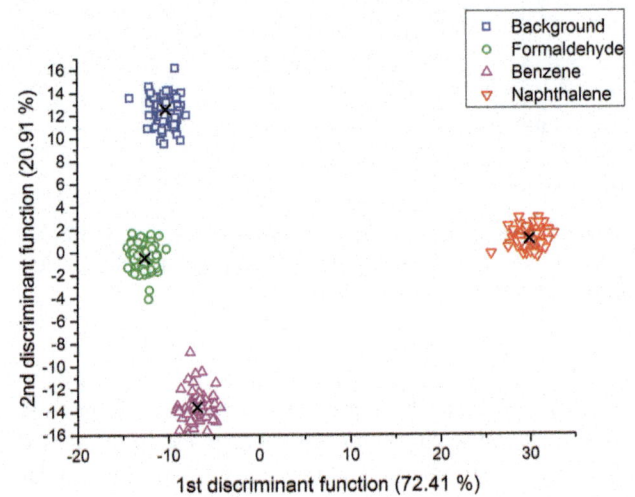

Figure 10. LDA plot based on data fusion of all three sensors.

Figure 8 shows the result of the LDA for the GGS 2330 sensor. Separation of the groups does not appear quite as distinct as for the GGS 1330 sensor, especially with formaldehyde and benzene having slightly more overlap. Leave-one-out cross-validation with kNN results in a correct classification of 96.6 % of all temperature cycles.

The result for the GGS 5330 sensor (Fig. 9) also shows a partial overlap of the groups, especially for formaldehyde and air; compared to the GGS 2230, however, the validation shows a slightly higher number of correct classifications at 98.4 %.

In addition to evaluating the single sensors, a combined processing of the data from the sensors is applied. In this sensor fusion, the feature vectors of two or three sensors are merged into a single data set for each temperature cycle, e.g., a 108-value vector for fusion of all three sensors. LDA

calculation with the combined data results in a much better separation of the groups, shown in Fig. 10 for the combination of all three sensor types. Now there is no overlap of the gas groups. Validation shows a classification accuracy of 100 %; all temperature cycles are classified correctly.

4.2 Field test sensor system characterization

For use in field tests, the sensors were integrated into field test electronics (Conrad et al., 2014). The systems are designed to operate two MOS gas sensors independently in temperature cycled operation, with different temperature cycles. Each sensor is mounted on a plug-in PCB (printed circuit board), which also contains an EEPROM (electronically erasable programmable read-only memory) for calibration data and LDA parameters of the individual sensor. With this setup, fast replacement of a sensor is possible without having

Figure 11. Exterior view of modular field test sensor system containing electronics (PCB) with two MOS gas sensors (Conrad et al., 2014).

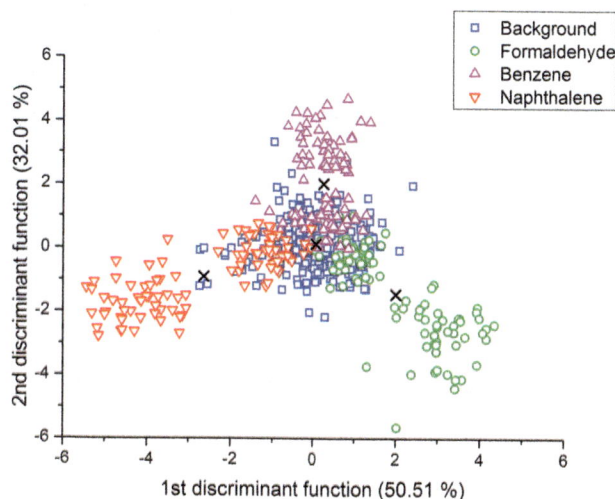

Figure 12. LDA plot of the lab characterization of a UST GGS 1330 sensor integrated in a field test system.

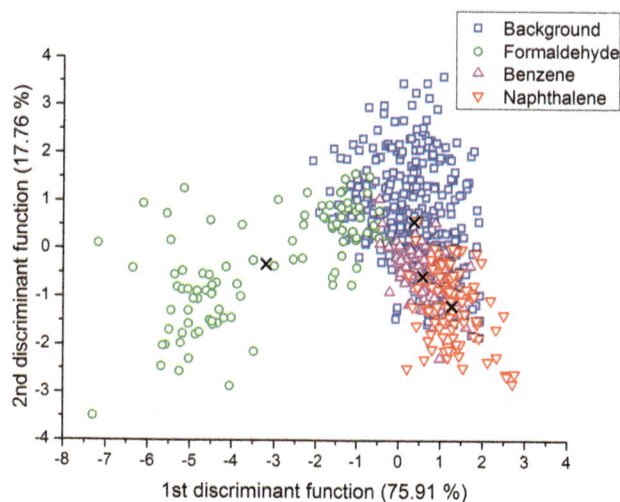

Figure 13. LDA plot of the lab characterization of a UST GGS 2330 sensor integrated in a field test system.

to perform a new calibration of the overall system. The sensor signals are acquired at a rate of up to 10 ksps and are stored on an SD memory card, which also contains general configuration data and the temperature cycle data sets. An on-board sensor measures air temperature and humidity; in addition, the system can be equipped with a dual-beam NDIR (nondispersive infrared) CO_2 sensor. Online preview of the measured data is possible via a selection of communication interfaces. The electronics are installed in a polymer housing (Fig. 11).

The performance of the systems was determined using the same test gas profile as for the sensor characterization in the stainless steel sensor chamber (Table 1). Three systems were characterized simultaneously, each equipped with two different UST gas sensor types with the temperature profiles identified during the lab optimization. A total of six MOS sensors were operated, two of every type; one sensor of every type was used for offline LDA signal processing. The systems were placed in a stainless steel measurement chamber with a volume of 3.5 L. The total gas flow was set to 800 mL min^{-1}, resulting in an air exchange rate of 13.7 ach (air changes per hour). Signal acquisition, pre-processing and feature extraction was performed identically to the characterization measurement of the sensors in the stainless steel sensor chamber.

The LDA result obtained with data from one of the GGS 1330 sensors is shown in Fig. 12.

Separation of the different gases is significantly less successful compared to the sensor characterization measurement (Fig. 7). Each VOC group is split into two sub-groups, reflecting the two tested VOC concentrations. While the higher concentrations are still discriminated from the background, the lower concentrations can no longer be separated from the background group. Using LOOCV, only 71.7 % of temperature cycles are now classified correctly, a significant reduction compared to the result of the sensor in the stainless steel sensor chamber which achieved 98.9 %.

Similar results were obtained for the other two sensor types. In the plot of the GGS 2330 LDA output (Fig. 13), the data groups of naphthalene and especially benzene are hardly separated from the background group and only the high formaldehyde concentration can be clearly discriminated. Only 66.1 % of the temperature cycles are assigned to the correct gas.

The GGS 5330 type sensor is much more sensitive to benzene and naphthalene compared to formaldehyde. This clearly shows in the LDA result (Fig. 14), where both formaldehyde concentrations are plotted overlapping with the background group. However, only the high concentrations of benzene and naphthalene are separated from the background, while the lower concentrations are not. The ratio of correct classifications is 62.0 %.

Table 2. List of leave-one-out cross-validation results with kNN-5 for the LDAs of the single sensors and sensor fusions

Sensor(s)	Correct classifications for the sensors in the stainless steel sensor chamber	Correct classifications for the sensors integrated in the field test systems
GGS 1330	98.9 %	71.7 %
GGS 2330	96.6 %	66.1 %
GGS 5330	98.4 %	62.0 %
GGS 1330 + GGS 2330	100 %	81.6 %
GGS 1330 + GGS 5330	100 %	76.5 %
GGS 2330 + GGS 5330	99.8 %	71.7 %
GGS 1330 + GGS 2330 + GGS 5330	100 %	83.4 %

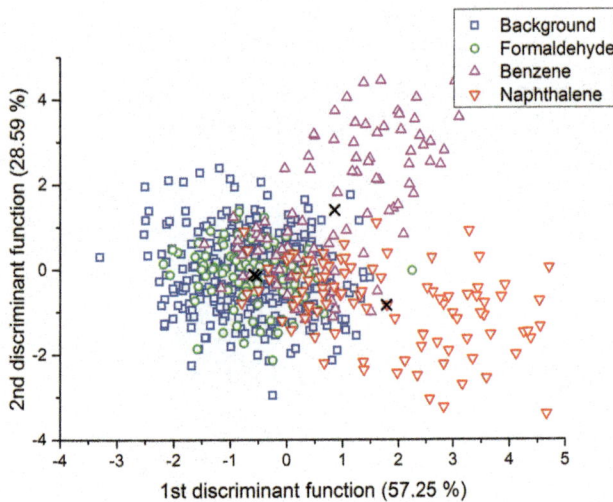

Figure 14. LDA result plot of the lab characterization of a UST GGS 5330 sensor integrated in a field test system.

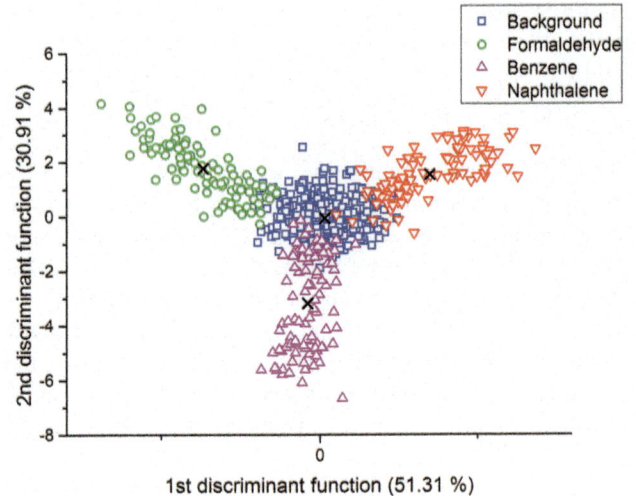

Figure 15. LDA plot based on data fusion of three sensors (one of every type) integrated in field test systems.

Data fusion was applied to the field test system sensor data as well; the resulting LDA plot for fusion one sensor of each of the three sensor types is shown in Fig. 15. As for sensor characterization setup, discrimination of the gases is significantly improved. Not only can the high concentrations of all three target gases be clearly discriminated, but now also the low concentrations are separated more clearly from the background compared to the results obtained with the individual sensors in the systems. LOOCV yields 83.4 % of all temperature cycles classified correctly, an improvement of 11.7 % over the best single sensor (71.7 % for the GGS 1330).

The results of the LDA validations for all the sensors and all possibilities of sensor fusion are listed in Table 2. For the sensors in the stainless steel sensor chamber, fusion of two sensors – GGS 1330 combined with any of the other two sensors – is already sufficient for reliable identification of the VOC. For the sensors integrated in the field test sensor systems, fusion of all three sensors is necessary for best selectivity.

Detailed LOOCVs of the LDA results of the sensors integrated in the systems are listed in Table 3. The different gas sensitivities of the three sensor types are clearly shown by the validation results for the different VOCs. While the GGS 1330 sensor has a similar sensitivity to all the gases, the GGS 2330 has an enhanced sensitivity to formaldehyde and a reduced sensitivity to the other two target VOCs. The GGS 5000 sensor is not very sensitive to formaldehyde but has higher numbers of correct classifications for naphthalene and especially benzene compared to the GGS 2330. These values show that the data from the different sensor types can reasonably be used in sensor fusion as the sensors complement each other in their responses to the target gases.

The reason for the reduced sensitivity of the sensors integrated in the field test systems was investigated further (Leidinger et al., 2014). As the main problem, gas emissions from the sensor system hardware components were determined. These emissions were identified and quantified using analytical methods, namely GC/MS VOC measurements according to the ISO 16000 standard.

Table 3. Detailed LOOCV results of the LDAs of the field test system MOS sensors; ratio of correct classifications for the single groups and overall.

Sensor(s)	Background	Formaldehyde	Benzene	Naphthalene	Overall
GGS 1330	82.8 %	60.8 %	52.6 %	61.5 %	71.75 %
GGS 2330	77.7 %	74.2 %	35.8 %	45.8 %	66.1 %
GGS 5330	84.0 %	11.3 %	48.4 %	46.9 %	62.0 %
GGS 1330 + GGS 2330	84.5 %	87.6 %	67.4 %	79.25 %	81.6 %
GGS 1330 + GGS 5330	84.5 %	66.0 %	61.1 %	72.9 %	76.5 %
GGS 2330 + GGS 5330	80.8 %	79.4 %	35.8 %	66.7 %	71.7 %
GGS 1330 + GGS 2330 + GGS 5330	87.7 %	85.6 %	68.4 %	80.2 %	83.4 %

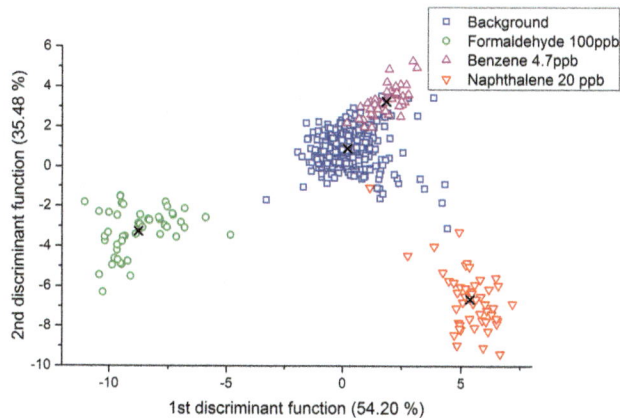

Figure 16. LDA result plot of the lab characterization of a UST GGS 1330 sensor integrated in a field test system, evaluated only for the high VOC concentrations.

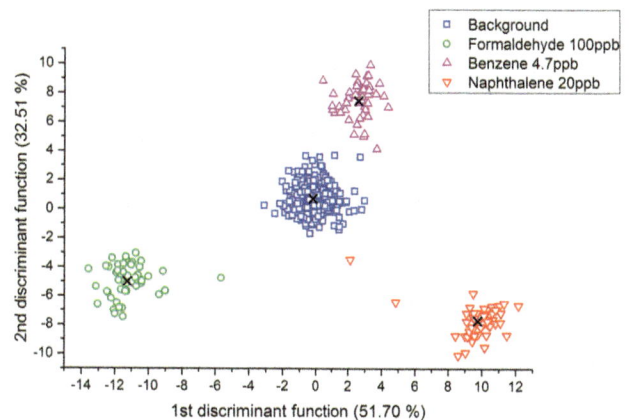

Figure 17. LDA plot based on data fusion of a GGS 1330 and a GGS 5330 integrated in field test systems, evaluated only for the high VOC concentrations.

For gas sampling, Tenax tubes were inserted into the outlet gas flow of the stainless steel measurement chamber containing three field test systems. Due to the requirements of this sampling method, air flow had to be reduced to 120 mL min^{-1} or 2.06 ach. The most significant results of the GC/MS analysis of the gas samples are listed in Table 4. The results obtained with the low flow rate were converted to the high flow rate of 13.7 ach used for the system characterization measurements, assuming that the gas emission rate from the systems is constant and independent of the gas flow at these air exchange rates. The conversion factor is 0.15, which is the ratio of the two gas flows (120 mL min^{-1} vs. 800 mL min^{-1}).

The TVOC value (last row Table 4) proves that there are significant VOC emissions from the systems, especially when heated up during operation. Measured TVOC emissions of three operating systems increase by a factor of approx. 20 compared to the unloaded test chamber and a factor of 12 compared to the systems being switched off and at room temperature. Thus, VOCs are produced by the systems, i.e., outgassing from either the PCB or the polymer housing (cf. Fig. 11). This is also confirmed by the reduced contamination observed after a heat treatment of the field test sensor systems (cf. last row in Table 4).

Looking at the specific gases, the amount of benzene measured is especially conspicuous. A concentration of 11.4 µg m^{-3} was determined, corresponding to 3.6 ppb. This strong benzene background, generated by the systems themselves, readily explains the reduced sensitivity to the applied benzene concentrations, especially the lower concentration of 0.5 ppb, compared to the single sensor measurements.

Naphthalene is not emitted from the systems in relevant amounts; the concentration measured with the systems operating is 0.2 µg m^{-3} or 0.04 ppb. Similarly, the concentration of formaldehyde was 1.2 ppb, or only 10 % of the lower test gas concentration of the calibration measurement. The most significant compound identified in the GC/MS analysis is 1,2-dimethoxyethane, with 168.8 µg m^{-3} or 45.8 ppb. The origin of this substance could not be determined.

Despite the high contamination levels, discrimination is still possible for the high concentrations of the target gases, as shown for the GGS 1330 sensor in Fig. 16. The high concentrations of formaldehyde and naphthalene can be mostly separated from the background. For benzene, discrimination does not seem as clear, but LOOCV shows that 94.5 % of temperature cycles are classified correctly. Sensor fusion further improves discrimination. Figure 17 shows the LDA

Table 4. Measured contaminations caused by outgassing of the field test system, converted from 2.06 ach to 13.7 ach; in µg m^{-3} according to the ISO 16000 standard (n.d.: not detectable; n/a: data not available).

Compound	Synthetic air 5.0	Measurement chamber, no systems	Systems OFF	Systems ON	Systems ON after heat treatment
Acetone	0.7	9.5	11.5	10.6	n/a
1,3-Dioxolane	0.3	0.4	1.9	24.8	12.8
1,2-Dimethoxyethane	0.1	1.9	12.3	168.8	84.2
Benzene	0.0	0.3	n.d.	11.4	5.7
Toluene	0.1	4.6	4.0	7.3	1.0
m/p-xylene	0.1	0.3	0.1	10.2	3.2
Naphthalene	n.d.	0.1	0.0	0.2	2.4
Formaldehyde	n/a	0.3	n/a	1.5	n/a
Acetaldehyde	n/a	n.d.	n/a	1.9	n/a
TVOC	1.74	14.1	22.2	270	164.6

Table 5. List of LOOCV results for the 2-D and 3-D LDAs of the single sensors and sensor fusions of the field test system sensors for the high VOC concentrations.

Sensor(s)	Correct LOOCV classifications with kNN-5, 2-D LDA	Correct LOOCV classifications with kNN-5, 3-D LDA
GGS 1330	94.5 %	97.8 %
GGS 2330	80.7 %	82.2 %
GGS 5330	87.4 %	87.0 %
GGS 1330 + GGS 2330	94.5 %	99.4 %
GGS 1330 + GGS 5330	98.6 %	99.2 %
GGS 2330 + GGS 5330	96.0 %	98.4 %
GGS 1330 + GGS 2330 + GGS 5330	95.3 %	99.6 %

plot for fusion data of a GGS 1330 sensor and a GGS 5330 sensor. With this sensor combination, 98.6 % of temperature cycles are classified in the correct group. The classification results for all sensor types in the field test systems and the combinations are listed in Table 5.

The results can be improved further by calculating 3-dimensional LDAs. Then the ratio of correct classifications reaches more than 99 % with sensor data fusion (Table 5, last column). One example of the 3-D LDA plot is given in Fig. 18.

A method to prevent or at least reduce gas emissions from the systems (PCB and housing) is heat treatment of the devices. This was performed in a climate chamber where the systems were kept for 13 h at 70 °C inside the stainless steel chamber while pure air was continuously flowing through the chamber in order to flush out all emissions from the systems. Afterwards, another gas sample was taken; see Table 4, last column. Obviously, VOC emissions have been reduced significantly by approx. 40 %, but are still more than 7 times higher compared to the unloaded test chamber. Thus, further heat treatment at higher temperature and/or different materials for the housing are required to achieve acceptable contamination levels of the integrated sensor systems.

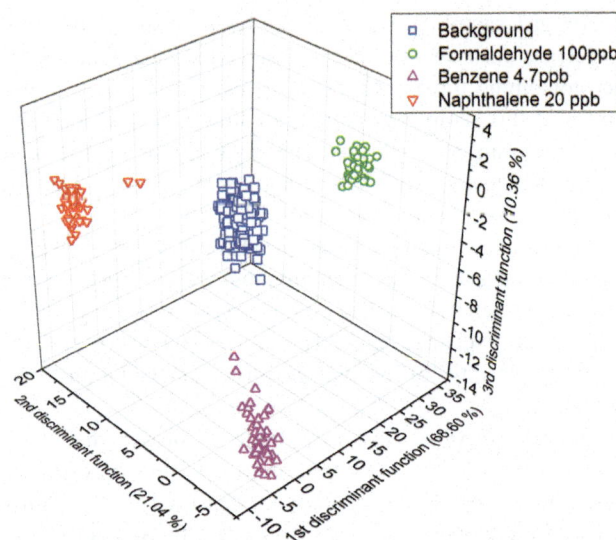

Figure 18. 3-D LDA plot based on data fusion of three sensors (one of every type) integrated in field test systems, evaluated only for the high VOC concentrations.

5 Conclusions and outlook

We have demonstrated that standard metal oxide semiconductor gas sensors operated in dynamic mode using TCO can detect and identify hazardous VOCs at ppb and sub-ppb levels, even in the presence of a much higher background concentration of ethanol (up to a factor of 4000 higher compared to the lower benzene concentration in the measurements).

In the sensor characterization measurements, when the sensors were installed in a stainless steel sensor chamber, the data sets from the sensor signals, containing several ethanol concentrations as well as gas humidities, could be assigned to the correct target gas with high reliability using a one-step LDA algorithm. The results of the data evaluation were improved significantly by sensor fusion, i.e., based on features obtained from two or three different sensors. For this measurement, 100 % of the temperature cycles were assigned to the correct gas by this method as verified by LOOCV. Further optimization of the sensor performance, e.g., using hierarchical data analysis (Schütze et al., 2004) or taking into account the information of further sensors, will be studied in the future.

For the integrated field test systems, however, the classification rate was reduced significantly compared to the sensor tests. Even with sensor fusion, only 83.4 % of the temperature cycles were classified correctly. This was attributed to VOC gas emissions from the system hardware, which have a profound effect on the performance of the individual sensors and the combined sensor array; the sensing capabilities are clearly impaired by the VOC emissions. Using only the high test gas concentrations for LDA processing, the ratio of correct classification rises to more than 95 % in a 2-D LDA and over 99 % in a 3-D LDA. These VOC concentrations, still in the ppb range, can be identified by the systems with a high success rate.

A first test of baking out the system showed promising results, as VOC emissions were significantly reduced. Separate heat treatment of the PCB and the housing would allow for application of a higher temperature to the PCB and should reduce gas emissions even further. The expected positive influence of reduced emissions on the sensing performance of the integrated sensor systems will be verified in future experiments. With these optimized integrated sensor systems field tests will be carried out in various typical indoor environments, e.g., offices and meeting rooms, to validate the performance of these systems for continuous monitoring of indoor air quality.

Acknowledgements. This work was in performed within the MNT-ERA.net project VOC-IDS. Funding by the German Ministry for Education and Research (BMBF, funding code 16SV5480K and 16SV5482) and the Portuguese Foundation for Science and Technology (FCT, ERA-MNT/0002/2010) is gratefully acknowledged.

Edited by: M. Penza
Reviewed by: two anonymous referees

References

Bernstein, J. A., Alexis, N., Bacchus, H., Leonard Bernstein., I, Fritz, P., Horner, E., Li, N., Mason, S., Nel, A., Oullette, J., Reijula, K., Reponen, T., Seltzer, J., Smith, A., and Tarlo, S. M.: The health effects of nonindustrial indoor air pollution, J. Allergy Clin. Immu., 121, 585–591, 2008.

Brinke, J. T., Selvin, S., Hodgson, A. T., Fisk, W. J., Mendell, M. J., Koshland, C. P., and Daisey, J. M.: Development of New Volatile Organic Compound (VOC) Exposure Metrics and their Relationship to "Sick Building Syndrome" Symptoms, Indoor Air, 8, 140–152, doi:10.1111/j.1600-0668.1998.t01-1-00002.x, 1998.

Bur, C., Bastuck, M., Lloyd Spetz, A., Andersson, M., and Schütze, A.: Selectivity enhancement of SiC-FET gas sensors by combining temperature and gate bias cycled operation using multivariate statistics, Sensor. Actuat. B-Chem., 193, 931–940, doi:10.1016/j.snb.2013.12.030, 2014.

Burge, P. S.: Sick building syndrome, Occup. Environ. Med., 61, 185–190, doi:10.1136/oem.2003.008813, 2004.

Conrad, T., Reimringer, W., and Rachel, T.: Modulare Systemplattform zur Bewertung der Luftqualität in Innenräumen basierend auf temperaturmodulierten Metalloxid-Gassensoren, ITG/GMA Fachtagung Sensoren und Messsysteme 2014, ITG Fachbericht 250, VDE Verlag, 2014.

French decree no. 2011-1727: Guideline values for indoor air for formaldehyde and benzene (Translation), original title: relatif aux valeurs-guides pour l'air intérieur pour le formaldéhyde et le benzene, 2011.

Fricke, T., Conrad, T., Sauerwald, T., and Schütze, A: Progress towards an Automatic T-Cycle Optimization Utilizing an Integrated Gas Sensor Testing and Evaluation Toolbox, IMCS2014 – the 15th International Meeting on Chemical Sensors (poster presentation); Buenos Aires, Argentina, 16–19 March, 2014.

Geiss, O., Giannopoulos, G., Tirendi, S., Barrero-Moreno, J., Larsen, B. R., and Kotzias, D.: The AIRMEX study – VOC measurements in public buildings and schools/kindergartens in eleven European cities: Statistical analysis of the data, Atmos. Environ., 45, 3676–3684, doi:10.1016/j.atmosenv.2011.04.037, 2011.

Gramm, A. and Schütze, A.: High performance solvent vapor identification with a two sensor array using temperature cycling and pattern classification, Sensor. Actuat. B-Chem., 95, 58–65, 2003.

Guo, H., Lee, S. C., Chan, L. Y., and Li, W. M.: Risk assessment of exposure to volatile organic compounds in different indoor environments, Environ. Res., 94, 57–66, doi:10.1016/S0013-9351(03)00035-5, 2004.

Gutierrez-Osuna, R.: Pattern Analysis for machine olfaction: A review, IEEE Sens. J., 2, 189–202, doi:10.1109/JSEN.2002.800688, 2002.

Helwig, N., Schüler, M., Bur, C., Schütze, A., and Sauerwald, T.: Gas mixing apparatus for automated gas sensor characterization, Meas. Sci. Technol., 25, 055903, doi:10.1088/0957-0233/25/5/055903, 2014.

Jones, A. P.: Indoor air quality and health, Atmos. Environ., 33, 4535–4564, 1999.

Koistinen, K., Kotzias, D., Kephalopoulos, S., Schlitt, C., Carrer, P., Jantunen, M., Kirchner, S., McLaughlin, J., Mølhave, L., Fernandes, E. O., and Seifert, B.: The INDEX project: executive summary of a European Union project on indoor air pollutants, Allergy, 63, 810–819, doi:10.1111/j.1398-9995.2008.01740.x, 2008.

Lee, A. P. and Reedy, B. J.: Temperature modulation in semiconductor gas sensing, Sensor. Actuat. B-Chem., 60, 35–42, 1999.

Leidinger, M., Sauerwald, T., Conrad, T., Reimringer, W., Ventura, G., and Schütze, A.: Selective Detection of Hazardous Indoor VOCs Using Metal Oxide Gas Sensors, Eurosensors XXVIII, 8–10 September 2014, Brescia, Italy, 2014.

Reimann, P. and Schütze, A.: Fire detection in coal mines based on semiconductor gas sensors, Sensor Rev., 32, 47–58, 2012.

Sagunski, H. and Heger, W.: Richtwerte für die Innenraumluft: Naphthalin, Bundesgesundheitsblatt – Gesundheitsforschung – Gesundheitsschutz, 47, 705–712, 2004.

Sauerwald, T., Baur, T., and Schütze, A.: Strategien zur Optimierung des temperaturzyklischen Betriebs von Halbleitergassensoren, Tagungsband zum XXVIII. Messtechnisches Symposium des Arbeitskreises der Hochschullehrer für Messtechnik, Shaker-Verlag, 2014.

Schüler, M., Helwig, N., Schütze, A., Sauerwald, T., and Ventura, G.: Detecting trace-level concentrations of volatile organic compounds with metal oxide gas sensors, IEEE SENSORS Conference (2013), 3–6 November 2013, doi:10.1109/ICSENS.2013.6688276, 2013.

Schütze, A., Gramm, A., and Rühl, T.: Identification of Organic Solvents by a Virtual Multisensor System with Hierarchical Classification, IEEE Sens. J., 4, 857–863, doi:10.1109/JSEN.2004.833514, 2004.

Umweltbundesamt: Beurteilung von Innenraumluftkontaminationen mittels Referenz- und Richtwerten, Bundesgesundheitsblatt – Gesundheitsforschung – Gesundheitsschutz, 50, 990–1005, 2007.

World Health Organization: WHO Guidelines for Indoor Air Quality: Selected Pollutants, Geneva, 2010.

Wu, C.-H., Feng, C.-T., Lo, Y.-S., Lin, T.-Y., and Lo, J.-G.: Determination of volatile organic compounds in workplace air by multisorbent adsorption/thermal desorption-GC/MS, Chemosphere, 56, 71–80, doi:10.1016/j.chemosphere.2004.02.003, 2004.

15

Catalytic and thermal characterisations of nanosized PdPt / Al$_2$O$_3$ for hydrogen detection

T. Mazingue[1], M. Lomello-Tafin[1], M. Passard[1], C. Hernandez-Rodriguez[1], L. Goujon[1], J.-L. Rousset[2], F. Morfin[2], and J.-F. Laithier[3]

[1]Univ. Savoie, SYMME, 74000 Annecy, France
[2]Institut de Recherches sur la Catalyse et l'Environnement de Lyon (IRCELYON, CNRS – University of Lyon), 2 avenue Albert Einstein, 69626 Villeurbanne CEDEX, France
[3]Comelec SA, Rue de la Paix 129 – 2301 La Chaux-de-Fonds, Switzerland

Correspondence to: T. Mazingue (thomas.mazingue@univ-savoie.fr)

Abstract. Palladium platinum (PdPt) has been intensively studied these last decades due to high conversion rate in hydrogen oxidation at room temperature with significant exothermic effects. These remarkable properties have been studied by measuring the temperature variations of alumina (Al$_2$O$_3$) supported nanosized PdPt nanoparticles exposed to different hydrogen concentrations in dry air. This catalyst is expected to be used as a sensing material for stable and reversible ultrasensitive hydrogen sensors working at room temperature (low power consumption). Structural and gas sensing characterisations and catalytic activity of PdPt / Al$_2$O$_3$ systems synthesised by co-impregnation will be presented. Catalytic characterisations show that the system is already active at room temperature and that this activity sharply increases with rise in temperature. Moreover, the increase of the PdPt proportion in the co-impregnation process improves the activity, and very high conversion can be reached even at room temperature. The thermal response (about 3 °C) of only 1 mg of PdPt / Al$_2$O$_3$ is reversible, and the time response is about 5 s. The integration of PdPt / Al$_2$O$_3$ powder on a flat substrate has been realised by the deposition onto the powder of a thin porous hydrophobic layer of parylene. The possibility of using PdPt in gas sensors will be discussed.

1 Introduction

Hydrogen is a promising non-polluting alternative to fossil fuels as an energy carrier. Its storage for domestic applications remains problematic, since the hydrogen molecule is very small (74×10^{-12} m), inducing a risk of leaks leading to explosion (lower explosion limit LEL $= 4\,\%$ of hydrogen in the air). There already exist many hydrogen sensors (Meixner and Lampe, 1996), based for example on electrolytic (Kroll and Smorchkov, 1996) or catalytic (Shin et al., 2009; Han et al., 2007) reactions, widely used in industrial plants. But most of these devices need important power supplies (for reaction activation and desorption process), and suffer from drift of measured values and poisoning (Barsan et al., 2007). We propose in this work to use palladium platinum (PdPt) nanoclusters supported by alumina (PdPt / Al$_2$O$_3$)

for H$_2$ detection. This bimetallic catalyst has been intensively studied these last decades for numerous remarkable properties, including good conversion rate in hydrogen oxidation at room temperature with high exothermal effects ($\Delta H_{25\,°C,\text{vap}} = -241.8\,\text{KJ mol}^{-1}$; Lide, 2001) and high resistance to poisoning (Rousset et al., 2001; Morfin et al., 2004; Rousset et al., 2001). We present here the PdPt / Al$_2$O$_3$ synthesis by co-impregnation, morphology, catalytic activity and gas sensing characterisations. Catalytic characterisations show that conversion starts below room temperature and increases quickly with temperature. The thermal response (about 3 °C) of only 1 mg of PdPt / Al$_2$O$_3$ is reversible at room temperature, and the time response is about 5 s. We conclude on the possibility of using PdPt in gas sensors.

2 Experiment

2.1 PdPt / Al$_2$O$_3$ synthesis

The PdPt / Al$_2$O$_3$ catalysts are prepared by co-impregnation of metallic precursors (Pd and Pt acetyl acetonates) on two Al$_2$O$_3$ powders with different grain sizes, crystalline phases and specific surface areas ($\alpha-$Al$_2$O$_3$: 0.3 μm, 7.7 m^2 g^{-1}, and $\gamma-$Al$_2$O$_3$: 2.6 μm, 56 m^2 g^{-1}) in a solution of toluene stirred for 24 h. Then the solution is dried (evaporation of the toluene under vacuum at 80 °C for a day). Finally, stages of decomposition (450 °C in Ar), calcination (350 °C in O$_2$) and reduction (450 °C in H$_2$) are performed. Three catalysts (A, B and C) have been synthesised on both sizes of alumina, and one with alpha alumina, with different concentrations of Pt and Pd. They all are in the form of a grey powder with a texture similar to flour.

2.2 Sample characterisations

The chemical composition of the catalysts synthesised is determined by ICP-OES (inductively coupled plasma – optic emission spectroscopy). Transmission electron microscopy (TEM) is used to characterise the size and the morphology of the metallic particles.

2.3 Catalytic activity

The oxidation of H$_2$ by synthesised PdPt / Al$_2$O$_3$ samples has then been carried out at atmospheric pressure in a continuous-flow fixed-bed reactor. The experimental set-up is described in detail in Rossignol et al. (2005). The catalyst (6 mg) is exposed to a reactive mixture consisting of 0.5 %H$_2$ + 10 %O$_2$ in He at a flow rate of 50 mL min^{-1}. The gas mixture is analysed at the output of the reactor and compared to the initial composition by a gas chromatograph Varian-Micro GC (CP2003) equipped with thermal conductivity detectors. The reactor is immersed in an isothermal water bath whose temperature was set between 0 and 70 °C. Catalytic conversion is measured after 20 min of stabilisation. A thermocouple located inside the catalytic bed is used to measure the catalyst temperature.

2.4 Gas sensitivity characterisations

The principle of the gas sensitivity measurement is to follow with an IR camera the temperature variation of PdPt / Al$_2$O$_3$ powder due to the exothermicity associated with the oxidation of H$_2$. This will provide a relation between the different H$_2$ concentrations and the variations of the temperature. This thermal response is considered as the gas sensitivity. The PdPt / Al$_2$O$_3$ samples are placed on a clean substrate in the gas chamber. The experimental set-up is described in more details in Mazingue et al. (2012). Before proceeding with the oxidation of H$_2$, a flow of dry air is sent into the

Table 1. Physical and chemical characterisations of PdPt / Al$_2$O$_3$ synthesised by co-impregnation.

Catalyst	A	B	C
Support	γ-Al$_2$O$_3$	α-Al$_2$O$_3$	γ-Al$_2$O$_3$
Pd (%wt)	0.7	0.6	5.4
Pt (%wt)	2.2	0.7	15.4
Φ (nm)	3.8	6.2	4–10

gas chamber for at least 10 min to evacuate eventual impurities and undesired uncontrolled humidity. The experimental protocol is then started with three alternate sequences of 30 s of dry air and a given concentration of H$_2$ (from 200 to 4000 ppm) in dry air under a flow rate of 500 mL min^{-1}. The repeatability and the linearity of the response can thus be evaluated. The thermal response is measured with an IR camera CEDIP JADE III MW under ALTAIR software, through a sapphire window above the gas chamber. The principle is to use each camera pixel as a small thermometer. A differential thermal average measurement is performed between a surface occupied by a determined mass of nanopowder and a free surface of the same area separated by a small distance on the same substrate. All electronic devices and pneumatic valves are monitored by a LabVIEW application, via a CompactRIO/FPGA.

3 Results and discussions

Catalysts A and B have been first synthesised as described above, respectively, on gamma and alpha alumina. Their chemical compositions are given in Table 1. It has been observed that much of platinum (70 %) introduced into the impregnating solution is lost during the synthesis of the catalyst on alpha alumina (sample B). This loss is much less pronounced for the palladium, and it is not present on the gamma alumina (sample A), which has a much larger specific surface area. The metallic proportion is therefore much higher for A (3.3 %wt) compared to B (0.9 %wt for B). As the efficiency of the catalyst increases with the metallic proportion, it has been chosen to synthesise sample C with higher quantities of metallic precursors on γ-Al$_2$O$_3$. This leads to powders with metallic concentrations up to 20.8 %wt for sample C. Figure 1 shows that the morphology of the catalyst remains in the form of nanoparticles deposited on alumina, independently of the metallic concentration.

Figure 2 shows that, without prior activation (a first cycle of heating and cooling), the catalytic conversion at room temperature is about 83 % for A but only 18 % for B. Full H$_2$ conversion is reached at about 60 °C for A and B, but A presents higher conversion rates for lower temperature. This can be explained by the greater amount of metal in the powder in A. Concerning sample C, 100 % of the H$_2$ is converted at room temperature. This value falls to about 68 % at 5 °C.

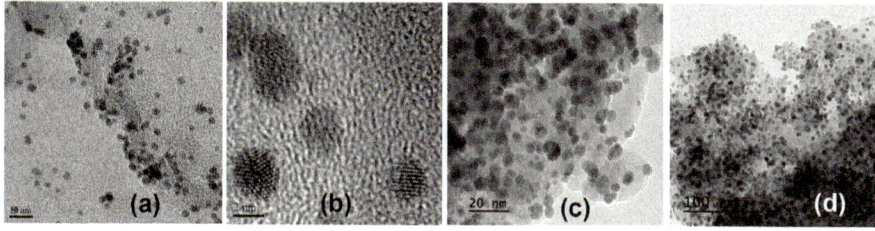

Figure 1. TEM micrographs of catalyst A: (a) and (b); catalyst C: (c) and (d).

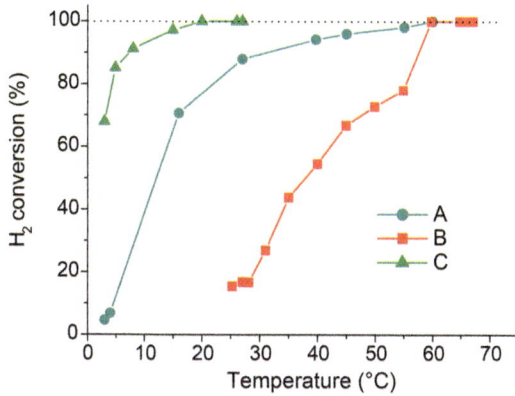

Figure 2. Dependence of H_2 conversion over PdPt / Al_2O_3 versus temperature in a mixture of $0.5\%H_2 +10\%O_2$ in He for 5.5 mg of synthesised catalysts A, B and C.

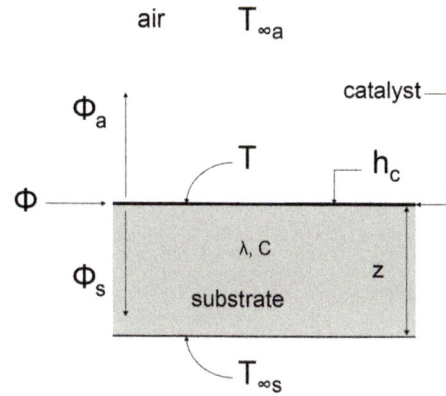

Figure 3. Modelling of the thermal problem.

The samples A and C exhibit very good activity even at room temperature and seem to be good candidates for easy integration into a transducing component, especially catalyst C. Considering these last results, only samples A and C have been taken into account for gas sensitivity characterisations.

We developed a simplified model to explain the thermal response of the catalyst. We consider that a mass m_c of catalyst with a negligible thickness is deposited on a surface S on a substrate with a mass m_s, a thickness z, a thermal conductivity λ and heat capacity C as shown in Fig. 3. The surrounding air temperature is $T_{\infty a}$ above the catalyst. A temperature $T_{\infty s}$ is imposed on the rear side of the substrate. We neglect at the moment the heat exchanges by radiation compared to those by convection ($h_r \ll h_c$). The experiment consists in measuring the temperature T at the surface of the catalyst. T is supposed to be also the temperature on the rear side of the catalyst.

The heat flux to which the system is subjected is as follows:

- $\Delta\Phi = m_s C \dfrac{\partial T}{\partial t}$ the variation of heat exchange of the system

- Φ the flux created at the surface of the catalyst during the catalytic reaction with the gas

- $\Phi_a = h_c S(T - T_{\infty a})$ the heat exchange with the surrounding air

- $\Phi_s = \lambda S/z(T - T_{\infty s})$ the heat exchange with the substrate.

The flow balance shows that the power variation of the system is equal to the power produced by the catalyst, minus those transferred to the environment, or $\Delta\Phi = \Phi - (\Phi_a + \Phi_s)$, which can also be written as follows:

$$m_s C \frac{\partial T}{\partial t} = \Phi - \left[h_c S(T - T_{\infty a}) + \lambda S/z(T - T_{\infty s})\right]. \quad (1)$$

This can be written in a reduced form:

$$\frac{\partial T}{\partial t} + \frac{T}{\tau} = \frac{T_f}{\tau}, \quad (2)$$

where T_f is the temperature reached by the catalyst once the thermal equilibrium established after a characteristic time τ. Those two last physical quantities are defined as follows:

$$\tau = \frac{m_s C/S}{h_c + \lambda/z}, \quad (3)$$

$$T_f = \frac{\Phi/S + h_c T_{\infty a} + (\lambda/z)T_{\infty s}}{h_c + \lambda/z}. \quad (4)$$

In absence of the target gas, no catalytic reaction occurs, $\Phi = 0$, and the temperature of the catalyst remains constant

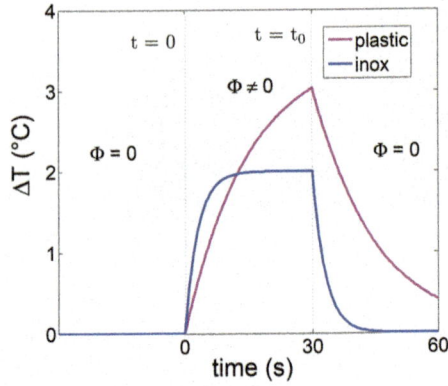

Figure 4. Modelling of the thermal response for a pulse of a given concentration of target gas for a plastic substrate ($\lambda_{plastic} = 0.15\,\mathrm{W\,m^{-1}\,K^{-1}}$, $C_{plastic} = 1800\,\mathrm{J\,K^{-1}\,kg^{-1}}$, pink curve) and an inox substrate ($\lambda_{inox} = 16\,\mathrm{W\,m^{-1}\,K^{-1}}$, $C_{inox} = 502\,\mathrm{J\,K^{-1}\,kg^{-1}}$, blue curve).

with temperature:

$$T_0 = \frac{h_c T_{\infty a} + (\lambda/z) T_{\infty s}}{h + \lambda/z}. \tag{5}$$

The resolution Eq. (1) describes the increase of the temperature as soon as the reaction takes place:

$$T(t) = (T_0 - T_f) e^{-t/\tau} + T_f. \tag{6}$$

The elevation of temperature is then

$$T_f - T_0 = \Delta T = \frac{\Phi/S}{h_c + \lambda/z}. \tag{7}$$

We consider now that, at $t < t_0$, the presence of the target gas has stabilised the temperature of the catalyst at $T = T_f$ thanks to $\Phi \neq 0$. At $t = t_0$, the gas inlet is suddenly cut off and Φ becomes zero. By injecting these conditions in Eq. (1), the expression of the temperature becomes, for $t > t_0$,

$$T(t) = (T_f - T_0) e^{-(t-t_0)/\tau} + T_0. \tag{8}$$

This very simplified model gives us important information:

1. ΔT depends on Φ, the thermal flux due to the catalytic reaction. Φ is directly linked to the concentration of the gas in the incoming air flow, and is therefore supposed to remain constant in the experimental conditions. ΔT increases as expected with $1/\lambda$ (homogenous to the thermal resistance) because a low substrate thermal conductivity prevents the heat from escaping on this side of the catalyst.

2. τ does not depend on Φ and therefore on gas concentration.

3. For a given value of Φ, the response time depends on the substrate physical properties. τ increases with the mass m and the heat capacity C of the substrate, and decreases with its thermal conduction λ.

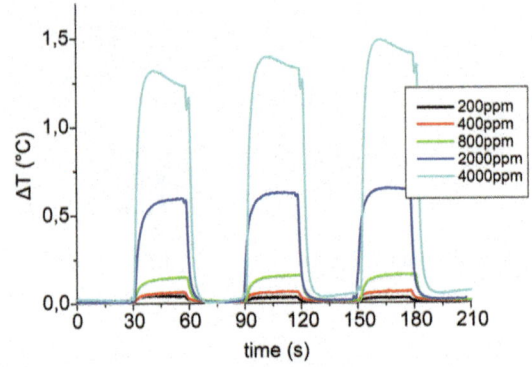

Figure 5. Gas sensitivity of catalyst A exposed to different H_2 concentrations in dry air at room temperature.

This means that the choice of the material used for the substrate is essential. A maximal value of ΔT obtained with a substrate with a low λ will induce a long response time. A compromise has to be found to get a good thermal response, stabilised as quickly as possible. Figure 4 shows the trends of the thermal response according to the model with two different material substrates: inox ($\lambda_{inox} = 16\,\mathrm{W\,m^{-1}\,K^{-1}}$, $C_{inox} = 502\,\mathrm{J\,K^{-1}\,kg^{-1}}$) and a plastic (polypropylene, $\lambda_{PP} = 0.15\,\mathrm{W\,m^{-1}\,K^{-1}}$, $C_{PP} = 1800\,\mathrm{J\,K^{-1}\,kg^{-1}}$). The shape is qualitative as physical data such as h_c or S are unknown. The influence of λ and C is clearly visible on ΔT and τ: in the case of an inox substrate (high thermal conductivity), the time response is low and the thermal equilibrium is quickly reached. In case of a plastic substrate, the response time is much longer and the temperature becomes higher, but the thermal equilibrium is not reached at the end of the H_2 exposure. However, the limits of the model are clearly highlighted by the absence of dependence of the thermal response ΔT with the mass m_c of the deposited catalyst. We neglect here the fact that the thermal contact between the catalyst and the substrate is far from perfect, and therefore that the temperature distribution in the catalytic layer is complex. Anyhow, this simplified model fits properly with the measurements, as seen in next paragraph.

Figure 5 shows the thermal response of 1 mg of sample A deposited onto an inox substrate under dry air mixtures with different concentrations of H_2 at room temperature. In accordance with the model described above, we observe that the catalyst exhibits a first-order response when exposed to H_2. Sample A presents a pronounced sensitivity to H_2 at room temperature: elevation of about $0.6\,^\circ\mathrm{C}$ for 2000 ppm and $1.2\,^\circ\mathrm{C}$ for 4000 ppm. The thermal response is rather fast (about 5 s), and we can note that the response time is not dependent on H_2 concentration, as predicted by our model. The thermal response is also reversible (quick return to the baseline in the absence of H_2), at room temperature. This means that no additional energy is needed to desorb H_2 on active sites where oxidation occurs. These results have to be

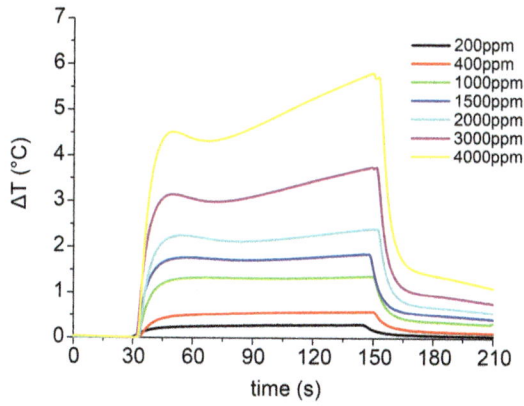

Figure 6. Stability of the thermal response of sample C to different concentrations of H_2 in dry air for 120 s.

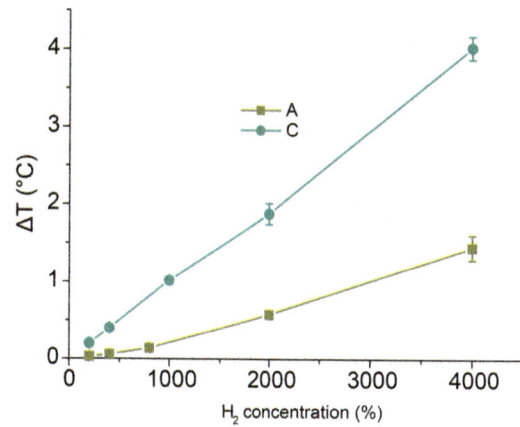

Figure 7. Linear response of the sensitivity of catalysts A and C.

compared with the 0.1 °C temperature variation obtained for the same mass of Au / CeO$_2$ under 500 ppm CO diluted in dry air (Mazingue et al., 2012). The shape of the plateau for an exposition of the catalyst to 4000 ppm shows that it can take time to get the thermal equilibrium. Indeed, the heat produced by the catalytic reaction increases the conversion rate of H_2, as observed in Fig. 2. With our model, we can deduce that Φ is then reaching a greater value, inducing a subsequent rise of temperature. This phenomenon is repeated as long as the external thermal conditions do not allow the thermal equilibrium for H_2 concentrations above 1500 ppm, as shown in Fig. 6. The stability of the thermal response decreases therefore when the H_2 concentration increases. However, the derivative of the expression of the temperature (Eq. 6) leads to the following formula:

$$\frac{dT}{dt} = \frac{\Delta T}{\tau} \exp^{-t/\tau} . \tag{9}$$

The slope of the curve at $t = 0$ is given by

$$\frac{\Delta T}{\tau} = \frac{\Phi}{m_s C} . \tag{10}$$

Since Φ is directly linked to the gas concentration, it is possible, by an adapted signal treatment on the derivative of the signal before the thermal equilibrium, to get a piece of information on the gas concentration.

The same tests have been performed with sample C and identical experimental conditions, giving the same kind of results but with higher thermal responses. Figure 7 gives a comparison between the performances of catalyst A and C: both exhibit linear and repeatable responses. Catalyst C is clearly the best candidate to be used as sensitive material in the H_2 sensor. The range of temperature variations can be easily detected by many kind of transducers, such as surface acoustic waves (SAWs) (Stelzer et al., 2008). These results also are promising for H_2 detection by optical means

(Mazingue et al., 2007) since recent studies showed that temperature variations in the °C range can be monitored (Wood et al., 2013).

The integration of the catalyst onto a transducer has to preserve the catalytic properties and supply a mechanical support. However, most of the used methods of deposition led to a loss of activity. Chemical methods such as dispersion into a porous sol-gel matrix induce an encapsulation of the catalyst. Incorporation of PdPt nanoclusters into thin mesoporous film such as TiO$_2$ or SiO$_2$ by co-impregnation (with the technique described above) gives a too high dispersion of the catalyst to allow a measurable thermal response. Physical methods such as chemical vapour deposition or pulsed laser deposition gave very flat catalyst layers, with limited reactive surface and therefore a low catalytic activity. We found an alternative by coating the catalyst with a thin layer of parylene, maintaining the powder onto a substrate without killing its catalytic activity. Parylene is a high-tech, ultrathin, transparent coating that is physically and chemically neutral, inert, biocompatible, and biostable. It insulates and protects, and can easily and accurately be applied in any thickness, from 50 nm to 100 µm. It is completely uniform, without pinholes, and suitable for application to very small technical components (protection against moisture, aggressive environment) and/or as an electrical insulator, resistance to all solvents, acids, and bases (Trantidou et al., 2013). The deposition process occurs in vacuum at room temperature. The coating ensures a hydrophobic protection to the catalyst, but is permeable enough to H_2.

A quantity of 1 mg of sample A deposited onto an oxidised Si wafer has been coated by different thicknesses of parylene (determined after deposition by interferometric measurements) and been exposed to increasing concentration of H_2 in dry air. Figure 8 shows that all the samples present a thermal response, but ΔT decreases with parylene thickness. This is not surprising, considering that parylene is a barrier for H_2 molecules to reach active sites of the catalyst. However, for a parylene thickness of 700 nm, a ΔT of 1 °C can

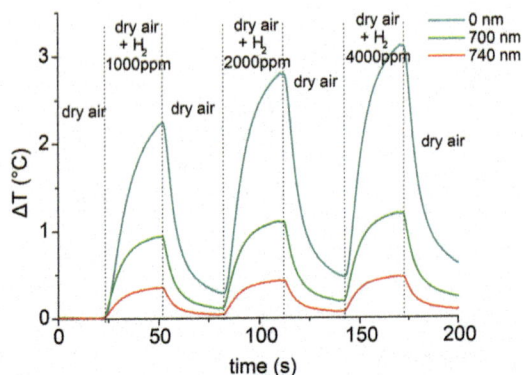

Figure 8. Thermal response of catalyst A coated by different thicknesses of parylene.

still be observed for a concentration of 4000 ppm. We can also notice that the shape of the thermal response is different from previous experiments because the substrate was in this case an oxidised Si wafer, with much lower thermal conductivity. As predicted by our simplified model and depicted in Fig. 6, the resulting thermal response is characterised by a higher ΔT and longer response time τ. A cumulative drift is then observed. This can be explained by the fact that the thermal equilibria cannot occur with exposure times as short as 30 s. We would observe no drift with exposure times much longer than response times. Moreover, the curves pictured in Fig. 8 have been performed after several tests according to the protocol described in the experimental part. During the first exposition of H_2, the samples coated with parylene gave a thermal response such as the ones in Fig. 8, whereas the uncoated sample exhibited no temperature rise. This can be explained by the fact that uncoated PdPt / Al$_2$O$_3$ must get rid of adsorbed water on active sites so that catalytic reaction can occur. This happens after several cycles of dry air and H_2 exposition, which would be problematic for a gas sensing application (hygrometry dependence). Coated catalysts do not have this problem, as the parylene deposition is performed in a vacuum at room temperature, inducing a water desorption of the active sites. Once coated, the latter are water free and interact directly with H_2 molecules, as explained in Appendix A. Parylene coating seems therefore to present all the required properties for efficient integration of PdPt / Al$_2$O$_3$ nanopowders on a transducer for gas sensing applications.

4 Conclusions

Catalytic activity and thermal response of PdPt / Al$_2$O$_3$ nanostructured catalysts to H_2 synthesised by the co-impregnation method have been characterised. It has been observed that the nature of the alumina support is relevant for the optimisation of the metallic proportion of the final system, leading to a greater catalytic activity. γ-PdPt / Al$_2$O$_3$ will therefore be preferred to α-PdPt / Al$_2$O$_3$ for gas sensing applications because of its higher specific surface area. Full H_2 conversion at room temperature can be reached by increasing the proportions of the metallic precursors in the synthesis process. The H_2 sensitivity tests showed that the thermal response of the catalysts is reversible with a time response of 5 s, at room temperature, and fits with a simplified model we developed to explain the shape of the curves. The temperature rise is in the °C range for only 1 mg of catalyst, and concentrations of H_2 in dry air ranging from 200 to 4000 ppm. However, the stability of thermal response is problematic for H_2 concentrations above 1500 ppm because of the establishment of the thermal equilibrium is harder to reach. Nevertheless, an integration method has been investigated by coating the catalyst with a parylene thin film. Sensitivity tests show that the thermal response decreases with the thickness of the coating, but still exhibits significant temperature rise (in the °C range). All these results are promising for integration of PdPt / Al$_2$O$_3$ nanostructured catalyst onto different transducers for fast-response, reversible, and room-temperature working gas-sensing systems. Alternative devices such as photonic (Bragg gratings and multimode interference couplers) or SAW transducers are under study to develop reversible remote controlled and passive components for the detection of H_2, working in a large range of temperatures in harsh environments, with no need for embedded energy.

Appendix A: Water permeability of the parylene

We presented in our article the possibility of using PdPt nanoclusters supported by alumina (PdPt / Al_2O_3) for H_2 detection. This bimetallic system is a catalyst for the following exothermal reaction (Munakata et al., 2011) :

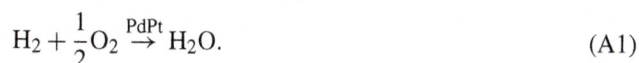

$$H_2 + \frac{1}{2}O_2 \xrightarrow{PdPt} H_2O. \tag{A1}$$

The heat generated during the reaction leads to a temperature rise of the catalyst that is directly linked to the H_2 concentration. The catalyst is maintained on a substrate with a thin film of parylene, a transparent and hydrophobic polymer. The aim of the following text is to show that the formed water during the catalytic reaction is in a gaseous state and induces no poisoning on the active sites of the catalyst.

The transmission of vapour through a membrane is measured at a temperature of 38 °C and 100 % of relative humidity, according to the norm ASTM F1249-06. The unit of measurement is $g\,mm\,m^{-2}\,d^{-1}$. This physical quantity can be considered as the mass of water vapour passing through a membrane 1 mm thick, per area unit (m^2) and per time unit (day).

As the molar mass of water is $M_{H_2} = 18\,g\,mol^{-1}$, we have, in a mass $m = 1\,g$ of water, $n = m/M_{H_2} = 0.056\,mol$. As the molar volume is $V_M = 24.79\,L\,mol^{-1}$ at $T = 25\,°C^{[1]}$, 1 g of water vapour occupies a volume of $V_M \times n = 24.79 \times 0.056 = 1.377\,L = 1377\,cm^3$.

The permeability of the parylene layer we used in this work is $0.1\,g\,mm\,m^{-2}\,d^{-1}$,[2]. The parylene thickness was $740\,nm = 740 \times 10^{-6}\,mm$. The permeability of the parylene film related to this thickness is therefore $135\,g\,m^{-2}\,d^{-1}$ or $1.86 \times 10^5\,cm^3\,m^{-2}\,d^{-1}$ (the molar volume of ideal gas is $V_M = 24.79\,L\,mol^{-1}$ at $T = 25\,°C^{[3]}$).

The gas permeability of a membrane is measured differently than for water. The used method is performed according to the norm ASTM D1434 F3985-05, at a temperature of 25 °C. The unit of measurement is the $cm^3\,mm\,m^{-2}\,d^{-1}\,bar^{-1}$. This physical quantity can be considered as the gas volume passing through a membrane 1 mm thick, per area unit (m^2), per pressure unit (bar) and per time unit (day).

The parylene we used in our work presents a H_2 permeability of $42\,cm^3\,mm\,m^{-2}\,d^{-1}\,bar^{-1}$,[2]. We had a parylene thickness of $740 \times 10^{-6}\,mm$. The permeability of the parylene film related to this thickness is therefore $42/740 \times 10^{-6} = 5.6 \times 10^4\,cm^3\,m^{-2}\,d^{-1}\,bar^{-1}$.

Since the gas mixture is injected in the test chamber at a pressure of 1 bar and a H_2 concentration of 4000 ppm, the partial H_2 pressure is $4 \times 10^{-3}\,bar$. The H_2 permeability of a thickness of 740 nm of parylene at this pressure is then $56 \times 10^3 \times 4 \times 10^{-3} = 227\,cm^3\,m^{-2}\,d^{-1}$. The parylene is therefore much more permeable to water vapour than to hydrogen.

In our conditions, the H_2 partial pressure $P_{H_2} = 4\,mbar$ is much lower than the saturated vapour pressure of water $P^0_{H_2O,25°C} = 31.7\,mbar^{[4]}$. Hence, even if we consider that the totality of the H_2 passes through the parylene and is oxidised into H_2O, the formed water will never condense since its partial pressure is much lower than the saturated vapour pressure. Moreover, we showed that the permeability of the parylene would be much higher for water vapour than for H_2. That means that vapour will be removed from the catalyst side faster than H_2 will be provided by our experiment. We can conclude that the formed water will not accumulate on the surface of the catalyst and thus will not poison it since it will stay in a gaseous state and will be removed quickly in the surrounding atmosphere.

---

[1] http://en.wikipedia.org/wiki/Standard_conditions_for_temperature_and_pressure

[2] http://www.comelec.ch/en/parylene_tableaux.php

[3] http://en.wikipedia.org/wiki/Standard_conditions_for_temperature_and_pressure

[4] http://www.thermexcel.com/english/tables/eau_atm.htm

Acknowledgements. Part of this work is funded by the French National Agency of the Research (ANR) within the framework of the 2009 P3N program (PEPS project) and Gravit Innovation Grenoble Alpes.

Edited by: A. Romano-Rodriguez
Reviewed by: two anonymous referees

References

Barsan, N., Koziej, D., and Weimar, U.: Metal oxide-based gas sensor research: How to?, Sens. Actuat. B-Chem., 121 , 18–35, 2007.

Han, C.-H., Hong, D.-W., Han, S.-D., Gwak, J., and Singh, K. C.: Catalytic combustion type hydrogen gas sensor using TiO_2 and UV-LED, Sens. Actuat. B-Chem., 125, 224–228, 2007.

Kroll, A. and Smorchkov, V.: Electrochemical solid-state microsensor for hydrogen determination, Sens. Actuat. B-Chem., 34, 462–465, 1996.

Lide, D. R.: Standard Thermodynamic Properties of Chemical Substances, CRC Handbook of Chemistry and Physics, 83rd Edn. CRC Press, Boca Raton, FL, 5-.4–5.60, 2001.

Mazingue, T., Kribich, R., Etienne, P., and Moreau, Y.: Simulations of refractive index variation in a multimode interference coupler: Application to gas sensing, Opt. Commun., 278, 312–316, 2007.

Mazingue, T., Lomello-Tafin, M., Gautier, G., Passard, M., Goujon, L., Hernandez-Rodriguez, C., Rousset, J.-L., and Morfin, F.: Characterizations of gold nanostructured catalysts for gas sensing applications, Proc. of AES-ATEMA, Advanced Engineering Solutions – Advances and Trends in Engineering Materials and their Applications 11th International Conference, Toronto, Canada, 6–10 August, 205–212, 2012.

Meixner, H. and Lampe, U.: Metal oxide sensors, Sens. Actuat. B-Chem., 33, 198–202, 1996.

Morfin, F., Sabroux, J.-C., and Renouprez, A.: Catalytic combustion of hydrogen for mitigating hydrogen risk in case of a severe accident in a nuclear power plant: study of catalysts poisoning in a representative atmosphere, Appl. Catal. B-Environ., 47, 47–58, 2004

Munakata, K., Wajima, T., Hara, K., Wada, K., Shinozaki, Y., Katekari, K., Mochizuki, K., Tanaka, M., and Uda, T.: Oxidation of hydrogen isotopes over honeycomb catalysts, J. Nucl. Mat., 417, 1170–1174, 2011.

Rossignol, C., Arrii, S., Morfin, F., Piccolo, L., Caps, V., and Rousset, J.-L.: Selective oxidation of CO over model gold-based catalysts in the presence of H_2, J. Catal., 230, 476–483, 2005.

Rousset, J.-L., Stievano, L., Cadete Santos Aires, F. J., Geantet, C., Renouprez, A., and Pellarin, M.: Hydrogenation of Toluene over γ-Al_2O_3-Supported Pt, Pd, and Pd-Pt Model Catalysts Obtained by Laser Vaporization of Bulk Metals, J. Catal., 197, 335–343, 2001.

Rousset, J.-L., Stievano, L., Cadete Santos Aires, F. J., Geantet, C., Renouprez, A., and Pellarin, M.: Hydrogenation of Tetralin in the Presence of Sulfur over γ-Al_2O_3-Supported Pt, Pd, and Pd-Pt Model Catalysts, J. Catal., 202, 163–168, 2001.

Shin, W., Nishibori, M., Ohashi, M., Izu, N., Itoh, T., and Matsubara, I.: Ceramic Catalyst Combustors of Pt-Loaded-Alumina on Microdevices, J. Ceram. Soc. Japan, 117, 659–6655, 2009.

Stelzer, A., Scheiblhofer, S., Schuster, S., and Teichmann, R.: Wireless sensor marking and temperature measurement with SAW-identification tags, Measurement, 41, 579–588, 2008.

Trantidou, T., Payne, D. J., Tsiligkiridis, V., Chang, Y.-C., Toumazou, C., and Prodromakis, T.: The dual role of parylene C in chemical sensing: Acting as an encapsulant and as a sensing membrane for pH monitoring applications, Sens. Actuat. B-Chem., 186, 1–8, 2013.

Wood, T., Le Rouzo, J., Flory, F., Kribich, R., Maulion, G., Signoret, P., Coudray, P., and Mazingue, T.: Study of the influence of temperature on the optical response of interferometric detector systems, Sens. Actuat. A-Phys., 203, 37–46, 2013.

16

Multi-channel IR sensor system for determination of oil degradation

T. Bley[1], **E. Pignanelli**[1], **and A. Schütze**[1,2]

[1]Centre for Mechatronics and Automation Technology (ZeMA), Saarbruecken, Germany
[2]Lab for Measurement Technology, Department of Mechatronics, Saarland University, Saarbruecken, Germany

Correspondence to: E. Pignanelli (eliseo.pignanelli@mechatronikzentrum.de)

Abstract. A miniaturized infrared (IR) multi-channel sensor system was realized to determine chemical oil degradation, e.g., oxidation, increasing water content. Different artificially aged oil samples (synthetic motor oil, mineral hydraulic oil and ester-based hydraulic fluid) were prepared by oxidative degradation at elevated temperatures or addition of water, and characteristic degradation features in the IR spectrum were detected using FTIR spectroscopy. In addition, the absorption behavior of water contaminated synthetic motor oil was analyzed with increasing temperature. To determine the influence of different degradation effects on the measurement results the sensor system was characterized with the various oil samples. The system uses a reference channel to suppress the effect of decreasing transmission over the entire spectrum caused, e.g., by increasing soot content in the oil or contamination of the optical path.

1 Introduction

1.1 Chemistry of oil deterioration

To determine the type and degree of oil degradation different chemical processes have to be taken into consideration, e.g., oxidation, nitration and sulfation, but also increased water content. The speed of oxidation and the specific species in the oil generated by oxidative processes mainly depend on temperature. Oxidation itself is an autocatalytic process. In the temperature range between 30 and 120 °C hydrocarbons react with metal catalysts, e.g., Co, Fe, Cr, Cu, V or Mn, to alkyl radicals, depending on the strength of the C–H bond. The alkyl radicals are converted via multiple reaction pathways to hydroperoxides and other alkyl radicals. Hydroperoxides are also generated by intramolecular propagation. These hydroperoxides break up into alkoxy and hydroxy radicals that take up hydrogen from the ambient. Secondary and tertiary alkoxy radicals react to aldehydes and ketones. The process terminates with a combination of radicals generating ketones and alcohols. At temperatures above 120 °C the reaction pathway is generally similar but reaction speed is increased and the selectivity of the generated oxygen-containing compounds is reduced. Carboxylic acids and esters are formed by reactions with alcohol. At high temperatures the viscosity of the hydrocarbon increases due to polycondensation of oxygenated products. Unsaturated aldehydes or ketones are formed by aldol condensation. Alkoxy radicals can start the polymerization or polycondensation. The oxidation process is increased by metal soaps reducing the activation energy for the decomposition of alkyl hydroperoxides (Mortier et al., 2010).

The presence of nitrogen or sulfur oxides in the oil, e.g., introduced through the blow-by effect in combustion engines (Pawlak, 2003), can lead to oil degradation by nitration or sulfation. In addition, these gases in combination with water contamination can lead to formation of acids, further intensifying the polycondensation process. Ester-based fluids, e.g., native or phosphate esters, which are often used as hydraulic fluids, can break up into alcohols and acid components (hydrolysis) when exposed to increasing water content, especially at high mechanical load or elevated temperatures (Totten et al., 2003). Comprehensive online condition monitoring would therefore require a sensor capable of analyzing the degree of degradation and also the relevant degradation processes to allow an estimation of the remaining lifetime of the oil in the specific application. Furthermore, such a sensor system would also allow identification of degrading operating conditions, i.e., could help to remedy the cause of excessive oil degradation that would otherwise lead to shortened oil change intervals.

1.2 State of the art

Determination of the chemical degradation is highly desirable for oils and lubricants in order to optimize oil change intervals and to prevent undue wear in combustions engines, gear boxes or hydraulic machinery. One primary driver today is the increasing demand for offshore wind turbines in which the main gear is a critical component. In these systems, a complex set-up including filters and coolers is used for oil conditioning. Online determination of the oil status is especially beneficial here, as access is limited and maintenance is costly. Current methods for determination of oil degradation can be divided into two categories: sampling and online measurement.

Sampling requires oil samples to be taken at regular intervals followed by analysis in dedicated laboratories. Here, spectroscopic analysis methods, especially IR spectrometry, are used as well as titration, viscosimetry, etc. The obtained analysis results are interpreted automatically with threshold limits adapted to and determined by the oil type and the application. Ideally, the analysis provides a profound insight into the actual oil condition, allowing an analysis of the degradation process and also of the machinery itself, e.g., by analyzing the particle contamination of the oil where metal particles can provide additional information about the location of system damage on the basis of the detected material. However, sampling can also lead to false analysis results, e.g., when sampling is done under different operating conditions or at unsuitable sample points. For example, taking samples at different levels of the oil sump can lead to false, often increased, values for the particle concentration (Booser, 1994). In addition, sampling and subsequent laboratory analysis typically require several days to obtain the results, which could lead to increased machine degradation if oil degradation has already passed a critical value. Finally, for some applications such as offshore wind turbines, sampling is simply too expensive for regular operation.

To determine the true oil status in real time, online monitoring directly in the oil and close to critical components is highly desirable. Different sensors for online determination of various, mostly physical, oil properties, e.g., viscosity, dielectric constant but also water content, have been developed. However, these values indicate the chemical status of the oil only indirectly and a change in the sensor signal can also be caused by factors independent of the actual oil deterioration, e.g., a refill with slightly different oil with a different base component mixture or other additives. Therefore these values are difficult to interpret and to relate to the results of laboratory analysis. Multi-sensor systems combining several "indirect" sensors to obtain a more accurate assessment of the oil condition have been reported (Duchowsky and Mannebach, 2006), but their uptake in real applications is slow due to the ambiguity of the measurement results.

Optical measurement techniques are promising candidates for monitoring the oil quality, especially in the infrared (IR) spectral region, i.e., multi-channel NDIR (non-diffractive infrared) absorption or IR spectroscopy. The latter method is also used in laboratories, thus allowing better correlation of online data and lab results. Wideband IR characterization of the oil can provide a wealth of information concerning changes in the molecular structure of the oil, e.g., for determining the effects of oxidation, nitration and sulfation, the concentration of water, the status of additives in the oil (antioxidants and anti-wear compounds), as well as soot content (ASTM International, 2007). A direct adaptation of laboratory methods into an online monitoring system is, however, difficult due to the complexity of the typical laboratory set-ups and, in addition, varying ambient conditions have to be taken into account. Several groups are currently working on the development of miniaturized optical set-ups for online monitoring of the oil condition to allow correlation with existing laboratory methods.

Agoston et al. (2004, 2008) demonstrated an experimental set-up using an infrared source, a CaF_2 fluid measurement cell and one infrared detector for evaluating the oxidation of engine oils during deterioration processes. To select different relevant spectral regions, various custom-made band-pass filters ($1710, 1970 \, cm^{-1}$) were placed in the IR path. Kudlaty et al. (2003) showed a measurement set-up for lubricating oils based on attenuated total reflection (ATR) with a pyroelectric detector and a chopper wheel with two integrated narrow band-pass IR filters. Endisch and Koch (2007) demonstrated a set-up with a movable gradient filter, a reference filter, two infrared detectors and a ZnSe fluid measurement cell with variable thickness allowing a low resolution spectral analysis of the IR transmission spectrum for lubricating oils. Later, Wiesent et al. (2010) demonstrated a similar set-up with a detector array combined with the gradient filter to avoid the need for movable parts. Wiesent et al. (2011) also studied a set-up based on two microstructured tunable Fabry–Perot filters for analysis of different IR spectral regions applied to condition monitoring of gear oils in offshore wind turbines. While spectral information is of course very helpful for data acquisition and comprehensive analysis, these systems are rather costly, thus preventing their widespread use. At the same time, the microcomponents used in these set-ups can only be manufactured at acceptable prices, if large volumes can be reached. Thus, these solutions while achieving promising results have not yet entered the market on a larger scale. Sumit et al. (2010) presented a system with an infrared source, a fluid cell consisting of two sapphire windows and a quadruple infrared detector equipped with different filter windows for analysis of phosphate ester hydraulic fluids, especially for aerospace applications.

In addition to approaches based on IR measurements for determining the oil quality, Schneidhofer and Dörr (2009) presented an oil monitoring system based on chemical corrosion sensors allowing an analysis of the acidic content of the oil. Finally, Wang (2001) used a set-up with platinum electrodes to determine oil changes in a car by measuring the consumption or transformation of additives. However, long-term stability and resolution are also a recurring concern for chemical sensor systems, especially in harsh environments such as gear boxes.

Taking into account the results obtained for the different methods, which clearly show the most promising results for multichannel NDIR measurements offering a good trade-off between information quality, stability and cost, an integrated sensor system based on MEMS components was developed, similar to the system demonstrated by Sumit et al. (2010). The modular system also uses a, in our case, commercially available microstructured infrared source and quadruple thermopile detector. To allow realization of a compact, low-cost sensor system with the focus on industrial applications, a fluidic cell based on silicon and realized using established microtechnologies is used. Compared to Sumit et al. (2010), the Si-based cell allows measurements over a wider spectral range (8000–1250 cm^{-1} corresponding to wavelengths of 1.25 to 8 μm), providing more information about the measured oil, because further relevant spectral changes can be determined. Si wafers with a thickness of 2 mm are used as base material for the fluid cell, allowing the realization of a sensor system able to monitor the mid-infrared spectrum up to a wavelength of 9 μm (Czochralski silicon) with a small and well-defined IR absorption path length of 200 μm. The channel is realized using pre-structured LTCC (low-temperature co-fired ceramic) spacers deposited on top of the Si wafer. Cells are realized using a batch process and mounted in a housing after dicing. The fluidic cell can be operated at pressures up to 100 bar (Bley et al., 2012b) due to the high bond strength between Si and the LTCC further optimized by nanostructuring of both materials as demonstrated by Günschmann et al. (2013). In addition, silicon oxide formed on the Si wafer surface is chemically inert, thus offering sufficient long-term stability against oil components and the reactive products of oil deterioration and contamination.

Some results obtained with the novel sensor system were presented previously (Bley and Schütze, 2011; Bley et al., 2012a, b, c). This paper is based on preliminary results presented in Bley et al. (2012a) and has been extended with measurements concerning the influence of temperature changes on the spectral transmission and especially reference measurements of acid number and viscosity. These were correlated with results of the measurement system for different oil types.

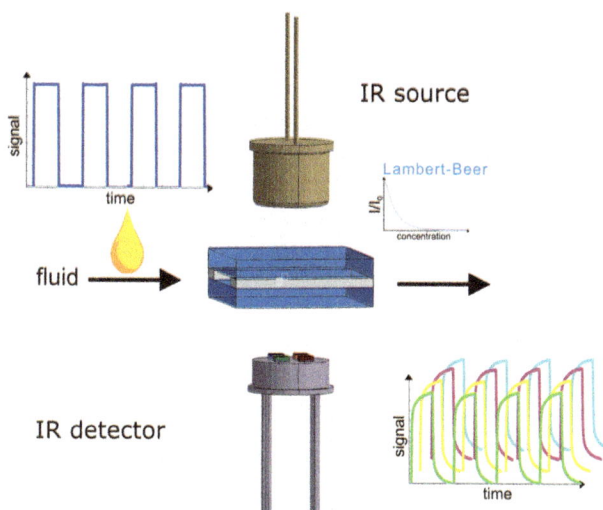

Figure 1. Schematic of the sensor system (Bley et al., 2012c).

2 Experimental

2.1 Sensor system

Based on spectral characterization of various oils and lubricants a miniaturized sensor system with an IR transparent silicon cell was developed for optical analysis of technical fluids in general and particularly for lubricating or hydraulic oils (Bley and Schütze, 2011). A schematic of the sensor system is shown in Fig. 1.

A commercial infrared source (Intex Inc., 2013) with a maximum permissible heater power of 980 mW, broad IR emission spectrum similar to a black body and a spectral time constant of 30 ms (heating) and 5 ms (cooling) is excited with a square wave signal at a frequency of 0.2 Hz. The small excitation frequency is used to ensure maximum signal intensity, i.e., allowing the IR source to reach its maximum temperature and also the thermopile detectors to reach steady state and to record signal amplitude and offset with a good signal-to-noise ratio. The electrical power used for the IR source in the on phase is approximately 715 mW, which achieves an increased lifetime (10 times) of the infrared source compared to operation at maximum power. The radiation of the IR source passes through the silicon cell in which a fluid channel with a height of 200 μm is defined using an LTCC spacer (Günschmann et al., 2012). The broadband spectrum of the IR source is attenuated at specific wavelengths depending on the specific spectral extinction $\varepsilon(\lambda)$ and concentration c of molecular species in the fluid as well as the film thickness d according to the Lambert–Beer law (Burns and Ciurczak, 2008):

$$I(\lambda) = I_0 \cdot e^{-\varepsilon(\lambda)\cdot c \cdot d}. \tag{1}$$

A modified commercial quadruple thermopile detector (Micro-Hybrid Electronic GmbH, 2013) with a thermal time constant of approximately 100 ms was used to detect the transmitted radiation. Each detector has a specific band-pass filter where the center wavelengths of three filters are matched to characteristic absorption features of the fluid, while the fourth is used as a reference channel. For reference a spectral region without characteristic absorption by the oil is selected to suppress the influence of broadband attenuation, e.g., soot contamination of the oil or degradation of the IR source or the optics.

In this paper oils based on hydrocarbons and on native esters are analyzed. During aging, both hydrocarbon oils and ester-based hydraulic fluids generate reaction products containing a carbonyl group. Thus, measuring the IR attenuation in the carbonyl-specific wavelength region in principle allows determination of the current oxidation condition of the oil. Ester-based hydraulic fluids, however, already contain carbonyl groups in their regular molecular structure, resulting in an almost complete absorption of the IR radiation in this region. Therefore, other specific regions in the IR spectrum, e.g., influenced by the hydroxyl group (OH), are required for determination of the oxidative degradation for ester-based hydraulic fluids. Consequently, the sensor set-up uses different, relevant regions of the IR absorption spectrum adapted to the general oil type (hydrocarbon oil or ester-based hydraulic fluid). Filter combination A is used for hydrocarbon-based oils, while filter combination B is used for ester-based hydraulic fluids; the filter parameters are listed in Tables 1 and 2, respectively. For each oil type the IR reference channel of the detector (AR, BR) is selected so that it is close to a one of the three measurement channels in order to minimize measurement errors resulting from, e.g., high soot content that can lead to a tilt of the IR spectrum (Mortier et al., 2010). Note that the filter selection can also be adapted to the expected dominating degradation processes, which are very different for combustion engines, gear boxes or hydraulic systems.

The raw detector signals, which are typically in the range of less than 5 mV, are amplified and electronically filtered with an active low-pass filter (72 Hz, 20 dB decade^{-1}). The electronic amplification stage has an overall gain of 341. The amplified signals are recorded with a sampling rate of 500 Hz using a 16 bit analog-to-digital converter with a real-time data acquisition system (CompactRIO System 9074, National Instruments) with a LabVIEW graphical user interface (GUI). The raw sensor signals of all four thermopile detectors are recorded over 12 consecutive on/off pulses of the IR source, i.e., over a total period of 60 s, and then evaluated in a post process in MATLAB using a DFT algorithm. The first harmonic DFT component of the detector signals, denoted below as DFT amplitude, is evaluated to improve the signal-to-noise ratio.

For evaluation of the sensor system with different oil samples a simple test set-up based on a membrane pump (KNF NF1.11) was realized. Each sample was pumped through the sensor system at a constant flow rate between 50 and

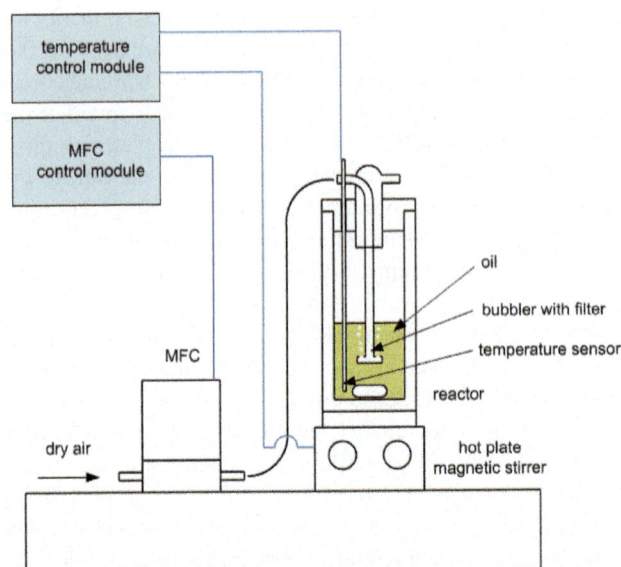

Figure 2. Experimental set-up for defined oil degradation by oxidation.

100 mL min^{-1}. After analysis of each sample the sensor system was cleaned with n-heptane and dry air to prevent carry-over from the different oil samples. The system also allowed testing under various pressure regimes and temperatures (Bley et al., 2012b).

2.2 Artificial oil degradation

To prepare different oil samples with well-defined and reproducible oxidative degradation a reactor set-up was realized as shown in Fig. 2. An oil sample (1500 mL) was placed in an aluminum reactor vessel that was heated to 170 °C for hydrocarbon oils and to 130 °C for ester-based hydraulic fluids using a hot plate to simulate extended operation of the oil in contact with ambient air. The temperature in the reactor was controlled with a temperature sensor placed directly in the oil to control the hot plate power. A mass flow controller (MFC) provided a constant flow of 50 mL dry air through the oil in the reactor. The MFC was connected to a bubbler with glass filter, which was dipped into the oil, generating finely distributed bubbles. The oil was continuously stirred using a magnetic stirrer. To analyze the degradation at different stages samples of 100 mL each were taken after predefined intervals; typically, four samples were taken over a period of several days. A volume of 100 mL per sample was chosen to minimize the effect of sampling on the degradation process on the one hand and to obtain sufficient material for analysis and testing on the other.

Samples with additional water contamination of 0.1, 0.2 and 0.4 weight percent (wt.%), respectively, were prepared by adding pure water (ASTM type II, cf. ASTM International, 2011a) to new and artificially aged oil samples. The samples were stirred with a magnetic stirrer at room

Table 1. Filter parameters of the band-pass filters for analysis of hydrocarbon-based oils, i.e., mineral and synthetic oils (filter combination A, hydrocarbon configuration).

	Filter A1	Filter A2	Filter A3	Filter AR
Center wavelength [cm^{-1}]	1722	3333	3532	2513
FWHM [cm^{-1}]	151	52	52	54
Absorption band (indicator)	Carbonyl group (oxidation)	Hydroxyl group (water)	Hydroxyl group (water)	Reference (soot, etc.)

Table 2. Filter parameters of the band-pass filters for analysis of ester-based hydraulic fluids (filter combination B, ester configuration).

	Filter B1	Filter B2	Filter B3	Filter BR
Center wavelength [cm^{-1}]	3333	3514	3634	3910
FWHM [cm^{-1}]	52	52	52	71
Absorption band (indicator)	Hydroxyl group (oxidation)	Hydroxyl group (oxidation)	Hydroxyl group (water)	Reference (soot, etc.)

temperature for several minutes after adding the water. Analysis was performed immediately afterwards to prevent segregation of oil and water mixture.

To measure the spectral influence of added water in combination with increasing temperature, oil samples with added water were filled into a closed fluid measurement cell (Omni Cell System, Specac). The samples were heated in the cell on a hotplate, controlled with a PT1000 sensor connected to the cell and then measured with the FTIR. Based on the negligible surface of oil and air during filling and the closed measurement cell evaporation of the water can be neglected.

For the determination of relevant spectral degradation features different oils were used:

– Hydrocarbon-based oils

  – Synthetic motor oil (Pegasus 1, Mobil)
  – Mineral hydraulic oil (Tellus Oil 46, Shell)

– Ester-based hydraulic fluid

  – Native ester (HETG 37, Meguin)

The synthetic motor oil and the mineral hydraulic oil are both hydrocarbon based and also contain various additives. The native ester fluid, on the other hand, is based on a molecular structure obtained by esterification of triglycerides, also with various additives. The two different molecular bases of the fluids cause different IR spectra with different regions of interest as shown in Sect. 3.1 below.

In addition to the artificially aged samples of the synthetic motor oil samples of the identical oil were taken from the gas engine of a combined heating and power unit (CHP) after 1025 and 1840 working hours, respectively. A standard laboratory analysis of these samples showed the oil to be marginal after 1025 h and strongly advised an immediate oil change after 1840 working hours. It was expected that these samples aged under actual operating conditions would show degradation effects in addition to oxidation due to, e.g., blow-by gases containing sulfur and nitrogen oxide and also water.

2.3 Reference IR spectra, acid number and viscosity

Each oil sample was analyzed using a high resolution FTIR spectrometer (Vertex 80V, Bruker Corporation, USA) to obtain reference spectra before testing with the multi-channel IR sensor system. In the spectrometer a fluid cell with 200 μm film thickness was used (Omni Cell with CaF$_2$ windows). The fluid cell was placed in the sample chamber of the FTIR spectrometer and spectra were recorded from 1000 to 4000 cm^{-1} at a resolution of 0.5 cm^{-1}. To analyze the influence of temperature on the IR spectra the oil samples were heated in the FTIR spectrometer using a hot plate and a fluid cell with integrated PT1000 temperature sensor.

In addition to the reference IR spectra the oil samples were further characterized by determining the acid number (AN) and the kinematic viscosity (η). The AN was determined by titration according to ASTM D974 guidelines (ASTM International, 2011b) using a Schott Titroline 6000. The kinematic viscosity was determined at 40 and 80 °C using an Ubbelohde viscometer.

3 Results

3.1 IR spectra

3.1.1 Synthetic motor oil

The spectrum of the synthetic motor oil shows multiple characteristic effects indicating the oil degradation under normal operating conditions as shown in Fig. 3. The oil aged over 1840 working hours shows a slight increase in broadband absorption caused by an increase in soot particles in the oil. Increased absorption is observed in the spectral region from 3100 to 3500 cm^{-1}. OH groups as found, e.g., in alcohols, and water and also amines can lead to an increased absorption in this spectral region. In oil samples, the water absorption peak is usually shifted toward higher wave numbers (Sumit et al., 2010); therefore we assume that the increased

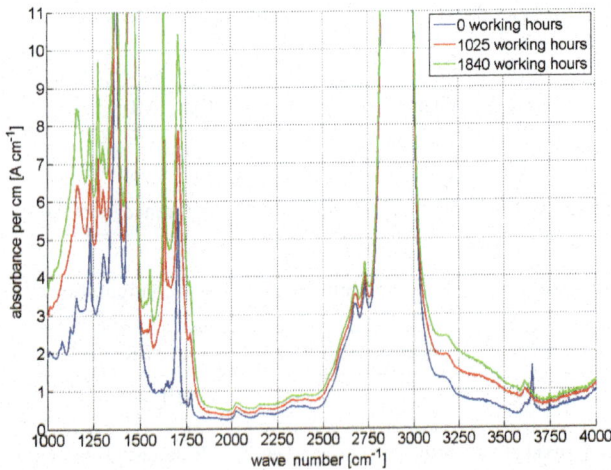

Figure 3. IR spectra of synthetic motor oil from a gas engine (combined heating and power unit) at different stages of aging under normal operating conditions.

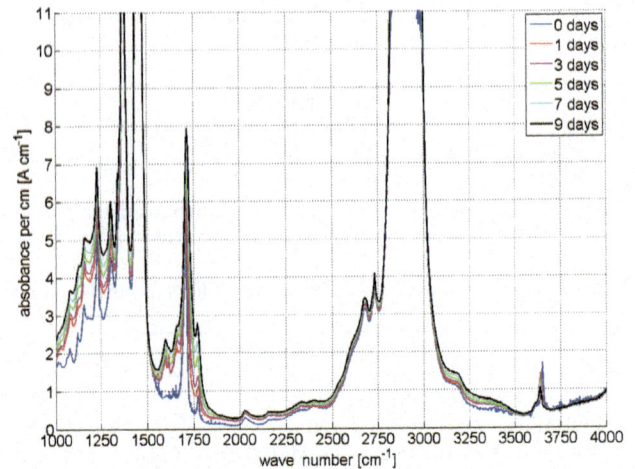

Figure 4. IR spectra of synthetic motor oil (1500 mL) artificially aged at 170 °C with 50 cm^3 min^{-1} air flow over 9 days.

absorption is caused by an increased concentration of alcohols due to oxidation.

The fresh oil shows a characteristic peak at 1700 cm^{-1}, which is probably due to an additive with a C＝O double bond, e.g., an ester. With increased aging a strong increase in this peak is observed caused by an increasing concentration of oxidized molecules with C＝O double bonds, e.g., aldehydes, ketones and carboxylic acids. At 1630 cm^{-1} a sharp absorption feature increases dramatically, which can be attributed to nitration of the oil due to nitrogen oxide carry-over caused by the blow-by effect in combustion engines (Pawlak, 2003).

The observed signal increase at 1150 cm^{-1} can be attributed to sulfation products while peaks at 1277 cm^{-1} and 1555 cm^{-1} suggest further reactions to nitrogen or oxygen containing compounds. At 3650 cm^{-1} the spectrum of the fresh oil shows a feature that is no longer evident in the aged samples. This absorption peak is probably due to a phenolic antioxidant additive (Pawlak, 2003; Pretsch et al., 2009), which is consumed under operating conditions and already used up after 1025 working hours.

It can be concluded that the spectrum of the synthetic motor oil shows multiple characteristic effects during aging. For an exact calibration of the sensor system the different degradation features in the IR absorption spectrum have to be analyzed separately to allow independent quantitative analysis of the various causes using the multi-channel system.

The spectrum of the synthetic motor oil artificially aged at 170 °C shows a characteristic increase in the C＝O double bond feature at 1700 cm^{-1} as shown in Fig. 4. After 9 days of artificial aging the peak height of this oxidation feature is comparable to that of the sample aged for 1025 working hours under operating conditions. In addition, a continuous increase in the OH feature at 3100–3500 cm^{-1} is observed,

which is similar in shape but somewhat smaller compared to the OH feature of the synthetic motor oil under operating conditions. This could indicate a higher water content of the samples from the CHP compared to the artificial aging in which no water could accumulate in the oil due to the high temperature during aging. However, while additional contamination of the synthetic motor oil with water also leads to increased absorption in this spectral region, as shown in Fig. 5, the resulting peak is shifted to larger wave numbers compared with the spectra shown in Figs. 3 and 4, respectively. The maximum absorption increase is observed at approximately 3400 cm^{-1} for increased water content compared to a maximum in the range of 3200 cm^{-1} observed both for aging under operating conditions as well as for artificial oxidation. This indicates that the larger OH absorption feature in the synthetic motor oil aged under operating conditions is caused by oxidation or other products caused by the reaction of oil and exhaust gases. Thus, the artificial aging employed here does not perfectly reflect the oxidation process under operating conditions.

The synthetic motor oil also shows that the concentration of different additives, e.g., with a carbonyl group (ester) or phenolic antioxidants, can be determined at a film thickness of 200 μm at wavelengths of 1700 and 3650 cm^{-1}, respectively. Very high additive concentrations could possibly attenuate the IR signal completely; thus, the system would have to be adapted by choosing an appropriate absorption path length, i.e., oil film thickness.

3.1.2　Mineral hydraulic oil

The spectra of the mineral hydraulic oil also artificially aged at 170 °C show a strong increase in the sharp C＝O double bond feature at 1700 cm^{-1} and a smaller increase in the broad OH feature at 3100–3500 cm^{-1} (Fig. 6) comparable

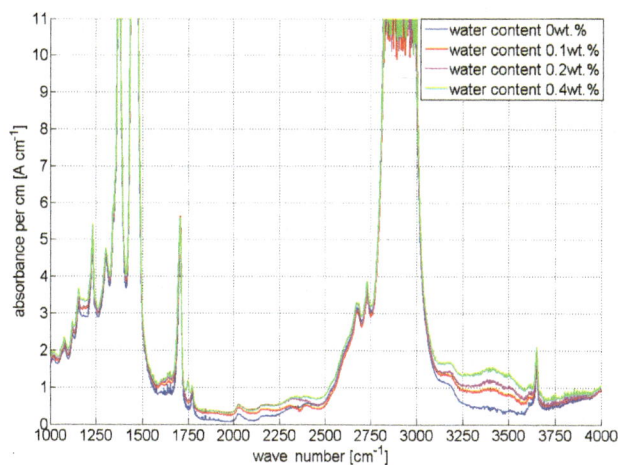

Figure 5. IR spectra of synthetic motor oil with added water (ASTM type II, 0, 0.1, 0.2 and 0.4 wt.%, respectively).

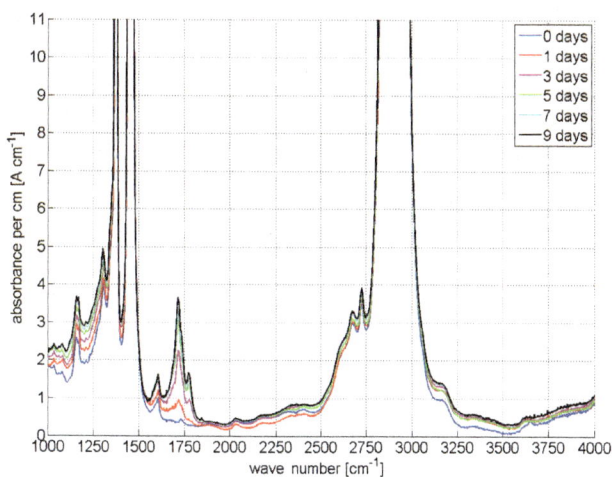

Figure 6. IR spectra of mineral hydraulic oil (1500 mL) artificially aged at $170\,^{\circ}C$ with $50\,cm^3\,min^{-1}$ air flow over 9 days.

to the results observed for the artificially aged synthetic oil. This similarity is due to the similar molecular base structure that is predominantly based on hydrocarbons for both oils. The main difference is the absence of the $C=O$ double bond feature for the fresh mineral hydraulic oil because it does not contain the ester additive as the synthetic motor oil. Also, the increase in the two relevant peaks seems slightly slower for the hydraulic oil, which could indicate a better oxidation stability. Comparison of the spectral features, however, indicates that the same spectral regions are relevant; thus, a multi-channel system with the same filter configuration could be used for both oils.

3.1.3 Ester-based hydraulic fluid

The spectra for native ester artificially aged at $130\,^{\circ}C$ show different oxidation features compared to the synthetic and mineral oils shown above. At $1700\,cm^{-1}$ the $C=O$ double bond feature due to an oxidation via hydroperoxides (Honary and Richter, 2011) overlaps with the $C=O$ double bond of the ester itself (Fig. 7) (Totten et al., 2003). Even for a film thickness of only $200\,\mu m$ the IR radiation is completely absorbed in the area of $1700\,cm^{-1}$ and, thus, cannot be evaluated for the determination of the oxidation status. Increasing absorption caused by increasing aging is observed in a broad spectral region around $3500\,cm^{-1}$ and, to a lesser extent, around 2500 and around $1600\,cm^{-1}$ on the shoulder of the saturated $C=O$ double bond feature. The main feature around $3500\,cm^{-1}$ is most suitable to obtain sufficient sensitivity for determination of the oil degradation by oxidation. However, comparing this result with the water contamination of the synthetic motor oil (cf. Fig. 4), an increasing water content would probably interfere with the oxidation in this spectral region (ASTM International, 2007) and thus, multiple channels would be required to be allow separate determination of oxidation and water contamination.

Table 3. Acid number (AN) and kinematic viscosity η for synthetic motor oil (Mobil Pegasus 1); actual samples taken from a gas engine in a combined heating and power unit.

	0 h	1025 h	1840 h	Limit
AN [mg KOH g^{-1}]	0.7	8.22	10.1	>5*
η [mm^2 s^{-1}]	92	123	163	±35 %

* Recommended limit values for mineral hydraulic fluids.

3.2 Acid number and viscosity

To complement the spectral features, additional measurements were performed to evaluate the quality and the degree of degradation of the different oils. The obtained results are compared with limit values proposed by Möller and Nassar (2002) for turbine and mineral hydraulic oils and Booser (1994) for engine oils (cf. Tables 3–6). The limit values for kinematic viscosity and AN given in the tables show the maximum acceptable deviation compared to fresh oil. For some specific applications limit values are not defined; these are instead compared to values of other applications and marked in the tables.

For synthetic motor oil, a similar height of the $C=O$ oxidation peak is reached after artificial aging for 9 days at $170\,^{\circ}C$ as for 1025 working hours under normal operating conditions (cf. Figs. 3 and 4). In comparison, the AN and viscosity of the sample aged under operating conditions show higher values compared to the artificial aging, see Tables 3 and 4. The higher AN results can be attributed to the increase in nitric acids in the oil, which also enhances polycondensation in the oil and increases the viscosity. However, the limit value taken from the literature would indicate that the oil is close to the limit after 1025 working hours, similar to the laboratory analysis, which indicates aging but still acceptable performance of the oil. The mineral hydraulic oil already

Table 4. Acid number (AN) and kinematic viscosity η for synthetic motor oil (Mobil 1 Pegasus); artificially aged samples, cf. Fig. 4.

	0 days	3 days	7 days	9 days	Limit
AN [mg KOH g^{-1}]	1.15	3.4	7	7.7	$\Delta > +2,5$
η [mm^2 s^{-1}]	90	94	101	110	$\pm 35\,\%$

* Limit value suggested by OelCheck (2013).

Table 5. Acid number (AN) and kinematic viscosity η for mineral hydraulic oil (Shell Tellus); artificially aged samples.

	0 days	3 days	7 days	Limit
AN [mg KOH g^{-1}]	0.54	2.01	2.31	> 5
η [mm^2 s^{-1}]	44	49	56	$\pm 10\,\%$

Table 6. Acid number (AN) and kinematic viscosity η for ester-based hydraulic fluid (Meguin HETG 37); artificially aged samples.

	0 days	2 days	4 days	6 days	Limit
AN [mg KOH g^{-1}]	0.67	3.0	4.95	6.7	$> 5^*$
η [mm^2 s^{-1}]	35.0	42.3	50.0	59.5	$\pm 10\,\%^*$

* Recommended limit values for mineral hydraulic fluids.

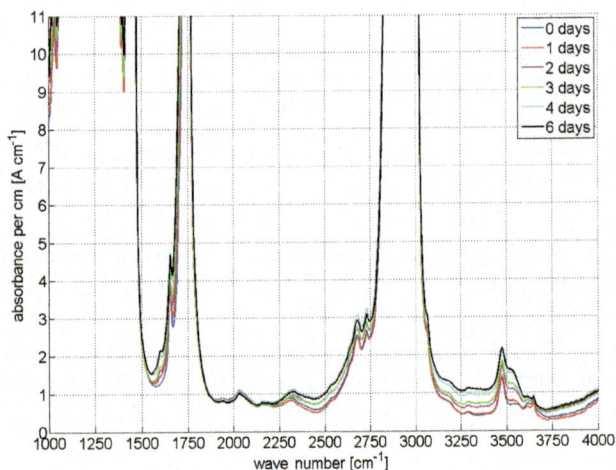

Figure 7. IR spectrum of ester-based hydraulic fluid (1500 mL) artificially aged at 130 °C with 50 cm^3 min^{-1} air flow over 6 days.

3.3   Multi-channel IR sensor system

The recorded IR spectra of the different oils have indicated specific spectral regions for determination of oxidation and water contamination for the hydrocarbon-based oils. On the other hand, IR analysis of the ester-based hydraulic fluid has shown an overlap of the spectral regions, indicating oxidation and water contamination. The same oil samples, fresh, artificially aged and, in the case of the synthetic motor oil, aged under operating conditions, were analyzed with the sensor system to evaluate the suitability of the system for determination of the two relevant conditions. For the two hydrocarbon-based oils the detector configuration A, Table 1, was used and for the native ester detector configuration B. The signal intensity is normalized, i.e., the values for fresh oils are set to one.

reaches the recommended threshold limit of the kinematic viscosity after only 3 days of artificial aging, see Table 5. The ester-based hydraulic fluid, finally, already exceeds the threshold limit of the viscosity after less than 2 days due to its lower resistance against oxidation (Bartz, 1993).

Figure 8 shows the IR absorption signal integrated over the characteristic oxidation region in the spectrum plotted vs. the measured AN of the three oil types for all artificially aged samples. All three show nearly linear correlation between IR absorption and AN. Note that due to the much smaller increase in the oxidation feature for the native ester the recorded slope is smaller than the slope of the synthetic motor oil and the mineral hydraulic oil. The different IR absorption values of the fresh samples for synthetic motor oil and mineral hydraulic oil are caused by the additive in the synthetic motor oil, which interferes with the measurement of the oxidation signal. It can be concluded that IR absorption measurement in a suitable spectral range allows an estimation of the AN, at least for thermal deterioration; however, the oil base type as well as the additives have to be taken into account, thus requiring individual calibration. When considering motor oils, on the other hand, estimation of the AN is more difficult due to additional acidic components generated by the combustion process and also blow-by gases. Therefore additional spectral ranges have to be taken into account.

The measurement results for the artificially aged samples of the synthetic motor oil (Fig. 9) show a continuous decrease in the reference channel AR due to an increase in soot particles in the oil. Therefore, the three measurement channels were weighted by dividing the normalized values through the reference channel signal to compensate for this effect. After normalization, a strong decrease in channel A1 corresponding to the C=O feature is observed for increasing aging as expected from the reference spectra (cf. Fig. 4). In addition, a slight decrease in channel A2 resulting from an increase in the OH feature can be observed. Channel A3 does not show any significant changes.

Figure 10 shows the effect of additional water contamination on a sample artificially aged for 9 days, also compared to the signal in fresh oil. Here, A1 shows a large difference between fresh and aged samples, but shows very little change with increasing water content. Channel A2 is only marginally affected and A3 is totally unaffected by the artificial aging, but both reflect the water contamination as expected from the reference spectra (cf. Fig. 5). Therefore, water content and oxidation can be independently determined well for the synthetic motor oil and presumably also for the

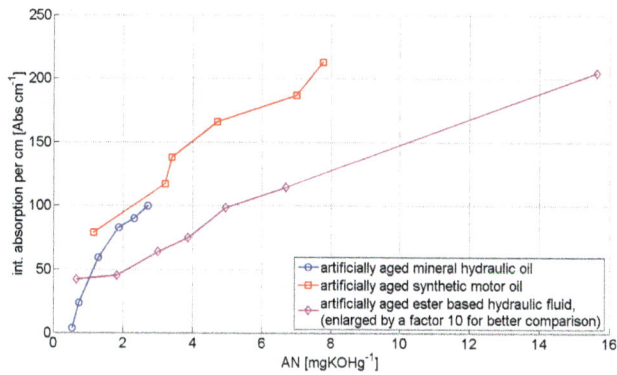

Figure 8. FTIR absorption integrated over the oxidation band plotted vs. acid number (AN) as determined by ASTM D974; evaluated spectral regions are $3333\,\mathrm{cm}^{-1}$ with FHWM $52\,\mathrm{cm}^{-1}$ for the artificially aged ester-based hydraulic fluid (corresponding to filter B1) and $1722\,\mathrm{cm}^{-1}$ with FWHM $151\,\mathrm{cm}^{-1}$, corresponding to filter A1, for the other artificially aged samples (synthetic motor oil, mineral hydraulic oil).

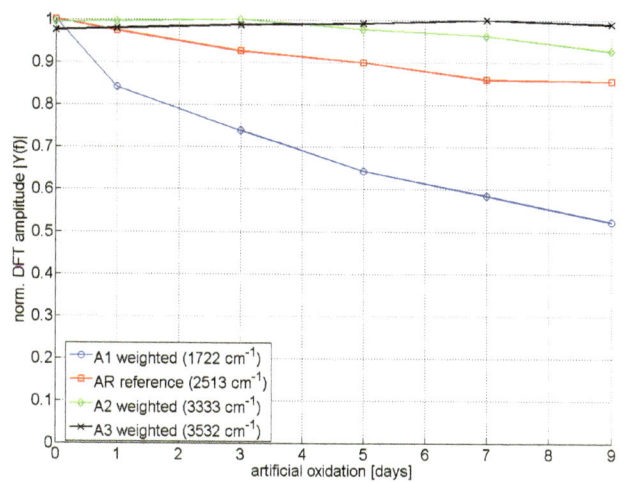

Figure 9. Measurement results for synthetic motor oil artificially aged at $170\,°\mathrm{C}$ with $50\,\mathrm{cm}^3\,\mathrm{min}^{-1}$ air flow over 9 days, measured with a detector in mineral oil configuration (reference AR).

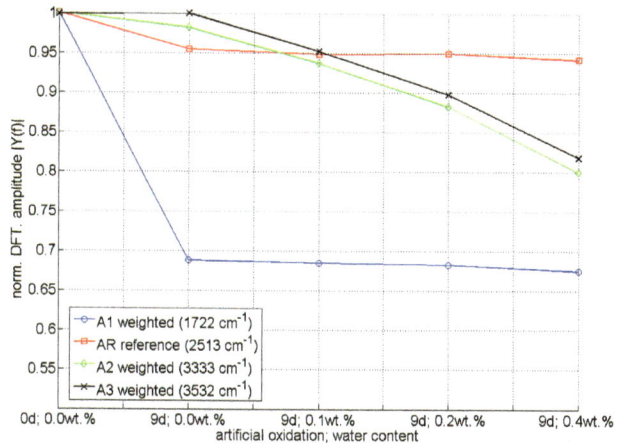

Figure 10. Comparison of aged synthetic motor oil (9 days) with water contaminations of 0, 0.1, 0.2, 0.4 wt.%, respectively, measured with a detector in mineral oil configuration (reference AR).

mineral hydraulic oil with the multiple measurement channels. In fact, the third channel could be tailored to indicate another degradation feature, e.g., to reflect the concentration of an additive.

For the artificially aged native ester, increased oxidation can be observed using measurement channels B1 and B2, both of which show a nearly linear decrease in the signal intensity with the duration of the artificial aging after normalization and weighting with the reference channel, Fig. 11. A slight decrease in the reference signal BR can also be observed, while channel B3 does not show a clear trend. Additional water contamination would thus be difficult to differentiate from the effect of oxidation due to the overlap of the spectral features. A correction was suggested by Sumit et al. (2010) using a differential algorithm. Analyzing the spectra of different ester-based hydraulic fluids, e.g., native hydraulic fluids, phosphor acid esters, etc., shows that usable wavelengths can be found (Bley, 2013). However, an adaption of the measurement channels, i.e., the filter wavelengths and possibly also the number of channels, for each fluid is necessary to obtain optimal results.

3.4 Influence of temperature variations

For monitoring of the oil degradation during normal operation of a machine the influence of varying ambient and, more importantly, oil temperatures has to be taken into account. Figure 12 (top) shows the reference spectra of synthetic motor oil artificially aged for 9 days and contaminated with 0.4 wt.% water measured with the FTIR spectrometer at different temperatures. Due to the temperature increase the density of the oil changes. According to the Lambert–Beer law the transmission depends on the concentration of radiation-absorbing molecules that decreases with decreasing density

of the fluid. At the same time, the extinction of the different molecules also changes with temperature as well as the absorption path length due to thermal expansion. Thus, the absorption spectrum will show complex changes when the oil temperature changes. The change of the absorption, $\Delta E(\lambda)$, is shown in Fig. 12 (bottom) for IR spectra measured at a temperature of 40.8, 50.8 and 57.9 °C, respectively, relative to the spectrum at 30.5 °C.

$$\Delta E(\lambda) = E(\lambda, T) - E(\lambda, 30.5\,°\mathrm{C}) \qquad (2)$$

$E(\lambda, T)$ absorption at temperature T, λ wavelength.

Changes in the spectrum caused only by the temperature can be observed. It is interesting to note that some spectral regions show increasing absorption with the temperature, e.g., below 1600 and above $3600\,\mathrm{cm}^{-1}$, while others

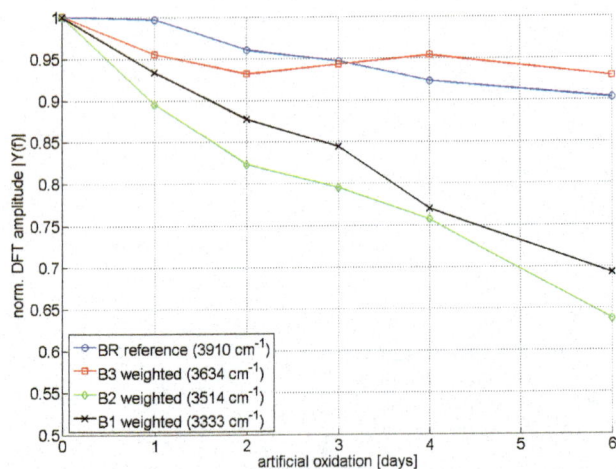

Figure 11. Measurement results for ester-based hydraulic fluid artificially aged at $130\,°C$ with $50\,cm^3\,min^{-1}$ air flow over 6 days, measured with a detector in ester configuration (reference BR).

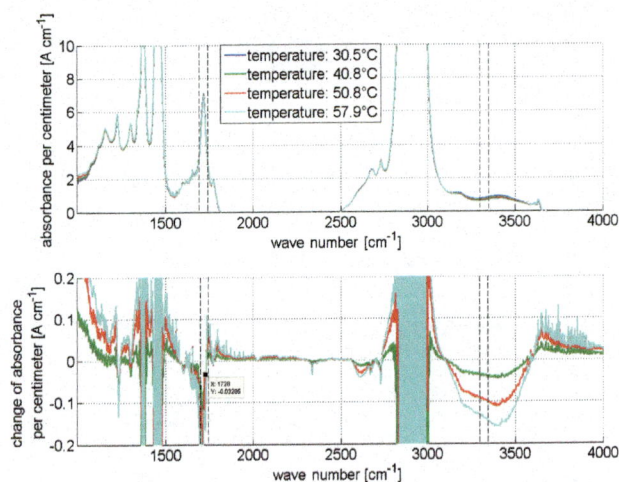

Figure 12. Temperature dependence of the FTIR spectra. Top: Transmission spectra for synthetic motor oil (artificially aged at $170\,°C$ with $50\,cm^3\,min^{-1}$ air flow over 9 days with 0.4 wt.% water contamination added after aging) at 30.5, 40.8, 50.8 and $57.9\,°C$. Bottom: Deviation of the transmission spectra at elevated temperatures relative to the transmission at $30.5\,°C$; areas for calculating the mean values around 1722 and $3333\,cm^{-1}$ shown in Fig. 13 are marked.

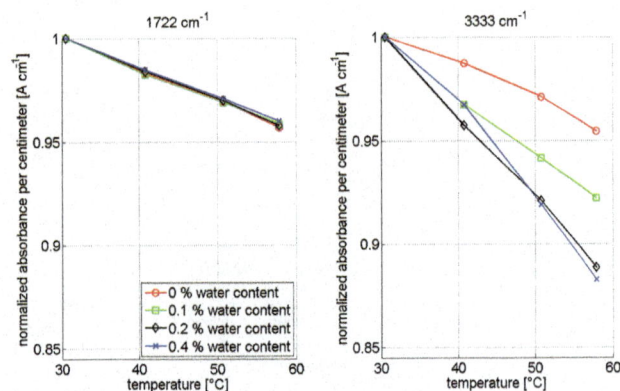

Figure 13. Change in absorbance at 1722 and $3333\,cm^{-1}$ for aged synthetic motor oil with increasing water content (0, 0.1, 0.2, 0.4 wt.%, respectively) and increasing temperature.

temperature. However, with increasing water content the change in absorption caused by the temperature increases in the IR region around $3333\,cm^{-1}$. Thus, the temperature has to be measured and taken into account to achieve a good resolution as the influence of changing temperature is on the same order as the effect of water contamination. For practical applications a calibration is required at different temperatures in order to be able to determine the water content correctly under varying operating conditions.

4 Conclusion and outlook

Hydrocarbon-based oils, i.e., a synthetic motor oil and a mineral hydraulic oil, as well as an ester-based hydraulic fluid, were analyzed using FTIR spectroscopy to identify the relevant spectral features caused by oxidation and water contamination. In addition, viscosity and acid number were determined using standard laboratory techniques. The FTIR analysis showed that for hydrocarbon-based oils the $C=O$ oxidation feature around $1700\,cm^{-1}$ can be used to evaluate the degree of oxidative degradation. However, the specific composition of the oil has to be taken into account, as additives can already lead to an increased absorption in this region even for fresh oil. For ester-based hydraulic fluids, on the other hand, this feature cannot be used as it already shows total extinction due to the base molecules. Alternatively, the OH feature can be used to evaluate the degree of oxidation. However, this shows a strong overlap with the effect of water contamination, thus requiring careful selection of the spectral filters and advanced signal analysis to allow independent determination of the oxidative degradation as well as the water content.

A miniaturized multi-channel IR sensor system was presented that is based on a Si flow cell, a micromachined IR source and four thermopiles with specific filters adapted to the application requirements. Two alternatives were presented, one suitable for hydrocarbon-based oils, the other for

show decreasing absorption, especially the water feature in the range from 3000 to $3600\,cm^{-1}$. Figure 13 shows the influence of the temperature at $1722\,cm^{-1}$ (no IR absorption by water molecules) and $3333\,cm^{-1}$ (IR absorption by water molecules) for different water contaminations. For these calculations a rectangular filter with a filter width of $52\,cm^{-1}$ was used. As expected, without water (0 wt.% H_2O) only a minimal change in the spectrum is observed at both wave numbers. This observed change is caused by the density decrease in the absorbing molecules with increasing ester-based hydraulic fluids. The sensor system is able to differentiate between increasing oxidation and increasing water contamination for synthetic motor oil and, due to the similarity of the base spectra, also for mineral hydraulic oil. The sensor system is also able to determine the oxidation status of ester-based hydraulic fluids; however, interference caused by water contamination requires advanced signal processing. The sensor system has to be equipped with an oil temperature sensor to compensate temperature dependent changes of the IR transmission spectrum and also requires a comprehensive calibration at different temperatures. On the other hand, the change of the recorded spectral information with temperature can be used to obtain additional information about the oil degradation or the concentration of water contamination.

This study was based on only one aging process, i.e., oxidative aging at elevated temperatures, and one contamination, i.e., with water. Other degradation processes and contaminations need to be studied to ensure that the approach will allow reliable results under various operating conditions. Future experiments will, for example, address the increase in nitric acid components caused by the blow-by effect in combustion engines. A specific IR filter will be implemented in the system for these compounds and controlled aging will be realized with an experimental set-up bubbling NO in N_2 through the oil at elevated temperatures, similar to the analysis of oxidative aging reported here. A combination of different well-controlled processes should also be performed to achieve degradation similar to that observed under normal operating conditions.

Acknowledgements. This work was part of the NaMiFlu project and the FluidSens project and was supported by the German Ministry for Education and Research (BMBF), the federal state of Saarland and the European Fund for Regional Development (EFRE).

Edited by: B. Jakoby
Reviewed by: two anonymous referees

References

Agoston, A., Ötsch, C., Zhuravleva, J., and Jakoby, B.: An IR-Absorption Sensor System for the Determination of Engine Oil Deterioration, Proceedings of IEEE Sensors Conference, 463–466, doi:10.1109/ICSENS.2004.1426200, 2004.

Agoston, A., Schneidhofer, C., Dörr, N., and Jakoby, B.: A concept of an infrared sensor system for oil condition monitoring, e & i Elektrotechnik und Informationstechnik, 125/3, 71–75, doi:10.1007/s00502-008-0506-3, 2008.

ASTM International: Standard Practice for Condition Monitoring of Used Lubricants by Trend Analysis Using Fourier Transform Infrared (FT-IR) Spectrometry, ASTM International, E2412-04, 2007.

ASTM International: Standard Specification for Reagent Water, ASTM International, D1193-06, 2011a.

ASTM International: Standard Test Method for Acid and Base Number by Color-Indicator Titration, ASTM International, D974-11, 2011b.

Bartz, W. J.: Biologisch schnell abbaubare Schmierstoffe und Arbeitsflüssigkeiten (Biologically rapid-degradable lubricants and working fluids), Expert Verlag, 43–44, ISBN 978-3-8169-0810-4, 1993.

Bley, T.: Integriertes Multisensorsystem zur Zustandsüberwachung von Schmierflüssigkeiten (Integrated Multisensor-System for Condition Monitoring of Lubricating Fluids), Dissertation, Saarland University, Lab for Measurement Technology, Shaker, Aachen, ISBN 978-3-8440-2198-1, 2013.

Bley, T. and Schütze, A.: A Multichannel IR Sensor System for Condition Monitoring of Technical Fluids, Proceedings IRS^2 2011 – 12th International Conference on Infrared Sensors and Systems, Nuremberg, 94–99, doi:10.5162/irs11/i4.3, 2011.

Bley, T., Pignanelli, E., and Schütze, A.: Multichannel IR Sensor System for Determination of Oil Degradation, Proc. IMCS 2012: The 14th International Meeting on Chemical Sensors, Nuremberg, Germany, 20–23 May, 2012a.

Bley, T., Pignanelli, E., Fischer, M., Günschmann, S., Müller, J., and Schütze, A.: IR-Optical Oil Quality Sensor System for High Pressure Applications, in: Mechatronics 2012 – The 13th Mechatronics Forum International Conference, Linz, Austria, 17–19 September 2012, edited by: Scheidl, R. and Jacoby, B., Proceedings Vol. 2/3, 351–358, 2012b.

Bley, T., Pignanelli, E., and Schütze, A.: COPS – Combined oil quality and Particle measurement System, ICST2012, Sixth International Conference on Sensing Technology, Kolkata, India, 18–21 December, 2012c.

Booser, E. R.: Handbook of Lubrication and Tribology, Vol. III CRC Press, 23–24, ISBN 978-1-4200-5045-5, 1994.

Burns, D. A. and Ciurczak, E. W.: Handbook of Near-Infrared Analysis, CRC Press, ISBN 978-1-4200-0737-4, 2008.

Duchowsky, J. K. and Mannebach, H.: A Novel Approach to Predictive Maintenance: A Portable, Multi-Component MEMS Sensor for On-Line Monitoring of Fluid Condition in Hydraulic and Lubricating Systems, Tribol. T., 49, 545–553, doi:10.1080/10402000600885183, 2006.

Endisch, P. and Koch, A.: In-situ-Infrarotsensor zur Ölzustandsanalyse (In-situ infrared sensor for oil condition monitoring), XXI, Messtechnisches Symposium des AHMT, Paderborn, 20–22 September 2007, Shaker Verlag, ISBN 978-3-8322-6539-7, 190–204, 2007.

Günschmann, S., Fischer, M., Bley, T., Käpplinger, I., Brode, W., Mannebach, H., Müller, J.: Fabrication of a Si-measurement cuvette using a new multifunctional bonding method, CICMT 2012 Ceramic Interconnect & Ceramic Microsystems Technologies, Erfurt, 2012.

Günschmann, S., Mannebach, H., Steffensky, J., Fischer, M., and Müller, J.: Herstellung einer Messküvette als Teil eines Messsensors für Echtzeit-Ölqualitätsmessungen (Fabrication of a measurement cuvette as part of a sensor for the realtime-oilquality-measurement), MST Congress, Aachen, ISBN 978-3-8007-3555-6, 2013.

Honary, L. and Richter, E.: Biobased Lubricants and Greases: Technology and Products, John Wiley and Sohns, 65–67, ISBN 978-0-470-74158-0, 2011.

Intex Inc.: INTX17-0900, available at: www.intexworld.com, last access: 26 June 2013.

Kudlaty, K., Purde, A., and Koch, A. W.: Development of an Infrared Sensor for On-line Analysis of Lubricant Deterioration; Proc. of IEEE Sensors 2003 Vol. 2, 903–908, Toronto, doi:10.1109/ICSENS.2003.1279074, 2003.

Micro-Hybrid Electronic GmbH: Four Channel Thermopile Detector, available at: www.micro-hybrid.de, last access: 26 June 2013.

Mortier, R. M., Fox, M. F., and Orszulik, S. T.: Chemistry and Technology of Lubricants, Springer, 107–115, ISBN 978-1-4020-8662-5, 2010.

Möller, U. J. and Nassar, J.: Schmierstoffe im Betrieb (Lubricating fluids in practical application), Springer, 662–687, ISBN 978-3-642-56379-9, 2002.

OelCheck: Limit values for gas engine oils, available at: http://www.oelcheck.de/en/home.html (last access: 3 July 2013), 2013.

Pawlak, Z.: Tribochemistry of Lubricating Oils, Elsevier, ISBN 978-0-444-51296-3, 2003.

Pretsch, E., Bühlmann, P., and Bardetscher, M.: Structure Determination of Organic Compounds, Springer, 287–288, ISBN 978-3-540-93810-1, 2009.

Schneidhofer, C. and Dörr, N.: Online-Ölzustandsüberwachung mit chemischen Korrosionssensoren (Online oil condition monitoring with chemical corrosion sensors), e & i Elektrotechnik & Informationstechnik, 126/1–2, 31–36, doi:10.1007/s00502-009-0606-8, 2009.

Sumit, P., Legner, W., Krenkow, A., Müller, G., Lemettais, T., Pradat, F., and Hertens, D.: Chemical Contamination Sensor for Phosphate Ester Hydraulic Fluids, Intern, J. Aerospace Eng., 2010, 156281, doi:10.1155/2010/156281, 2010.

Totten, G. E., Westbrook, S. R., and Shah, R. J.: Fuels and Lubricants Handbook, ASTM International, 787–791, ISBN 978-0-8031-2096-9, 2003.

Wang, S. S.: Road tests of oil condition sensor and sensing technique, Sensors und Actuators, B73, 106–111, doi:10.1016/S0925-4005(00)00660-2, 2001.

Wiesent, B. R., Dorigo, D. G., and Koch, A. W.: Limits of IR-spectrometers based on linear variable filters and detector arrays, Proc. SPIE 7767, Instrumentation, Metrology, and Standards for Nanomanufacturing IV, 77670L (24 August, 2010); doi:10.1117/12.860532, 2010.

Wiesent, B. R., Dorigo, D. D., and Koch, A. W.: Suitability of tunable Fabry-Perot spectrometers for condition monitoring purposes of gear oils in offshore wind turbines, Proceedings IRS[2] 2011 – 12th International Conference on Infrared Sensors and Systems, Nuremberg, 82–87, doi:10.5162/irs11/i4.2, 2011.

17

Development of a portable active long-path differential optical absorption spectroscopy system for volcanic gas measurements

F. Vita[1], C. Kern[2], and S. Inguaggiato[1]

[1]Istituto Nazionale di Geofisica e Vulcanologia – Sezione di Palermo – Via Ugo La Malfa, 153, 90146 Palermo, Italy
[2]U.S. Geological Survey, Cascades Volcano Observatory, 1300 SE Cardinal Ct S100, Vancouver, Washington 98683, USA

Correspondence to: F. Vita (fabio.vita@ingv.it)

Abstract. Active long-path differential optical absorption spectroscopy (LP-DOAS) has been an effective tool for measuring atmospheric trace gases for several decades. However, instruments were large, heavy and power-inefficient, making their application to remote environments extremely challenging. Recent developments in fibre-coupling telescope technology and the availability of ultraviolet light emitting diodes (UV-LEDS) have now allowed us to design and construct a lightweight, portable, low-power LP-DOAS instrument for use at remote locations and specifically for measuring degassing from active volcanic systems. The LP-DOAS was used to measure sulfur dioxide (SO_2) emissions from La Fossa crater, Vulcano, Italy, where column densities of up to 1.2×10^{18} molec cm^{-2} (~ 500 ppmm) were detected along open paths of up to 400 m in total length. The instrument's SO_2 detection limit was determined to be 2×10^{16} molec cm^{-2} (~ 8 ppmm), thereby making quantitative detection of even trace amounts of SO_2 possible. The instrument is capable of measuring other volcanic volatile species as well. Though the spectral evaluation of the recorded data showed that chlorine monoxide (ClO) and carbon disulfide (CS_2) were both below the instrument's detection limits during the experiment, the upper limits for the X / SO_2 ratio (X = ClO, CS_2) could be derived, and yielded 2×10^{-3} and 0.1, respectively. The robust design and versatility of the instrument make it a promising tool for monitoring of volcanic degassing and understanding processes in a range of volcanic systems.

1 Introduction

The emission of volatile species from volcanic magmas is linked to geochemical processes occurring at depth. Volatiles are exsolved as a result of recharge, decompression, crystallisation or mixing of magmas. The composition and flux of gases being emitted from a volcanic vent can provide insights into geophysical and geochemical parameters such as magma supply, pressure, temperature and degassing depth – parameters that can used by volcanologists to forecast eruptions.

The major volatile species emitted from volcanic vents are water vapour (H_2O), carbon dioxide (CO_2), sulfur dioxide (SO_2), hydrogen sulfide (H_2S) and hydrogen halides (HX with X = Fl, Cl, Br, I). Of these, CO_2 has the lowest solubility in most magma types, and is typically emitted first as a magmatic system is recharged (Werner et al., 2013; Burton et al., 2013). Therefore, developing methods for accurately monitoring CO_2 emissions from volcanoes has been a major focus of the volcanology community in recent years.

Unfortunately, volcanic CO_2 emission rates are considerably more difficult to measure than those of some of the other species, the main reason being the significant background concentration of CO_2 in the Earth's atmosphere. Unless measurements are made very close to the vent where high magmatic/atmospheric gas ratios can be sampled, the enhancement of CO_2 in a volcanic plume is oftentimes only a few parts per million (ppm) over the background of approximately 400 ppm. Passive remote sensing instruments that use scattered solar radiation as a light source and therefore make an integrated measurement over the entire atmospheric column attempt to detect an even smaller enhancement. The relative increase in the total CO_2 column is of the order of only 10^{-3} for large emitters and can be significantly less at volcanoes with moderate to small emission rates. For this reason, CO_2 emissions from volcanoes have not yet been detected from satellite platforms, and direct measurements of CO_2 emission rates from volcanic vents can only be made by laborious and costly aircraft traverse techniques (Gerlach et al., 1997).

On the other hand, SO_2 has a negligible atmospheric background concentration and is much more accessible to remote sensing observations, with routine measurements made both from ground-based and satellite-based instrumentation (Galle et al., 2010; Oppenheimer et al., 2011; Carn et al., 2013). In fact, arguably the most robust method currently available for measuring volcanic CO_2 emission rates is by determining the molecular ratio of CO_2 to SO_2 in a volcanic plume and multiplying this by the SO_2 molecular emission rate obtained from remote sensing instrumentation (Gerlach et al., 1998; Burton et al., 2013; Pering et al., 2014). CO_2 / SO_2 molecular ratios can either be determined by collecting gas in Giggenbach bottles (Giggenbach and Goguel, 1989) for analysis in the laboratory or using in situ chemical and optical sensors in a multicomponent gas analyzer system (Multi-GAS, Aiuppa et al., 2005; Shinohara, 2005). While the sampling of fumaroles with Giggenbach bottles is limited in its applicability for continuous monitoring, these relatively novel Multi-GAS instruments now allow real-time measurements of volcanic gas compositions suitable for eruption forecasting (Aiuppa et al., 2007).

Despite their relative versatility, the Multi-GAS instruments also have some drawbacks, and determining the gas speciation representative of the bulk plume at a volcanic vent remains challenging. For one, the instruments typically rely on an optical CO_2 sensor and multiple electrochemical sensors for other gas species (Aiuppa et al., 2005). The different sensors tend to have significantly different response times, with the optical sensor being quicker to respond to a change in composition than the electrochemical sensors. This can lead to errors and uncertainties in the derived gas ratios if not properly taken into account (Roberts et al., 2014). Also, the electrochemical sensors have cross-sensitivities to other gases – oftentimes gases found in great abundance in volcanic plumes (Roberts et al., 2012). A typical electrochemical H_2S sensor, for example, can have a cross-sensitivity to SO_2 of up to 20 % (Roberts et al., 2014). Finally, while in situ sampling lends itself to deriving the gas composition at a specific location, problems arise if the volatile speciation is heterogeneous in space in the area around the volcanic vent. Such heterogeneities are quite common, and can be caused by a complex plumbing system and interaction of volatiles with a hydrothermal system, or result from gas and

fluid temperature variations in the proximity of the vent. In such cases, the gas composition measured by an in situ measurement may not be representative of the bulk plume.

In light of these challenges, improved methods for measuring plume compositions at volcanic vents are still being sought. One very sophisticated method is to use Fourier transform infrared spectroscopy (FTIR) to measure the absorption or emission of infrared (IR) radiation by volcanic gases (Oppenheimer et al., 1998; Stremme et al., 2012). FTIR is sensitive to a large number of species, including both CO_2 and SO_2, and can therefore be used in combination with SO_2 emission rate measurements to derive CO_2 emission rates (Burton et al., 2000). When used in an open-path configuration with the light path penetrating the bulk plume, representative gas compositions can be obtained. However, FTIR instruments are very costly and limited in their portability, as they rely on a moving interferometer to record spectral information. Also, deployment can be challenging in the absence of accessible hot lava, as they then require either a bulky thermal IR emitter requiring tens of watts of power to be deployed on the opposite side of the vent, or must be located such that the plume is between the instrument and the Sun, which is then used as an IR source.

Near-infrared (NIR, 800 nm < λ < 2500 nm) tuneable diode laser absorption spectroscopy (TDLAS) instruments represent a somewhat simpler alternative (Schiff et al., 1994; Carapezza et al., 2011; Pedone et al., 2014). Although typically only sensitive to a single gas species according to which the laser wavelength is chosen, today's NIR-TDLAS units are small and portable, typically weighing ~ 2.5 kg, and can run for extended periods of time on battery power. When set up in an open-path configuration, these instruments only require that a retroreflector be deployed on the opposite side of the desired optical path, as both the light source and the detector are usually located in the same unit. Reflectors are considerably simpler to deploy than active light sources, as they do not require power, do not need to be carefully adjusted and can withstand harsh environmental conditions more easily. NIR-TDLAS instruments measuring CO_2, H_2S, HF or CH_4 along open paths are commercially available for purchase.

Measuring SO_2 in the infrared absorption is also possible, but due to the lack of near-infrared (NIR) absorption bands, measurements are made in the thermal infrared regions (3.5 μm < λ < 20 μm). Though the technology of thermal infrared lasers is constantly evolving (Tittel et al., 2012), they remain considerably more complex than NIR diode lasers, and generally require significantly more power. Depending on which thermal IR band is chosen, additional complications can arise from interference by pressure-broadened absorption lines of other atmospheric compounds (mainly CO_2 and H_2O). Therefore, IR laser absorption measurements of SO_2 are usually performed in cells at reduced barometric pressures (Richter et al., 2002; Rawlins et al., 2005), and no IR laser absorption measurements of SO_2 along open atmospheric paths are known to date. In the absence of an SO_2 measurement, emission rates of gases detected using NIR-TDLAS cannot easily be determined using the ratio method described earlier.

For this reason, implementation of a portable, active open-path measurement of SO_2 is highly desirable. In this study, we describe how a miniature active long-path differential optical absorption spectroscopy (LP-DOAS) instrument was designed to measure SO_2 and potentially also other trace gases with a high degree of accuracy and precision along well-defined open paths around volcanic vents. Emphasis was placed on ensuring portability of the instrument on foot to remote locations, a sufficient degree of environmental hardening, low power consumption that can easily be supplied by battery, and simple set-up and removal at the target site. After the instrument was constructed, tests were performed at the island of Vulcano (Aeolian arc, Italy) in February 2010, the results of which are described below.

2 Design of a portable LP-DOAS instrument

The first measurements of atmospheric trace gases by LP-DOAS were performed in the late 1970s (Platt et al., 1979). The technique has since become widely used in the atmospheric sciences community, particularly because the contact-free nature of the measurement allows the detection of highly reactive radical species (Platt and Stutz, 2008). More recently, active differential absorption measurements have also been applied to volcanic gases (O'Dwyer, 2003; Kern et al., 2008), but deployment of active LP-DOAS systems has been severely limited by the size (a typically longer than 1 m focal length telescope) and weight (typically tens to hundreds of kg) of such systems and by the power consumption of the ultra-violet light sources required (typically tens to hundreds of watts).

Two recent technological advances have now made it possible to design a lightweight portable LP-DOAS instrument. For one, the size and weight of the telescope unit can be dramatically reduced by replacing the conventional Newtonian set-up with a fibre-coupling coaxial design. In this set-up, first introduced by Merten et al. (2011), the light source is coupled into the telescope using a fibre bundle, one end of which is placed in the focal plane of the spherical primary mirror. Light from the desired light source is coupled into several of the fibres in the bundle, reflected on the primary mirror, and transmitted in a nearly parallel beam into the atmosphere to an array of retroreflectors located tens or hundreds of metres away. These reflect the light back to the primary mirror of the telescope, where, partly due to spherical aberration of the mirror, some of it enters the receiving fibres that couple it into a moderate resolution ultraviolet (UV) spectrometer for analysis. The use of this fibre-coupling design eliminates the need for any secondary mirrors and al-

lows separation of the light source from the telescope, thus making a lighter, portable telescope design possible.

In our design, the fibre bundle consists of seven $100\,\mu m$ diameter fused silica fibres, six of which are arranged in a circular pattern around a central seventh fibre. As suggested by Merten et al. (2011), we use the six outer fibres of the bundle for transmission and the central fibre to receive radiation that has passed through the optical system. However, we have modified the original design by additionally splitting the transmission fibres into three groups of two (Fig. 1). This allows the coupling of three different light sources into the optical path, a feature that is unique to our system and allows a high degree of flexibility with regard to measurement wavelength (see below).

In this coaxial set-up, the telescope primary mirror acts both as a sending and receiving unit. A three-dimensional fibre positioner is used to mount the end of the fibre bundle in the focal plane of the mirror (Fig. 1), which is protected from environmental conditions by an MgF_2 coating. Compared to previously described LP-DOAS systems, we use a significantly smaller primary mirror diameter d of only 150 mm with a focal length f of 400 mm to reduce the overall instrument size and weight. In the paraxial approximation, and considering the fibre bundle diameter of about $300\,\mu m$, this set-up leads to a beam spread angle α of approximately 0.04 degrees. The diameter D of the retroreflector array required to reflect the entire emitted radiation back to the telescope is then given by

$$D = d + 2L \tan\left(\frac{\alpha}{2}\right) \approx d + L \tan\alpha. \qquad (1)$$

Here, L is the distance between the telescope and reflector array, which is also exactly half of the optical path length. In our set-up, an array diameter of 19 cm is needed at a distance of 50 m. For a path length of 200 m, an array diameter of 29 cm is required to reflect all transmitted radiation. A larger telescope would reduce beam spread, but the compact design chosen here makes the instrument portable and allows measurements to be made in otherwise inaccessible remote locations. Path lengths of up to a few hundred metres are sufficient for making measurements at most volcanic vents. For the test measurements, we used an array of ten 52 mm diameter fused silica corner cubes densely packed in a triangular frame mounted on a tripod. This array had a total diameter of approximately 40 cm, thus allowing one-way paths of up to about 350 m without significant loss of radiative energy. We successfully made measurements along paths of up to 200 m with the set-up (see Sect. 3).

For one-way paths longer than 350 m, one might expect a significant drop in the measurement signal-to-noise ratio (SNR). Because the fraction of reflected light is proportional to D^{-2} and therefore proportional to L^{-2} for long light paths (Eq. 1), photon statistics yield an SNR of the measured spectrum that is proportional to the square root of the intensity, or L^{-1}. However, since the column density is proportional to

(a)

(b)

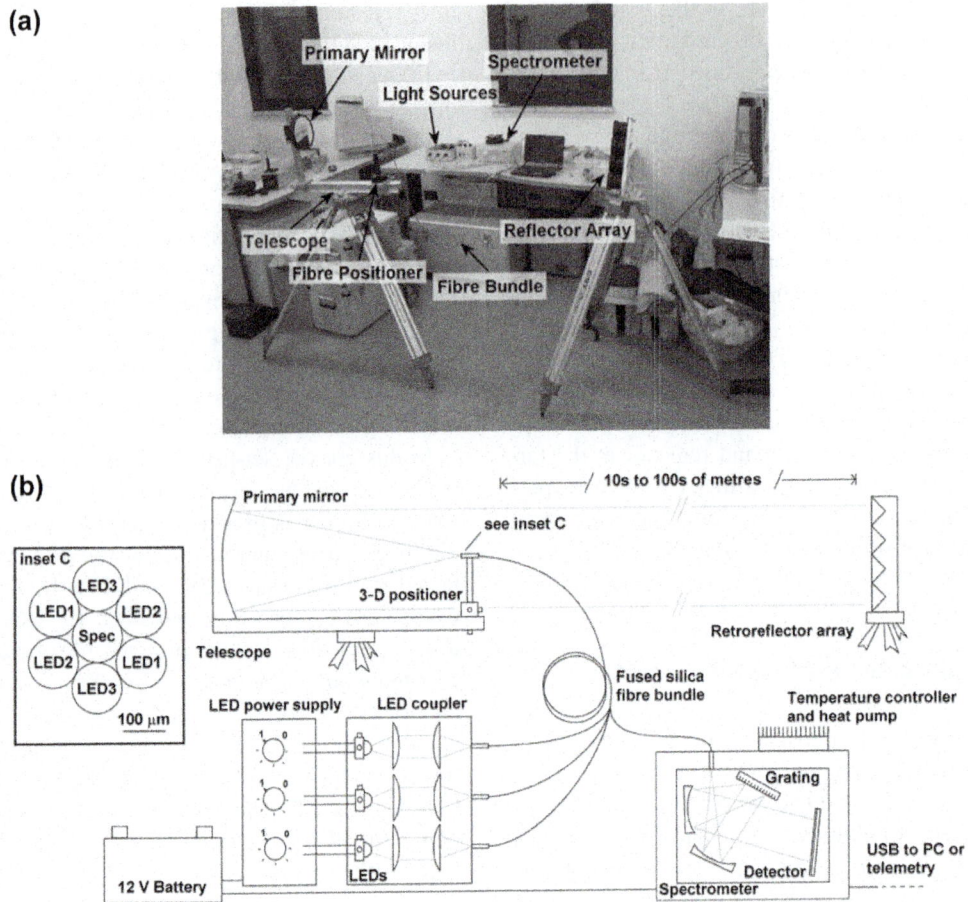

Figure 1. Photograph **(a)** and schematic **(b)** of the portable active long-path DOAS instrument. Light from three UV LEDs is coupled to a fibre bundle, collimated to a parallel beam and sent to an array of reflectors. Upon returning, it is coupled to a moderate-resolution UV spectrometer for analysis of absorption features.

the light path length (if the plume fills the optical path), the measured optical depth will be proportional to L, thus yielding a measurement SNR that is, in first order, independent of the light path length. However, if opaque volcanic plumes are measured, scattering of light out of the optical path by aerosols and water droplets will further decrease the SNR, particularly for longer light paths (see Sect. 4.2).

In addition to allowing for a decoupling of the light source from the telescope, the fibre design allows for fine adjustments of the beam direction without the need to adjust the pointing direction of the telescope (Merten et al., 2011). After aiming the telescope in the vicinity of the reflectors, slight adjustments to the fibre position can be used to fine-tune the pointing direction of the instrument. This technique, however, requires that the $f/\#$ of the telescope be smaller than that of the spectrometer. In other words, the area from which light is collected by the spectrometer must be smaller than the primary mirror. If this is not the case, light that has not passed along the optical path can enter the system from adjacent to the telescope primary mirror when the fibre bundle is moved out of the focal point. We avoid this

problem by using an $f/2.7$ telescope and an $f/4$ spectrometer. The $f/4$ spectrometer collects light from a circular 10 cm diameter subsection of the primary mirror, so the fibre bundle can be moved up to about 2.5 cm to each side of the focal point without collecting light from beside the mirror. This allows adjustment of the telescope pointing direction by ± 3.6 degrees in the left/right and up/down directions without moving the telescope itself.

The second technological advance which allowed us to design a portable LP-DOAS instrument is the availability of an efficient UV light source. Besides the weight of the instrument itself, the application of LP-DOAS systems in remote locations has also been severely limited by the high power consumption of conventional UV light sources. Typical light sources such as xenon arc and deuterium lamps only emit a fraction of their consumed energy as optical output – by far the majority of the energy is lost to wavelengths that are not relevant for the measurement, especially in the thermal IR. The advent of UV-LEDs and their applicability to LP-DOAS instruments has now revolutionised the achievable power efficiency of such systems (Kern et al., 2006; Sihler et al., 2009). By utilising three UV LEDs with maximum emission wavelengths at 285, 310 and 315 nm instead of a gas discharge lamp, the power consumption of the light source was reduced from what was typically between 50 and 500 W to less than 1 W without significantly affecting the optical output power at the required wavelengths. We used Sensor Electronic Technology UVTOP285, UVTOP310 and UVTOP315 flat window LEDs with maximum emission wavelengths of 285, 310 and 315 nm, respectively. Each consuming approximately 180 mW of electrical power, these LEDs are rated as emitting between 600 and 800 µW of optical power at wavelengths within approximately ±15 nm (12 nm FWHM) around their peak emission (Sensor Electronic Technology, 2014).

As shown in Fig. 2, the UV LED wavelengths were chosen to overlap with the differential absorption features of SO_2, chlorine monoxide (ClO) and carbon disulfide (CS_2) (Bogumil et al., 2003; Vandaele et al., 2009). However, other LEDs can be chosen to target other species (e.g. bromine monoxide (BrO) at ~ 340 nm, chlorine dioxide (OClO) at ~ 350 nm (Kern et al., 2008) or nitrogen dioxide (NO_2) at ~ 350–450 nm, Kern et al., 2006; Chan et al., 2012), or three LEDs of the same wavelength can be coupled to the system to enhance the SNR of the measurement in one specific wavelength region. The UV LEDs are each mounted on a 2-D positioner and coupled to the fibre bundle using two fused-silica plano-convex lenses (Fig. 1). The lens positioning along the optical axis is individually adjusted so as to provide maximum coupling efficiency for a given LED wavelength, thus avoiding chromatic aberration.

After the radiation has passed along the open optical path, the spectrum is analysed using an Ocean Optics QE65000 spectrometer with a cooled, back-illuminated CCD detector. The installed grating provided a wavelength range of 260 to 340 nm at a resolution of 0.5 nm. The spectrometer was placed in an insulated enclosure and the temperature of the optical bench was stabilised using a pulse-width-modulated thermoelectric cooling/heating unit to avoid fluctuations in calibration or other optical properties of the system. Excluding the notebook computer that was used for data collection in these test measurements, the power consumption of the entire system was below 25 W during continuous operation. This represents a considerable reduction in power use when compared to any previously reported LP-DOAS set-up, and allowed us to make battery-operated measurements.

The majority of the consumed power is actually used for temperature stabilisation of the spectrometer optical bench. The achieved temperature stability of better than ±0.1 °C is required when measuring weak absorption features of trace gases in the optical path of smaller than ~ 0.01 optical depth (see next section). If only gas species such as SO_2 with relatively strong absorption bands are targeted, this stabilisation is not required, and the power consumption of the system can be reduced to as low as 5 W (depending on the CCD chip

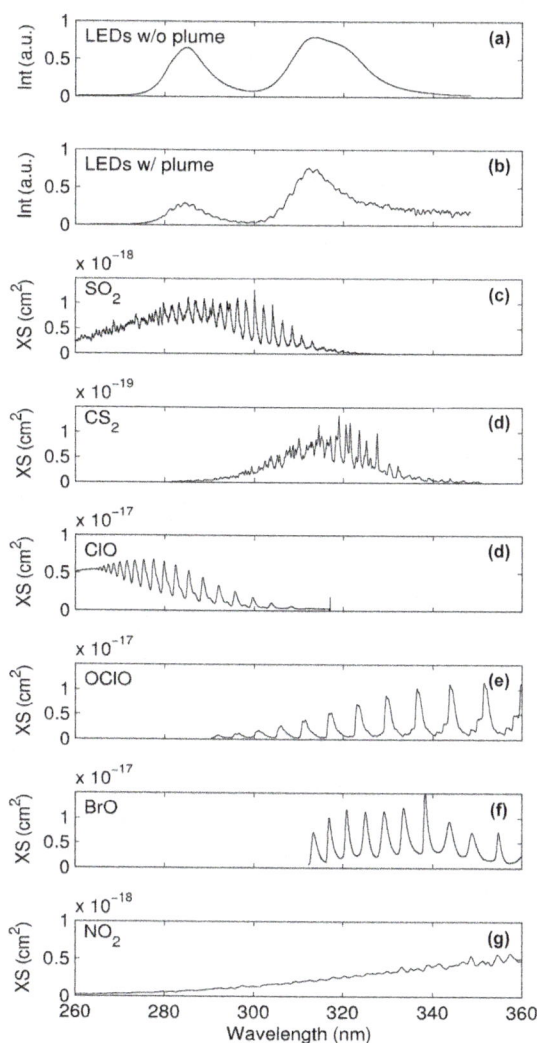

Figure 2. (a) Relative intensity of the three UV LEDs recorded when no plume was in the instrument's path. (b) Light intensity recorded when a plume drifted into the open path of the instrument. (c–f) Absorption cross sections (Wahner et al., 1988; Bogumil et al., 2003; Vandaele et al., 2009) of various trace gases that can be measured with the portable LP-DOAS system. Note that the LED emission spectrum overlaps the cross sections of SO_2, CS_2 and ClO. The LEDs selected for these measurements were chosen specifically to provide sensitivity to these species.

temperature). During our test measurements described in the next section, we used a laptop computer for data acquisition. This added an additional ~ 15 W to the total power consumption, and we chose to use a 45 Ah, 12 V lead acid battery to run the measurements for the entire day. However, a set-up for making continuous measurements could either use a low-power integrated PC or make use of a USB device server (~ 2 W) instead of the laptop, and include a radio telemetry link (~ 5 W) to the observatory where the spectral acquisition and processing would occur. Without temperature stabilisation (e.g. if only measuring SO_2), the system could be run continuously in this set-up with a total power consumption of approximately 12 W. Use of a solar power system for supplying the instrument power is therefore possible.

3   Configuration of measurements at Vulcano

On 25 February 2010, the portable LP-DOAS was tested at the island of Vulcano (Aeolian arc, Italy). Measurements were performed using two different light paths. During the first measurement period from approximately 12:00 to 12:45 local time, the array of retroreflectors was positioned behind a low-temperature fumarole about 100 m from the telescope (see Fig. 3a). During this time, the gases being emitted by this fumarole were measured. Little or no disturbance from other fumaroles is expected in this data set, as favourable winds were blowing the gases from other fumaroles in the area away from the light path. Depending on the wind, the light path length through the fumarolic gases was between 0 and 15 m one way, so a total light path of between 0 and 30 m resulted.

During the second measurement period, the reflectors were moved to a position farther down in the crater (approximately 200 m from the telescope) to enable the measurement of gases being emitted from the higher-temperature fumaroles located on the edge of the crater floor (Fig. 3b). The light path length inside the fumarolic gases varied considerably due to variations in wind direction as well as variations in the degassing strength of the fumaroles themselves.

4   Evaluation of spectral data and measurement results

4.1   Spectral retrieval of SO_2

Due to its importance as a tracer for other volcanic gases (see Sect. 1), SO_2 is the primary target species for the LP-DOAS instrument. SO_2 has several distinct absorption features in the spectral region between 260 and 320 nm, the strongest of which lie around 300 nm (Vandaele et al., 2009, Fig. 2). However, we chose to measure the differential absorption of SO_2 between 280 and 290 nm, because the relatively high column densities generally encountered in the plumes of active volcanoes make absolute sensitivity a secondary concern. More importantly, the ability to measure at shorter wavelengths eliminates any interference from scattered solar radiation entering the spectrometer, as the ozone layer effectively absorbs all incident solar radiation at wavelengths shorter than 300 nm (Hartley absorption bands). A UVTOP LED with a 285 nm peak transmittance wavelength was selected for this purpose.

The SO_2 column density was derived from each measured spectrum using a DOAS retrieval. First, each spectrum was corrected for the electronic offset of the spectrometer. This offset is added to each measured spectrum by the spectrometer electronics prior to digitisation in order to avoid nega-

Figure 3. **(a)** Location of the retroreflector array during the first experiment on the island of Vulcano on 25 February 2010. The telescope was situated approximately 100 m off to the left of this photograph, but only the last 15 m of the one-way optical path were filled intermittently with gas from the low-temperature fumaroles shown here, depending on the wind direction. **(b)** Configuration of the second experiment. Here, the retroreflector array (indicated by an arrow) was located approximately 200 m from the telescope and farther down in the crater of the volcano. Gases originating from the higher-temperature fumaroles at the base of the crater would intermittently blow into the optical path.

tive intensity values. It is best measured by co-adding a large number of exposures at very short exposure time and with no incident light. Since the electronic offset is added to each exposure, the co-added spectrum is then divided by the number of exposures, multiplied by the number of exposures in the measurement spectrum, and then subtracted from it.

In the next step, each measurement spectrum is corrected for the detector dark current. This contribution stems from thermal excitation of electrons in the CCD detector, and is proportional to the exposure time. It is therefore best characterised by taking a long exposure of the CCD with no incident radiation. After subtracting the offset, the dark current spectrum is divided by the exposure time, multiplied by the exposure time of the measurement spectrum, and then subtracted from it. Note that both the offset and dark current are typically dependent on the temperature of the spectrometer. However, since the spectrometer and the CCD detector were both kept at a stable temperature during our measurements, characterisation of the offset and dark current could be performed before and after the measurements were made.

In order to evaluate only the differential absorption features in a measured spectrum, the next step is to derive the differential optical density τ'. τ' is defined as the negative logarithm of the ratio of measured intensity I to transmitted intensity I_0 which, according to the Beer–Lambert–Bouguer law of absorption, is proportional to the column density S of the absorber in the light path.

$$\tau' = -\ln\left(\frac{I}{I_0}\right) = \sigma \cdot S \tag{2}$$

Here, σ is the absorption cross section of the absorbing gas. The initial intensity I_0 is the spectrum of the light sources before the radiation has passed through the atmosphere. Since I_0 can depend on the temperature of the LEDs (Kern et al., 2006; Sihler et al., 2009), it is best measured in close temporal proximity to the measurement spectra. In LP-DOAS instruments, characterisation of the light sources can be performed using a "shortcut" system which redirects light emitted from the light sources directly into the spectrometer without passing along the optical path. Since we use a fibre assembly to couple radiation into and out of the telescope, a shortcut system can be implemented by simply placing a reflective diffusor plate in front of the fibre bundle in the focal point of the mirror (Merten et al., 2011). Radiation emitted from the fibre ring is scattered on this plate and, in part, enters the central collection fibre for analysis. In between measurements, we manually inserted a sand-blasted aluminium plate a few cm from the fibre bundle and measured I_0. However, owing to the variable wind directions encountered within the La Fossa crater, we also intermittently obtained measurement spectra with no volcanic gas in the light path of the instrument. These were later found to be the best measure of I_0, because exactly the same optical path through the instrumentation is used as during the measurements, only in the absence of volcanic gas. In our evaluations, we co-added 60 such spectra and used these as an LED reference (shown in Fig. 4).

In a differential optical absorption spectroscopy retrieval (Platt and Stutz, 2008), the column density S of one or multiple absorbers i (in this initial case, just SO_2) in the instrument's light path is derived by fitting the laboratory absorption cross sections σ_i to the measured optical density, along with a polynomial P_λ (we use a fifth order) to account for broadband spectral features. We varied this approach slightly, and instead fit the SO_2 absorption cross section σ_{SO_2} and the

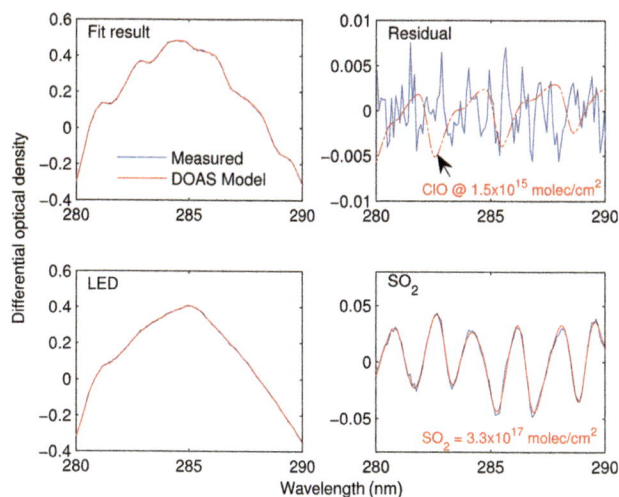

Figure 4. Evaluation of SO_2 absorption in the measured spectra. The absorption cross section of SO_2 was fitted to the differential optical depth. In this example evaluation, a column density of 1.5×10^{17} molec cm^{-2} was retrieved. The detection limit of ClO was derived by comparing the fit residual with the absorption features of a hypothetical column density of 1.5×10^{15} ClO molec cm^{-2}. See text for details.

logarithm of the shortcut spectrum I_0 to the logarithm of the measurement spectrum I, as shown in Eq. (3).

$$\ln I = \ln I_0 - \sum_i \sigma_i S_i + P_\lambda \tag{3}$$

In this manner, a slight shift in wavelength calibration of the spectrometer between shortcut and measurement, should this occur, can be accounted for in the fit. Since the spectral resolution of the instrument does not match that of the laboratory reference (we use the SO_2 cross section measured by Vandaele et al., 2009), the reference cross section is convolved with the instrument line shape before the fit is performed. The line shape or so-called "slit function" was characterised by recording the spectrum of a mercury vapour lamp and measuring the shape of a single atomic emission line at the spectrometer's resolution. The convolved cross section was then used in the fit.

An example of the SO_2 fit is shown in Fig. 4. In this example, an SO_2 column density of 1.5×10^{17} molec cm^{-2} of SO_2 was retrieved. The residual, shown at the top right of the figure and representing the difference between the measured spectrum and the fit result, is about 1 % peak to peak. When compared to the SO_2 absorption cross section, this noise level corresponds to an SO_2 detection limit of approximately 2×10^{16} molec cm^{-2} or 8 parts per million × metres (ppmm). In other words, concentrations below 1 ppm can be detected along a 10 m total path (5 m one way).

The results of the SO_2 evaluation for the spectra recorded during the first measurement period at the low-temperature fumarole are shown in Fig. 5a. The time series is characterised by a strong fluctuation of SO_2 column densities in time which is mainly caused by wind blowing the fumarolic gas into or out of the light path of the instrument. At times, the retrieved SO_2 column density reached up to 2.8×10^{17} molec cm^{-2} (~ 110 ppmm). Assuming a total optical path length of up to 30 m in the fumarolic gas (see Fig. 3a), this would correspond to an average SO_2 mixing ratio of 3.7 ppm.

The second optical path that was chosen was more favourable in that it did not pass through as dense a gas plume as in the first measurement period (Fig. 3b). This led to a lower opacity and slower fluctuations in SO_2 column density (Fig. 5b). Up to 1.2×10^{18} molec cm^{-2} (~ 480 ppmm) of SO_2 were measured along this 200 m (one-way) path. We estimate that the volcanic gas filled about 1/4 of the optical path for this measurement. Under this assumption, the measured column density corresponds to an average mixing ratio of about 4.8 ppm. When further taking into account that this plume was considerably more diluted than the plume measured at the first position, this second plume appears to have a significantly higher SO_2 fraction. This finding is in agreement with the results of Aiuppa et al. (2005), who also found a significantly higher SO_2 fraction in the plume originating from the inner-crater fumaroles. The combination of slower column density fluctuations and improved integration time (typically shorter than 300 ms) also led to smaller errors in the retrieval for this time period (see next section).

4.2 Rapid variation of gas and aerosol concentrations in the optical path

The errors in the results obtained for the first measurement site (depicted in Fig. 5a) are significant, particularly for individual measurements of high column densities preceded and followed by lower values. The relatively large error in these measurements therefore appears to be associated with a change in gas and/or aerosol concentration in the instrument's light path during a measurement. Here we discuss the instrument's sensitivity to such fluctuations.

Two effects can be separated. First, the concentration of aerosols and water droplets in the light path can change over time. A portion of the light emitted by the LP-DOAS instrument is scattered onto particles and droplets while passing along the open optical path. In the narrow beam approximation (Platt and Stutz, 2008), light that has been scattered once will not re-enter the light path, and is therefore removed from the system. This leads to a reduction in the light intensity received by the telescope and inherently decreases the SNR of the measurement according to photon statistics.

However, the DOAS approach is not susceptible to a systematic error when the intensity changes during an exposure. Since a polynomial P_λ is included in the fit to account for

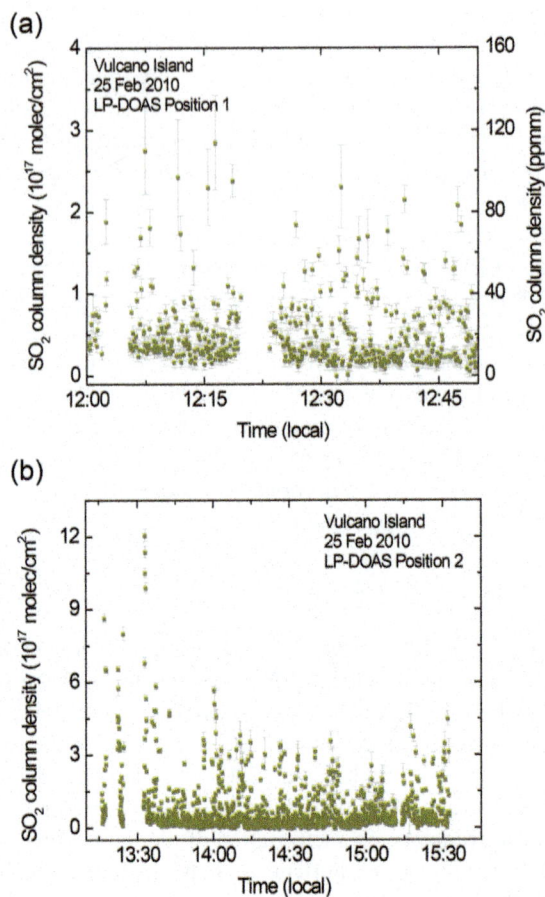

(a)

(b)

Figure 5. Time series of SO_2 column densities measured in the two experiments at the island of Vulcano on 25 February 2010. **(a)** depicts the results obtained during the first measurement period at a low-temperature fumarole on the crater rim, and **(b)** depicts the results obtained while measuring gas stemming from the fumaroles along the base of the crater during the second measurement period. Rapid fluctuations in column density arise from various amounts of gas being blown into the optical path of the LP-DOAS system.

broadband spectral features such as those associated with particle scattering (see Eq. 3), the column density S is retrieved solely from the differential depth of the narrowband absorption lines of the respective trace gas. If the broadband intensity of radiation returning from the atmospheric path changes without a change in the relative depth of the absorption lines, an exposure integrated over this change will still exhibit the differential optical depth associated with the encountered gas column density. This can easily be shown for two spectra with different on-band intensities I_1 and I_2, and off-band intensities $I_{0,1}$ and $I_{0,2}$. If we require that

$$\frac{I_1}{I_{0,1}} = \frac{I_2}{I_{0,2}}, \tag{4}$$

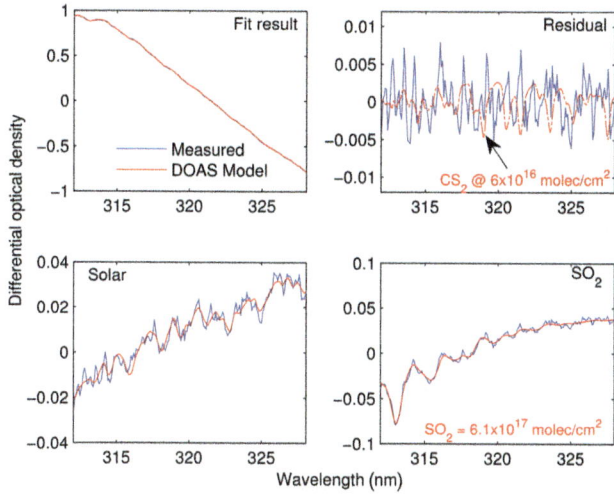

Figure 6. Evaluation of the measured spectra for CS_2 absorption. Here, a solar spectrum was included in the fit to remove features associated with scattered solar radiation entering the instrument's light path (Ring spectrum not shown for clarity). CS_2 was not detected during the experiments, but an upper limit could be determined from the fit residual. Here, the residual is compared to the absorption features associated with a hypothetical column density of 6×10^{16} CS_2 molec cm^{-2}.

in other words, that the amount of gas in the light path be constant (see Eq. 2), then

$$\frac{I}{I_0} = \frac{I_1 + I_2}{I_{0,1} + I_{0,2}} = \frac{(I_1 + I_2) I_{0,2}}{(I_{0,1} + I_{0,2}) I_{0,2}} = \frac{I_1 \cdot I_{0,2} + I_2 \cdot I_{0,2}}{I_{0,1} \cdot I_{0,2} + I_{0,2}^2} \quad (5)$$

$$\overset{\text{Eq. (4)}}{=} \frac{I_2 \cdot I_{0,1} + I_2 \cdot I_{0,2}}{I_{0,1} \cdot I_{0,2} + I_{0,2}^2} = \frac{I_2 (I_{0,1} + I_{0,2})}{I_{0,2} (I_{0,1} + I_{0,2})} = \frac{I_2}{I_{0,2}} = \frac{I_1}{I_{0,1}}.$$

Therefore, the optical depth τ will remain an accurate measure of the column density S of gas in the instrument's well-defined light path (Eq. 2) if only the aerosol concentration changes. This stands in contrast to the effect that aerosols have on passive DOAS measurements which use scattered solar radiation to measure volcanic gases. For these, any change in the aerosol concentration along the viewing direction will directly influence the atmospheric light path along which the measurement is being conducted, thus also potentially changing the measured optical depth τ and greatly impacting the accuracy of the measurement (Kern et al., 2009).

Particles and water droplets in the LP–DOAS optical path can also scatter solar radiation into the receiving optics of the instrument. This leads to a "background" solar spectrum that appears superimposed on the spectrum of the active light sources (shown in Fig. 2a). This contribution to the measurement needs to be corrected during the spectral retrieval, as is discussed in the next section. Correction is not necessary when operating at wavelengths shorter than

300 nm (as we do here), because solar radiation is negligible in this region.

The second effect that can occur during a measurement is a rapid change in the column density S of the measured trace gas. This effect is significantly more problematic than a change in the aerosol concentration. When measuring volcanic plumes, the two effects will often occur simultaneously as SO_2 and aerosol-rich volcanic gas is blown into and out of the instrument light path. If we assume a constant off-band intensity I_0, then a change in trace gas column density will lead to a change in the on-band intensity of a measurement from I_1 to I_2. If these two successive measurements are co-added (or integrated during a single exposure of the CCD), the optical depth τ can be calculated from the sum of the individual intensities:

$$\tau = \ln \frac{I_1 + I_2}{2 I_0} = \ln \frac{I_1 \left(1 + \frac{I_2}{I_1}\right)}{2 I_0} = \ln \frac{I_1}{I_0} + \ln \frac{1 + \frac{I_2}{I_1}}{2} \quad (6)$$

$$= \tau_1 + \ln \frac{1 + \frac{I_2}{I_1}}{2} \neq \frac{1}{2} (\tau_1 + \tau_2)$$

Equation 6 shows that, owing to the nonlinearity of the Beer–Lambert–Bouguer law (Eq. 2), τ is not equal to the average of the two optical depths τ_1 and τ_2 associated with the gas column densities at time 1 and time 2. Instead, there is a systematic shift towards lower optical depths. This effect will also influence the shape of the absorption lines in the measured spectra, and the fit of the absorption cross sections according to Eq. (3) will result in a systematic residual structure. Accurate measurements therefore require that the exposure time be short compared to the timescales over which the column density of the absorbers changes in the light path.

During our test measurements at the first site, individual spectra were recorded at exposure times varying between approximately 150 and 600 ms, depending on the opacity of the plume in the light path at any given time. Because the retroreflector was positioned in direct proximity to the fumaroles, high opacities were encountered fairly often, and exposure times of ~ 600 ms were not uncommon. As shown above, changes in plume opacity alone will not lead to systematic errors, but in this case, such changes were likely accompanied by changes in the gas column in the instrument's light path, as the fumarole's plume drifted in and out. This often led to a misfit between measured optical depth and the absorption cross section and, thus, resulted in the relatively large errors (Fig. 5a). During the measurements conducted at the second site, generally lower plume opacities led to shorter exposure times. Thus, the assumption of a constant SO_2 column density in the light path is valid, and the errors are significantly reduced (Fig. 5b). To avoid systematic errors in the future, care must be taken to ensure that the exposure time is always shorter than the timescale of variations in the trace gas column density, even if this leads to undersaturated exposures and, associated with this, slightly decreased SNRs for the measurement.

4.3   ClO and CS$_2$ detection limits

Besides SO$_2$, CS$_2$ and ClO both have characteristic absorption features in the 260–360 nm wavelength region (Fig. 2). The absorption cross section of ClO was originally included in the SO$_2$ fit described above, but no differential features associated with ClO absorption were detected during the evaluation, so it was later omitted from the SO$_2$ retrieval. In an attempt to improve the SNR of the measurement and thereby improve the ClO detection limit, the evaluated spectra were sorted by their SO$_2$ column density, binned in groups of 10 to 20 such that all spectra in a single group had comparable SO$_2$ columns, and co-added to improve the statistics of the measurement.

These spectra were again evaluated for ClO absorption, but none was found. Due to their high solubility in water, halogen gases are very efficiently scrubbed when passing through large-scale hydrothermal systems (Symonds et al., 2001). Considering the significant hydrothermal activity at the island of Vulcano, it is therefore not surprising that halogen emissions appear negligible. However, an upper limit could be derived from the spectral retrieval. For this, the magnitude of the fit residual was analysed.

The fit residual represents the difference between the measured optical depth and the optical depth derived from the DOAS fit. Absorption structures of gases not accounted for by the fit would appear in the residual. For a given measurement spectrum, the detection limit of a gas species not included in the DOAS model can therefore be derived by comparing the fit residual with the absorption cross section of that species multiplied by a hypothetical column density. This is demonstrated in Fig. 4. Here, the magnitude of the fit residual is compared to the absorption features associated with a hypothetical column density of 1.5×10^{15} ClO molec cm^{-2}. By making the conservative assumption that any ClO absorption features larger than these could not be masked by the noise in the fit residual, a column of 1.5×10^{15} ClO molec cm^{-2} represents the upper limit for this particular spectrum. This corresponds to a path-averaged mixing ratio of about 1.5 ppb (note that the volcanic gas only fills part of the 400 m total light path) and an upper limit for the ClO/SO$_2$ ratio of about 5×10^{-3} for this example. When analysing the entire collected data set, and particularly the co-added spectra containing larger amounts of SO$_2$ than in the given example, an upper limit for the ClO/SO$_2$ molar ratio of about 2×10^{-3} was obtained for the emissions from the fumaroles on the edge of the crater floor. Considering an average SO$_2$ emission rate from the fumarolic area of about 12 t d^{-1} (as reported by Vita et al. (2012) for the 2008–2010 period), this corresponds to an approximate upper limit for the ClO flux of 20 kg d^{-1}.

Finally, the recorded spectra were also evaluated with regards to the potential absorption of CS$_2$. CS$_2$ was first measured by DOAS in the urban atmosphere above Shanghai by Yu et al. (2004). Though it has never been detected by DOAS in a volcanic environment, an attempt was made here, because reduced sulfur species (particularly H$_2$S, Aiuppa et al., 2006) are known to be present in abundance in the fumaroles at La Fossa crater on the island of Vulcano, where these measurements were made. The differential absorption features of CS$_2$ in the 310–330 nm wavelength range (Fig. 2) do make it potentially accessible to our LP-DOAS instrument. A separate CS$_2$ evaluation was performed in which the wavelength region between 312 and 328 nm was analysed. At shorter wavelengths, strong SO$_2$ absorption interfered with the measurement. At longer wavelengths, the intensity of the available LED was insufficient. In addition to the cross section of CS$_2$, the cross section of SO$_2$ was included in the fit, as well as a reference spectrum compiled from 60 spectra without SO$_2$ absorption. A fifth-order polynomial was again used to account for any broadband structures in the retrieval. Because this retrieval was performed at wavelengths longer than 300 nm, the spectrum of solar radiation scattered into the instrument's light path by condensed water droplets and volcanogenic aerosols was found superimposed onto the LED spectra. This is clearly demonstrated by the non-zero baseline at 350 nm in the example spectrum shown in Fig. 2b. To correct for this effect, a Fraunhofer solar reference spectrum (measured with the same geometry but with the LED switched off) and a Ring spectrum (used to correct for inelastic scattering in the atmosphere, Grainger and Ring, 1962) were also included in the fit.

As was the case for ClO, CS$_2$ concentrations were below the instrument's detection limit throughout the measurement period. However, the same approach was used to derive an upper limit for the CS$_2$/SO$_2$ molecular ratio in the fumarolic emissions. CS$_2$ was omitted from the DOAS model and, after the fit accounted for SO$_2$ absorption and interference from solar radiation scattered into the optical path, the fit residual was compared to the absorption features of hypothetical columns of CS$_2$. Figure 6 shows an example in which an upper limit of 6×10^{16} CS$_2$ molec cm^{-2} (~ 24 ppmm) was obtained. This corresponds to an upper limit for the path-integrated average mixing ratio of about 60 ppb, a CS$_2$/SO$_2$ molecular ratio of less than 0.1 and an average CS$_2$ flux of less than about 1.4 t d^{-1} (again assuming an average SO$_2$ emission rate of 12 t d^{-1} as reported by Vita et al., 2012).

5   Conclusions and outlook

Recent technological advances in fibre-coupled telescopes and UV-LED technology have made design and construction of a mobile active LP-DOAS for use in remote environments possible. Here, we built an instrument specifically for measuring volcanic gas emissions. The tests conducted at La Fossa crater, Vulcano, showed that in the current configuration, the instrument is highly sensitive to SO$_2$, with open-path measurements of SO$_2$ column densities as low as about

2×10^{16} molec cm^{-2} (~ 8 ppmm) possible. For the 400 m total light path length used here, this corresponds to a detection limit of about 20 ppb SO$_2$ averaged along the light path, but even higher sensitivities could potentially be achieved by using longer light paths.

Collocation of the LP-DOAS with laser absorption-based open-path CO$_2$ instruments would enable the quantification of CO$_2$ / SO$_2$ ratios in the bulk plume, not just at a single location within the plume, as stationary in situ techniques currently allow. In fact, both instruments can use the same retroreflector as long as a fused-silica prism assembly is chosen. At the same time, the low power consumption and lack of moving parts make the instrument significantly more portable and durable than FTIR spectrometers, which previously represented the only remote sensing approach to obtaining this geochemical parameter. Despite the difficulties involved with making measurements in the past, monitoring the CO$_2$ / SO$_2$ ratio at active volcanoes throughout the world has already led to significant insights into volcanic processes (Duffell et al., 2003; Allard et al., 2005; Aiuppa et al., 2007; Burton et al., 2007; Shinohara et al., 2008, 2011; Ohba et al., 2011; Werner et al., 2012, 2013) and allows the estimation of CO$_2$ input into the atmosphere from global volcanism (Burton et al., 2013). In the future, mobile LP-DOAS instruments could help quantify this parameter at hitherto inaccessible volcanic systems. The fact that the LP-DOAS instruments need not be located within the volcanic plume itself also makes them potentially attractive for long-term continuous monitoring, as the highly corrosive gases in volcanic plumes make permanent installation of equipment in their midst extremely difficult. The retroreflectors are made of plastic and fused silica and can be located in corrosive plumes without fear of damaging them (although periodic cleaning is necessary). This allows the light path length to be adapted to the expected plume opacity, with short light paths chosen for very dense plumes. The technical difficulties and safety concerns associated with approaching an active volcanic vent probably represent the main limitations to instrument applicability at a given location.

Aside from SO$_2$ measurements, there is also significant potential for the quantitative detection of other volatile species in volcanic plumes using the described LP-DOAS instrumentation. Though neither ClO nor CS$_2$ column densities were found above the instrument's detection limits in the fumarolic gases at La Fossa crater, the instrument sensitivity would be sufficient to detect ClO in volcanic environments in which it is present in higher abundances. For example, Lee et al. (2005) report a ClO / SO$_2$ ratio of 5×10^{-2} in the plume of Sakurajima volcano (Japan), a value that is more than 1 order of magnitude above the detection limit of our LP-DOAS. On the other hand, little is known about the actual magnitude of CS$_2$ emissions from volcanic systems, so it is unclear whether the achieved detection limit will make the LP-DOAS a valuable tool for assessing this trace species. Globally, CS$_2$ is only a very minor component (Halmer et al., 2002) when compared to other sulfur-bearing species (mainly SO$_2$ and H$_2$S), but trace amounts have been reported both in explosive eruption plumes (Rasmussen et al., 1982) and in fumarolic gases (Stoiber et al., 1971), so its exact role remains uncertain, and abundances may vary widely from system to system.

The described LP-DOAS system can also be used to measure other volcanic volatiles (Fig. 2), most notably perhaps the halogen oxide compounds OClO and BrO. The design of the fibre-coupling system allows the addition of LEDs at other wavelengths if targeting species with absorption cross sections outside the current range. Alternatively, three LEDs emitting at the same wavelength can be used to improve the light throughput and SNR of the system at that wavelength. Another envisioned modification that will likely further improve the sensitivity of the instrument is the implementation of a pulsed LED power supply. The optical output power of LEDs is limited by the heat generated by the dissipated electrical input power. Therefore, LEDs can be pulsed at significantly higher input currents than they can sustain in continuous operation. Such a pulsed operation does not significantly improve the time-integrated optical output power of the LEDs but, if synchronised with the spectrometer read-out electronics, the system could measure the background spectrum during periods when the LEDs are off, thus providing a more or less contemporaneous measurement of the background solar radiation scattered into the light path. Currently, accurate correction of the scattered solar contribution appears to be one of the main limitations of the system, particularly when applying it to opaque, optically thick plumes, and a pulsed acquisition stands to improve this correction significantly.

The flexibility of the system allows application to a wide range of volcanic system types, from active high-temperature vents to hydrothermal fumarole fields. The active light sources allow measurements to be made both during the day and at night, which can provide insights into the photochemistry taking place in volcanic plumes, as for example shown by Kern et al. (2008). The instrument's light weight, rugged design and low power consumption allow active DOAS measurements to be made at remote locations that were previously inaccessible. It therefore has the potential to make significant improvements to our ability to monitor and understand geochemical processes in volcanic systems around the world.

Acknowledgements. The authors would like to acknowledge Tom Pering, Tamar Elias and two anonymous reviewers for their thoughtful reviews of the manuscript. Any use of trade, firm, or product names is for descriptive purposes only and does not imply endorsement by the US Government.

Edited by: A. Schütze
Reviewed by: three anonymous referees

References

Aiuppa, A., Federicao, C., Giudice, G., and Gurrieri, S.: Chemical mapping of a fumarolic field: La Fossa Crater, Vulcano Island (Aeolian Islands, Italy), Geophys. Res. Lett., 32, L13309, doi:10.1029/2005GL023207, 2005.

Aiuppa, A., Federico, C., Giudice, G., Gurrieri, S., and Valenza, M.: Hydrothermal buffering of the SO_2/H_2S ratio in volcanic gases: Evidence from La Fossa Crater fumarolic field, Vulcano Island, Geophys. Res. Lett., 33, L21315, doi:10.1029/2006GL027730, 2006.

Aiuppa, A., Moretti, R., Federico, C., Giudice, G., Gurrieri, S., Liuzzo, M., Papale, P., Shinohara, H., and Valenza, M.: Forecasting Etna eruptions by real-time observation of volcanic gas composition, Geology, 35, 1115, doi:10.1130/G24149A.1, 2007.

Allard, P., Burton, M., and Muré, F.: Spectroscopic evidence for a lava fountain driven by previously accumulated magmatic gas, Nature, 433, 407–410, doi:10.1038/nature03246, 2005.

Bogumil, K., Orphal, J., Homann, T., Voigt, S., Spietz, P., Fleischmann, O. C., Vogel, A., Hartmann, M., Bovensmann, H., Frerick, J., and Burrows, J. P.: Measurementsof molecular absorption spectra with the SCIAMACHY pre-flight model: Instrument characterization and reference data for atmospheric remote sensing in the 230—2380 nm region, J. Photoch. Photobio. A, 157, 167–184, 2003.

Burton, M. R., Oppenheimer, C. M., Horrocks, L. A., and Francis, P. W.: Remote sensing of CO_2 and H_2O emission rates from Masaya volcano, Nicaragua, Geology, 28, 915–918, 2000.

Burton, M. R., Sawyer, G. M., and Granieri, D.: Deep Carbon Emissions from Volcanoes, Rev. Mineral. Geochem., 75, 323–354, doi:10.2138/rmg.2013.75.11, 2013.

Burton, M., Allard, P., Muré, F., and La Spina, A.: Magmatic gas composition reveals the source depth of slug-driven strombolian explosive activity., Science, 80, 227–230, doi:10.1126/science.1141900, 2007.

Carapezza, M. L., Barberi, F., Ranaldi, M., Ricci, T., Tarchini, L., Barrancos, J., Fischer, C., Perez, N., Weber, K., Di Piazza, A., and Gattuso, A.: Diffuse CO_2 soil degassing and CO_2 and H_2S concentrations in air and related hazards at Vulcano Island (Aeolian arc, Italy), J. Volcanol. Geoth. Res., 207, 130–144, doi:10.1016/j.jvolgeores.2011.06.010, 2011.

Carn, S. A., Krotkov, N. A., Yang, K., and Krueger, A. J.: Measuring global volcanic degassing with the Ozone Monitoring Instrument (OMI), in: Remote Sensing of Volcanoes and Volcanic Processes: Integrating Observations and Modelling, edited by: Pyle, D. M., Mather, T. A., and Biggs, J., Geological Society, London, 2013.

Chan, K. L., Pöhler, D., Kuhlmann, G., Hartl, A., Platt, U., and Wenig, M. O.: NO_2 measurements in Hong Kong using LED based long path differential optical absorption spectroscopy, Atmos. Meas. Tech., 5, 901–912, doi:10.5194/amt-5-901-2012, 2012.

Duffell, H. J., Oppenheimer, C., Pyle, D. M., Galle, B., McGonigle, A. J., and Burton, M. R.: Changes in gas composition prior to a minor explosive eruption at Masaya volcano, Nicaragua, J. Volcanol. Geoth. Res., 126, 327–339, doi:10.1016/S0377-0273(03)00156-2, 2003.

Galle, B., Johansson, M., Rivera, C., Zhang, Y., Kihlman, M., Kern, C., Lehmann, T., Platt, U., Arellano, S., and Hidalgo, S.: Network for Observation of Volcanic and Atmospheric Change (NOVAC) – A global network for volcanic gas monitoring: Network layout and instrument description, J. Geophys. Res., 115, D05304, doi:10.1029/2009JD011823, 2010.

Gerlach, T. M., Delgado, H., Mcgee, K. A., Doukas, M. P., Venegas, J. J., and Cardenas, L.: Application of the LI-COR CO_2 analyzer to volcanic plumes: A case study, volcan Popocatepetl, Mexico, 7 and 10 June, 1995, J. Geophys. Res., 102, 8005–8019, 1997.

Gerlach, T. M., Mcgee, K. A., Sutton, A. J., and Elias, T.: Rates of volcanic CO_2 degassing from airborne determinations of SO_2 emission rates and plume CO_2/SO_2: Test study at Pu'u 'O'o Cone, Kilauea volcano, Hawaii, Geophys. Res. Lett., 25, 2675–2678, 1998.

Giggenbach, W. F. and Goguel, R. L.: Methods for collection and analysis of geothermal and volcanic water and gas samples, Dep. Sci. Ind. Res. Chem. Div., Report 240, 1989.

Grainger, J. F. and Ring, J.: Anomalous Fraunhofer Line Profiles, Nature, 193, 762 pp., 1962.

Halmer, M. M., Schmincke, H.-U., and Graf, H.-F., The annual volcanic gas input into the atmosphere, in particular into the stratosphere: a global data set for the past 100 years, J. Volcanol. Geoth. Res., 115, 511–528, doi:10.1016/S0377-0273(01)00318-3, 2002.

Kern, C., Trick, S., Rippel, B., and Platt, U.: Applicability of light-emitting diodes as light sources for active differential optical absorption spectroscopy measurements, Appl. Optics, 45, 2077–2088, 2006.

Kern, C., Sihler, H., Vogel, L., Rivera, C., Herrera, M., and Platt, U.: Halogen oxide measurements at Masaya Volcano, Nicaragua using active long path differential optical absorption spectroscopy, B. Volcanol., 71, 659–670, doi:10.1007/s00445-008-0252-8, 2008.

Kern, C., Deutschmann, T., Vogel, L., Wöhrbach, M., Wagner, T., and Platt, U.: Radiative transfer corrections for accurate spectroscopic measurements of volcanic gas emissions, B. Volcanol., 72, 233–247, doi:10.1007/s00445-009-0313-7, 2009.

Lee, C., Kim, Y. J., Tanimoto, H., Bobrowski, N., Platt, U., Mori, T., Yamamoto, K., and Hong, C. S.: High ClO and ozone depletion observed in the plume of Sakurajima volcano, Japan, Geophys. Res. Lett., 32, 10–13, doi:10.1029/2005GL023785, 2005.

Merten, A., Tschritter, J., and Platt, U.: Design of differential optical absorption spectroscopy long-path telescopes based on fiber optics, Appl. Optics, 50, 738–54, 2011.

O'Dwyer, M.: Real-time measurement of volcanic H_2S and SO_2 concentrations by UV spectroscopy, Geophys. Res. Lett., 30, 12–15, doi:10.1029/2003GL017246, 2003.

Ohba, T., Daita, Y., Sawa, T., Taira, N., and Kakuage, Y.: Coseismic changes in the chemical composition of volcanic gases from the Owakudani geothermal area on Hakone volcano, Japan, B. Volcanol., 73, 457–469, doi:10.1007/s00445-010-0445-9, 2011.

Oppenheimer, C., Francis, P., Burton, M., Maciejewski, A. J. H., and Boardman, L.: Remote measurement of volcanic gases by Fourier transform infrared spectroscopy, Appl. Phys. B, 67, 505–515, 1998.

Oppenheimer, C., Scaillet, B., and Martin, R. S.: Sulfur Degassing From Volcanoes: Source Conditions, Surveillance, Plume Chemistry and Earth System Impacts, Rev. Mineral. Geochem., 73, 363–421, doi:10.2138/rmg.2011.73.13, 2011.

Pedone, M., Aiuppa, A., Giudice, G., Grassa, F., Cardellini, C., Chiodini, G., and Valenza, M.: Volcanic CO_2 flux measurement at Campi Flegrei by tunable diode laser absorption spectroscopy, B. Volcanol., 76, 812, doi:10.1007/s00445-014-0812-z, 2014.

Pering, T. D., Tamburello, G., McGonigle, A. J. S., Aiuppa, A., Cannata, A., Giudice, G., and Patanè, D.: High time resolution fluctuations in volcanic carbon dioxide degassing from Mount Etna, J. Volcanol. Geoth. Res., 270, 115–121, doi:10.1016/j.jvolgeores.2013.11.014, 2014.

Platt, U. and Stutz, J.: Differential Optical Absorption Spectroscopy, Springer, Berlin, Heidelberg, 2008.

Platt, U., Perner, D., and Patz, H. W.: Simultaneous Measurement of Atmospheric CH_2O, 0_3, and NO_2 by Differential Optical Absorption, J. Geophys. Res., 84, 6329–6335, 1979.

Rasmussen, R. A., Khalil, M. A., Dalluge, R. W., Penkett, S. A., and Jones, B.: Carbonyl sulfide and carbon disulfide from the eruptions of mount st. Helens., Science, 215, 665–667, doi:10.1126/science.215.4533.665, 1982.

Rawlins, W. T., Hensley, J. M., Sonnenfroh, D. M., Oakes, D. B., and Allen, M. G.: Quantum cascade laser sensor for SO_2 and SO_3 for application to combustor exhaust streams, Appl. Optics, 44, 6635–6643, 2005.

Richter, D., Erdelyi, M., Curl, R. F., Tittel, F. K., Oppenheimer, C., Duffell, H. J., and Burton, M.: Field measurements of volcanic gases using tunable diode laser based mid-infrared and Fourier transform infrared spectrometers, Opt. Laser Eng., 37, 171–186, doi:10.1016/S0143-8166(01)00094-X, 2002.

Roberts, T. J., Braban, C. F., Oppenheimer, C., Martin, R. S., Freshwater, R. A., Dawson, D. H., Griffiths, P. T., Cox, R. A., Saffell, J. R., and Jones, R. L.: Electrochemical sensing of volcanic gases, Chem. Geol., 332–333, 74–91, doi:10.1016/j.chemgeo.2012.08.027, 2012.

Roberts, T. J., Saffell, J. R., Oppenheimer, C., and Lurton, T.: Electrochemical sensors applied to pollution monitoring: Measurement error and gas ratio bias — A volcano plume case study, J. Volcanol. Geoth. Res., 281, 85–96, doi:10.1016/j.jvolgeores.2014.02.023, 2014.

Schiff, H. I., Mackay, G. I., and Bechara, J.: The use of tunable diode laser absorption spectroscopy for atmospheric measurements, in: Air Monitoring by Spectroscopic Techniques, edited: by M. W. Sigrist, 239–333, John Wiley and Sons, Inc., New York, Chister, Brisbane, Toronto, Singapore, 1994.

Sensor Electronic Technology, I., UVTOP Deep UV LED Technical Catalogue, 1–40, 2014.

Shinohara, H.: A new technique to estimate volcanic gas composition: plume measurements with a portable multi-sensor system, J. Volcanol. Geoth. Res., 143, 319–333, doi:10.1016/j.jvolgeores.2004.12.004, 2005.

Shinohara, H., Aiuppa, A., Giudice, G., Gurrieri, S., and Liuzzo, M.: Variation of H_2O/CO_2 and CO_2/SO_2 ratios of volcanic gases discharged by continuous degassing of Mount Etna volcano, Italy, J. Geophys. Res., 113, B09203, doi:10.1029/2007JB005185, 2008.

Shinohara, H., Matsushima, N., Kazahaya, K., and Ohwada, M.: Magma-hydrothermal system interaction inferred from volcanic gas measurements obtained during 2003–2008 at Meakandake volcano, Hokkaido, Japan, B. Volcanol., 73, 409–421, doi:10.1007/s00445-011-0463-2, 2011.

Sihler, H., Kern, C., Pöhler, D., and Platt, U.: Applying light-emitting diodes with narrowband emission features in differential spectroscopy., Opt. Lett., 34, 3716–3718, 2009.

Stoiber, R. E., Leggett, D. C., Jenkins, T. F., Murrmann, R. P., and Rose, W. I.: Organic Compounds in Volcanic Gas from Santiaguito Volcano, Guatemala, Geol. Soc. Am. Bull., 82, 2299–2302, 1971.

Stremme, W., Krueger, A., Harig, R., and Grutter, M.: Volcanic SO_2 and SiF4 visualization using 2-D thermal emission spectroscopy – Part 1: Slant-columns and their ratios, Atmos. Meas. Tech., 5, 275–288, doi:10.5194/amt-5-275-2012, 2012.

Symonds, R. B., Gerlach, T. M., and Reed, M. H.: Magmatic gas scrubbing: implications for volcano monitoring, J. Volcanol. Geoth. Res., 108, 303–341, doi:10.1016/S0377-0273(00)00292-4, 2001.

Tittel, F. K., Lewicki, R., Lascola, R., and McWhorter, S.: Emerging Infrared Laser Absorption Spectroscopic Techniques for Gas Analysis, in: Trace Analysis of Specialty and Electronic Gases, edited by: Geiger, W. M. and Raynor, M. W., 71–110, John Wiley and Sons, Inc., New York, Chister, Brisbane, Toronto, Singapore, 2012.

Vandaele, A. C., Hermans, C., and Fally, S.: Fourier transform measurements of SO_2 absorption cross sections: II. Temperature dependence in the 29000-44000 cm^{-1} (345–420 nm) region, J. Quant. Spectrosc. Ra., 110, 2115–2126, 2009.

Vita, F., Inguaggiato, S., Bobrowski, N., Calderone, L., Galle, B., and Parello, F.: Continuous SO_2 flux measurements for Vulcano Island, Italy, Ann. Geophys., 55, 301–308, doi:10.4401/ag-5759, 2012.

Wahner, A., Ravishankara, A. R., Sander, S. P., and Friedl, R. R.: Absorption cross section of BrO between 312 and 385 nm at 298 and 223 K, Chem. Phys. Lett., 152, 507–512, 1988.

Werner, C., Evans, W. C., Kelly, P. J., McGimsey, R., Pfeffer, M., Doukas, M., and Neal, C.: Deep magmatic degassing versus scrubbing: Elevated CO_2 emissions and C/S in the lead-up to the 2009 eruption of Redoubt Volcano, Alaska, Geochem. Geophy. Geosyst., 13, Q03015, doi:10.1029/2011GC003794, 2012.

Werner, C., Kelly, P. J., Doukas, M., Lopez, T., Pfeffer, M., McGimsey, R., and Neal, C.: Degassing of CO_2, SO_2, and H_2S associated with the 2009 eruption of Redoubt Volcano, Alaska, J. Volcanol. Geotherm. Res., 259, 270–284, doi:10.1016/j.jvolgeores.2012.04.012, 2013.

Yu, Y., Geyer, A., Xie, P., Galle, B., Chen, L., and Platt, U.: Observations of carbon disulfide by differential optical absorption spectroscopy in Shanghai, Geophys. Res. Lett., 31, L11107, doi:10.1029/2004GL019543, 2004.

18

Aerosol-assisted CVD synthesis, characterisation and gas-sensing application of gold-functionalised tungsten oxide

F. Di Maggio[1], M. Ling[1], A. Tsang[1], J. Covington[2], J. Saffell[3], and C. Blackman[1]

[1]Department of Chemistry, University College London, 20 Gordon Street, London WC1H 0AJ, UK
[2]School of Engineering, University of Warwick, Coventry CV4 7AL, UK
[3]Alphasense Ltd, 300 Avenue West, Skyline 120, Great Notley, Essex CM77 7AA, UK

Correspondence to: C. Blackman (c.blackman@ucl.ac.uk)

Abstract. Tungsten oxide nanoneedles (NNs) functionalised with gold nanoparticles (NPs) have been integrated with alumina gas-sensor platforms using a simple and effective co-deposition method via aerosol-assisted chemical vapour deposition (AACVD) utilising a novel gold precursor, $(NH_4)AuCl_4$. The gas-sensing results show that gold NP functionalisation of tungsten oxide NNs improves the sensitivity of response to ethanol, with sensitivity increasing and response time decreasing with increasing amount of gold.

1 Introduction

Tungsten oxide functionalised with gold nanoparticles (NPs) has found application in photocatalysis (Xi et al., 2012) and gas sensing (Vallejos et al., 2011; Annanouch et al., 2013), and recently Vallejos et al. (2011) demonstrated a surfactant- and polymer-free, single-step aerosol-assisted chemical vapour deposition (AACVD) method to synthesise thin films of tungsten oxide nanoneedles (NNs) decorated with gold NPs, which provides for direct integration of the nanomaterial with gas-sensing platforms (Vallejos et al., 2011). The observation that chemoresistive gas sensors fabricated from metal oxides functionalised with catalytic metal NPs show enhanced sensor response towards certain analytes has been explained using three mechanisms: spillover (enrichment of the surface of the metal oxide with analyte), Fermi level control (analyte changes the chemical state of the metal NPs leading to a large change in electronic interaction between oxide and metal) and the presence of oxidised states on the metal NPs forming an associated bulk effect and local/surface site on the oxide which interacts strongly with the analyte (Cuenya, 2010; Hübner et al., 2011). Only relatively small amounts of metal NPs need to be loaded on tungsten oxide NNs to show dramatically enhanced sensing properties (Vallejos et al., 2013), but the limit of the enhancement under variable loading is rarely reported. This paper reports on the successful fabrication of various loadings of gold NPs, utilising a novel gold AACVD precursor, $(NH_4)AuCl_4$, supported on tungsten oxide NNs, deposited directly on alumina gas-sensor substrates, and their gas-sensing properties with respect to ethanol and the correlation of sensitivity with gold loading.

2 Experimental section

Film deposition: gold NPs supported on tungsten oxide NNs were co-deposited at 375 °C via AACVD of tungsten hexaphenoxide $(W(OPh)_6)$ (Cross et al., 2003) and ammonium tetrachloroaurate(III) hydrate $(NH_4AuCl_4.xH_2O)$ (Alfa Aesar, 99.9 %) dissolved in acetone (10 cm³, Sigma-Aldrich, $\geq$ 99.6%) and methanol (5 cm³, Sigma-Aldrich, $\geq$ 99.6 %) (Table 1) using a custom-built reactor.

A Johnson Matthey Liquifog 2 operating at 1.6 MHz was used to generate an aerosol from a solution containing both the tungsten and gold precursors, with the aerosol droplets transported to the heated substrate by a nitrogen (oxygenfree, BOC) gas flow. Annealing of the samples was carried out in air at 500 °C for 2 h. Alumina (A493 Kyocera) or gas-sensor platforms (as shown schematically in Fig. 1), on which both the heater and sensor are printed on the same

Table 1. Details of samples.

	W(OPh)$_6$ (g)	NH$_4$ AuCl$_4$ (g)	Solvent (acetone : methanol 2 : 1) (cm^3)	Au at. %	Temperature (°C)
Sample 1	0.10	0.010	15	4.5	375
Sample 2	0.10	0.003	15	2.1	375
Sample 3	0.10	0	15	0	375

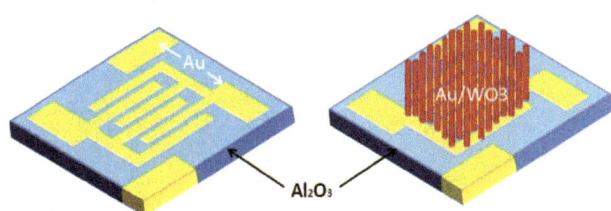

Figure 1. Schematic fabrication processes of the micro-sensor based on Au NPs supported on tungsten oxide NNs.

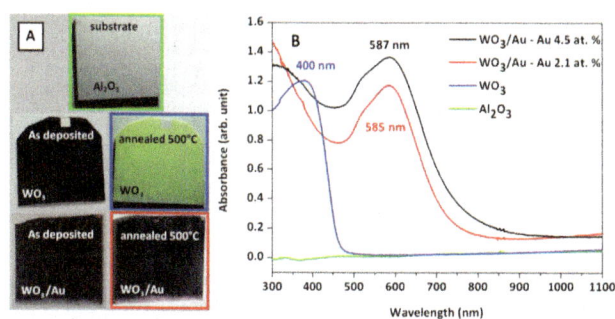

Figure 2. (a) Pictures of films deposited onto alumina substrate, before and after heating treatment at 500 °C. **(b)** UV–Vis analyses of Au-functionalised WO$_3$ (in black, 4.5 at. %; red, 2.1 at. %) and non-functionalised WO$_3$ (in blue).

side of an alumina tile, were used as substrates. Both were cleaned with acetone and isopropyl alcohol prior to use.

Film analysis: the crystalline structures of samples were determined via X-ray diffraction (XRD) (Bruker, LinxEye D8-Discover, using Cu Kα radiation operated at 40 kV and 40 mA) with glancing incident angle (tube at 1°), 0.05° per step and 1 s per step, and 2θ from 10 to 66°. The microstructure of the films was examined with scanning electron microscopy (SEM) (Jeol 6310F, 5 kV) and high-resolution transmission electron microscopy (HR-TEM) (Jeol-2100, 200 kV), which was also used to obtain the information of lattice structure. The chemical composition measurement was determined via energy dispersive X-ray spectroscopy (EDX) (20 kV) (Jeol 6310F). Chemical and electronic states of the elements in the thin films were examined by means of X-ray photoelectron spectroscopy (XPS) (Thermo Scientific K-Alpha), using monochromatic Al Kα radiation (0.6 eV) and charge compensation by means of dual-beam charge neutralisation with an electron gun (1 eV) and argon-ion gun (≤ 10 eV), calibrated by the C 1s peak at 284.7 eV. UV–Vis spectroscopy was performed using a double-monochromated Perkin Elmer Lambda 950 UV–Vis–NIR spectrophotometer in the 300 to 1100 nm range.

Gas-sensing tests: gas sensors were exposed to various concentrations of ethanol at 0.5, 1, 2, 3 and 4 ppm respectively, at an operating temperature of 300 °C and relative humidity of 50 %. The sensor response is defined as $R = R_a/R_g$, where R_a is the resistance of the sensor in air and R_g is the resistance in ethanol. Sensors were exposed to ethanol vapours for 4 min and afterwards to air for 4 min.

3 Results and discussion

Figure 2a shows pictures of Au-functionalised and non-functionalised tungsten oxide films deposited on alumina

substrates before and after heat treatment in air at 500 °C. The pre-annealed films were characterised by a dark colour, as expected in the presence of partially reduced tungsten species, and after heating in air, non-functionalised tungsten oxide turned bright yellow, as expected for stoichiometric WO$_3$, whilst Au-functionalised tungsten oxide was dark purple in colour, indicative of the presence of gold nanoparticles.

UV–Vis analyses (Fig. 2b) were performed collecting reflectance data (R) and then converting into absorbance (A) using the following formula (Viscarra Rossel et al., 2006):

$$A = \log_{10}(1/R), \tag{1}$$

analogous to absorbance units, $\log(1/T)$, for transmission measurements.

For the non-functionalised tungsten oxide film, a peak is observed at 400 nm, corresponding to the WO$_3$ band edge (E_{gap} WO$_3$ ~ 2.7 eV). The Au-functionalised WO$_3$ films show an additional absorbance at 585 nm, corresponding to the surface plasmon resonance (SPR) peak of gold nanoparticles (El-Brolossy et al., 2008). Increasing the amount of gold precursor led to an increase in the absorbance of the gold plasmon resonance peak, suggesting a greater incorporation of gold particles.

XRD analysis (Fig. 3) of films deposited on Kyocera alumina substrates, after annealing, revealed the presence of monoclinic-phase WO$_3$ (P2$_1$/n space group, $a = 7.306$ Å, $b = 7.540$ Å, $c = 792$ Å, and $\beta = 90.88$°; ICCD card no. 72-0677; Loopstra and Rietveld, 1969). Alumina peaks, exhibited in all patterns and depicted with dashed vertical lines, come from the alumina substrate. Both the WO$_3$ / Au and the

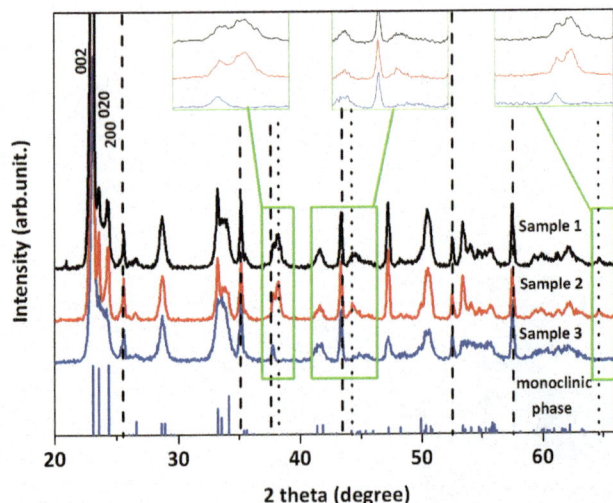

Figure 3. XRD patterns of Au-functionalised WO_3 (in black, 4.5 at.%; red 2.1 at.%) and non-functionalised WO_3 (in blue). Monoclinic-phase (blue), Al_2O_3 trigonal (- - -) and Au FCC (• • •) structures are also presented.

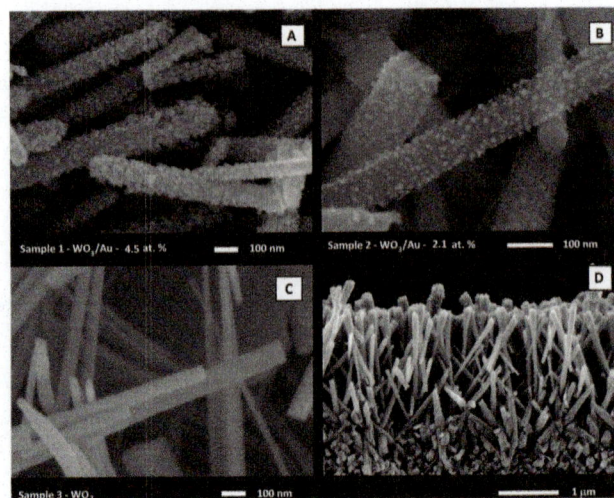

Figure 4. SEM images of **(a)** sample 1 (WO_3/Au 4.5 at.%), **(b)** sample 2 (WO_3/Au 2.1 at.%), **(c)** sample 3 (WO_3) and **(d)** cross section of sample 3.

WO_3 patterns showed preferred orientation in the [001] direction. In particular, the peak (002) at 23.11° 2θ is very intense relative to the (020) peak. Gold FCC peaks (dotted vertical lines) at 38.25° 2θ, 44.30° 2θ and 64.55° 2θ were observed only in the WO_3/Au pattern (zoomed patterns). This result confirms the co-deposition of tungsten oxide and gold metal. The mean grain size of Au nanoparticles was calculated to be about 26 nm based on the Scherrer formula:

$$D = \frac{0.89\lambda}{\beta\cos\theta}, \qquad (2)$$

where D is the mean size of the crystalline domains, λ is the Cu Kα wavelength, β is the line broadening at half the maximum intensity (FWHM) in radians and θ is half of the Bragg angle.

No shifts in the WO_3 peaks position were observed, demonstrating that the monoclinic crystal structure of tungsten oxide was not changed after the addition of gold. Similar results for WO_3 monoclinic phase were obtained by Vallejos et al. (2013) when using $HAuCl_4$ as the gold precursor, but no Au peaks were observed in their XRD patterns, likely due to the lower amount of gold in the samples (0.09 Au at.%).

Images from EDX analysis of the Au-functionalised and non-functionalised samples are shown in Table 2. The measured Au atomic percentage composition (Au at.% on Au–WO_3) in the deposited films was 4.5 and 2.1 for sample 1 and sample 2 respectively, much lower than the theoretical values based on the composition of the precursor solution, which were 10.4 and 3.4 at.% respectively. The incorporation efficiency was 44% for sample 1 and 63% for sample 2, much higher than the one determined by Vallejos et al. (2013), which had 5–10% efficiency.

Table 2. Precursors theoretical and EDX experimental atomic percentage composition for gold element.

	Theoretical Au at.%	Experimental Au at.%
Sample 1	10.4	4.5
Sample 2	3.4	2.1
Sample 3	0	0

SEM images are depicted in Fig. 4, clearly showing the increase in the amount of decorating nanoparticles on increasing the amount of gold precursor, with sample 1 (Fig. 4a) displaying a larger number of surface (gold) particles compared to sample 2 (Fig. 4b). No decorating particles were observed in the non-functionalised sample 3 (Fig. 4). The decorating particles were well dispersed all along the needles, demonstrating the good efficiency of the synthesis in terms of composition homogeneity. The thickness of the films, visible in a cross-section image in Fig. 4d, was roughly 1.5 μm in all cases, and the tungsten oxide "needles" (ca. 0.1–0.2 μm diameter and ca. 1–2 μm length), although not perfectly vertically aligned, were predominantly perpendicular to the substrate.

HR-TEM was also used to characterise the crystalline habit of the nanoneedles and decorating nanoparticles (Fig. 5). The crystal planes in the long axis of the needles were separated by 0.378 nm (Fig. 5c), corresponding to the (020) plane of monoclinic WO_3 monoclinic, and the decorating NP (Fig. 5d) exhibited planes separated by 0.23 nm, consistent with (111) atomic planes of FCC gold. TEM examination showed that the gold nanoparticles were polydispersed both in shape (triangle, trapezoid, rod, sphere) and size (ca. 18 and 62 nm diameter), with an average size of 30 nm. This result matches with the gold nanoparticles' grain

Table 3. Au 4f and W 4f scans fitting parameters: peak, FWHM and area.

			Peak 1 (4f$_{5/2}$)			Peak 2 (4f$_{7/2}$)		
			Position	FWHM	Area	Position	FWHM	Area
Sample 1	Au 4.5 at. %	W 4f	37.84	1.098	51 429	35.65	1.098	68 586
Sample 2	Au 2.1 at. %	W 4f	37.79	1.115	52 690	35.60	1.115	70 267
Sample 3	Au 0 at. %	W 4f	37.84	1.165	86 442	35.65	1.165	115 284
Sample 1	Au 4.5 at. %	Au 4f	87.68	0.982	36 077	83.90	0.982	48 120
Sample 2	Au 2.1 at. %	Au 4f	87.68	0.997	37 438	83.90	0.997	49 935

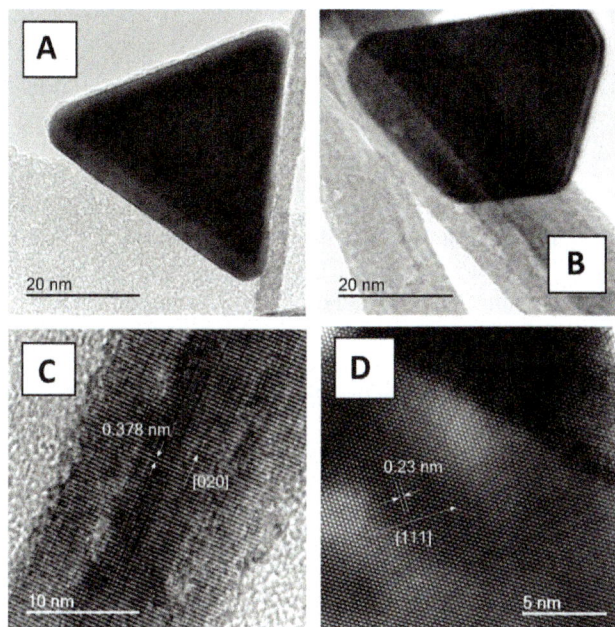

Figure 5. (a) and **(b)** TEM images of WO$_3$ / Au sample. HR-TEM for **(c)** WO$_3$ nanoneedle and **(d)** gold nanoparticles.

Table 4. XPS Au atomic percentage composition. Oxygen : tungsten ratio.

	Au at. % EDX	Au at. % XPS	O / W XPS
Sample 1	4.5	9.5	2.7
Sample 2	2.1	7.1	2.8
Sample 3	0	0	2.8

Table 5. Response time and recovery time variation to the different gold loadings.

	Response time (s)	Recovery time (s)
Sample 1	2.8	14.0
Sample 2	5.3	9.1
Sample 3	4.5	8.5

size, 26 nm, observed by using the Scherrer formula on the XRD pattern.

This is different to that observed previously using HAuCl$_4$ as a gold precursor (Vallejos et al., 2013). In that case the gold NP displayed a spherical shape with a size between 4 and 11 nm. This may be due to the different nature of the precursor or the different gold loading.

XPS high-resolution spectra for WO$_3$ and WO$_3$ / Au, after annealing, are shown in Fig. 6 and the relative fitting parameters are listed in Table 3.

The W 4f spectra in Fig. 6a show a single tungsten environment for all the samples, with the 4f$_{7/2}$ peak centred at 35.6 eV, which corresponds to the W^{6+} oxidation state (Fleisch and Mains, 1982). The Au 4f peaks for the Au-functionalised WO$_3$ samples were fitted using single Gaussian–Lorentzian functions and the Au 4f$_{7/2}$ binding energy was found to be 83.9 eV (Fig. 6b), corresponding to metallic gold (Dückers and Bonzel, 1989). Therefore W(VI)

and Au(0) were the only oxidation states found for these elements.

XPS analysis of annealed samples indicated a W / O ratio of 2.8 for samples 2 and 3 and a ratio of 2.7 for sample 1 (Table 4). This value is much higher compared to the 2.4 ratio obtained by Vallejos et al. (2013) in both W and Au / W samples. The gold atomic percentage (Au at. % on Au–WO$_3$) obtained via XPS (Table 4) is much higher than the one obtained via EDX (Table 2). This can be explained by taking into account the morphology of the WO$_3$ / Au system and the area analysed; SEM images (Fig. 4) show the Au nanoparticles are dispersed on the surface of the tungsten oxide nanoneedle, and therefore, with XPS being a surface analysis technique, the Au at. % appears to be greater than EDX, which gives bulk information.

4 Gas-sensing properties

Gas-sensing properties of gold-functionalised and non-functionalised WO$_3$ nanoneedles were tested towards ethanol vapours. Figure 7 shows the sensor response towards different amounts of analyte at an operating temperature of 300 °C in the presence of 50 % of relative humidity. Gas sensors based on non-functionalised WO$_3$ (sample 3) showed a very low response towards 0.5 ppm of ethanol, with sensitiv-

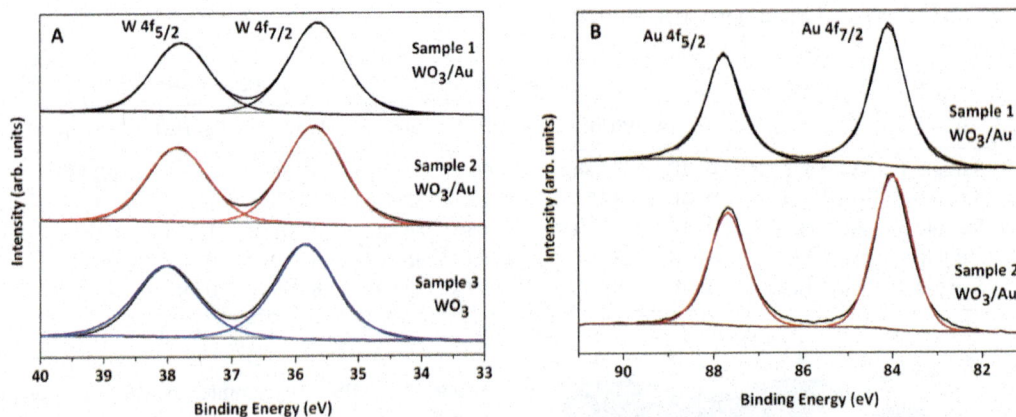

Figure 6. XPS spectra for sample 1 (WO$_3$ / Au 4.5 at. %), sample 2 (WO$_3$ / Au 2.1 at. %) and sample 3 (WO$_3$) films on alumina. **(a)** W 4f peaks and **(b)** Au 4f, fitted with parameters in Table 3.

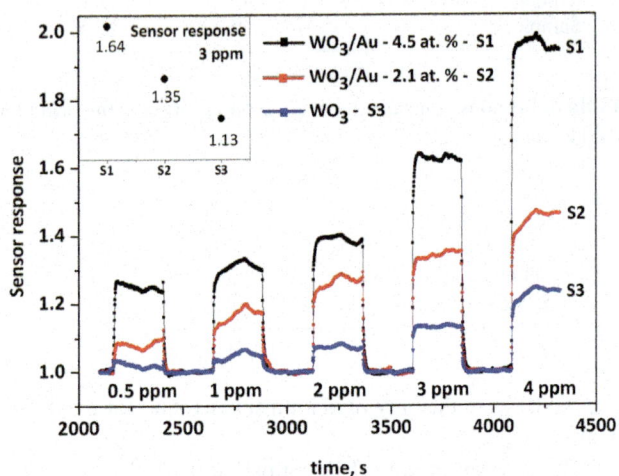

Figure 7. Gas-sensing response towards different amount of ethanol at 300 °C as a function of increasing gold loading.

ity increasing in the presence of higher amounts of analyte as expected, up to a highest value of 1.2 when the concentration of ethanol was 4 ppm.

Sample 2 (Au 2.1 at. %) and sample 1 (Au 4.5 at. %) showed respectively 2-fold and 5-fold higher sensing response than non-functionalised gas sensors (sample 3), with increased response even towards low concentrations of analyte, close to 1.3 towards 0.5 ppm of ethanol for sample 1 (4.5 % Au). These results demonstrate an enhancement in gas-sensing response by increasing the amount of gold loading. The highest loading amount used in this study was 4.5 %; however, as has been seen in catalysis studies (Epling and Hoflund, 1999), increasing the loading percentage does not necessarily lead to a linear improvement in catalytic performance. Therefore, further experiments will be conducted in order to determine the optimum loading of gold beyond which the activity is stable or decreases.

All sensors displayed an n-type response with decreasing resistance in the presence of the reducing gas (Fig. 7). The response and recovery times calculated for 3 ppm ethanol (Table 5) exhibit approximately the same values for sample 2 (2.1 % Au) and 3 (non-functionalised), a response time of 5 s and recovery time of 9 s. Sample 1 (4.5 % Au) had the fastest response time, 2.8 s, but the slowest recovery time, 14 s.

5   Conclusions

Tungsten oxide NNs functionalised with Au NPs integrated have been synthesised via AACVD using a single-step method and integrated onto alumina gas-sensing platforms to produce gas sensors. The average of Au NPs diameter is 30 nm on the tungsten oxide NNs (ca. 0.1–0.2 μm diameter and ca. 1–2 μm length). An increase in the amount of Au precursor (NH$_4$AuCl$_4$) in the initial solution produced a higher loading of Au NPs on the tungsten oxide NNs. Gas-sensing tests towards ethanol vapours showed an increase in sensor response in samples with higher amounts of gold nanoparticles, with increases in response towards 3 ppm of ethanol of 2-fold and 5-fold over non-functionalised WO$_3$ as the gold loading was raised from 2.1 to 4.5 % respectively.

Acknowledgements. This work was supported by UCL through its Impact Studentship Programme.

Edited by: A. Lloyd Spetz
Reviewed by: two anonymous referees

References

Annanouch, F. E., Vallejos, S., Stoycheva, T., Blackman, C., and Llobet, E.: Aerosol assisted chemical vapour deposition of gas-sensitive nanomaterials, Thin Solid Films, 548, 703–709, 2013.

Cross, W. B., Parkin, I. P., O'Neill, S. A., Williams, P. A., Mahon, M. F., and Molloy, K. C.: Tungsten oxide coatings from the aerosol-assisted chemical vapor deposition of W (OAr) 6 (Ar = C6H5, C6H4F-4, C6H3F2-3, 4); photocatalytically active γ-WO$_3$ films, Chem. Mater., 15, 2786–2796, 2003.

Cuenya, B. R.: Synthesis and catalytic properties of metal nanoparticles: Size, shape, support, composition, and oxidation state effects, Thin Solid Films, 518, 3127–3150, 2010.

Dückers, K. and Bonzel, H. P.: Core and valence level spectroscopy with Y Mζ radiation: CO and K on (110) surfaces of Ir, Pt and Au, Surf. Sci., 213, 25–48, 1989.

El-Brolossy, T. A., Abdallah, T., Mohamed, M. B., Abdallah, S., Easawi, K., Negm, S., and Talaat, H.: Shape and size dependence of the surface plasmon resonance of gold nanoparticles studied by Photoacoustic technique, Eur. Phys. J.-Spec. Top., 153, 361–364, 2008.

Epling, W. S. and Hoflund G. B.: Catalytic oxidation of methane over ZrO2-supported Pd catalysts, J. Catal., 182, 5–12, 1999.

Fleisch, T. H. and Mains, G. J.: An XPS study of the UV reduction and photochromism of MoO3 and WO$_3$, J. Chem. Phys., 76, 780–786, 1982.

Hübner, M., Koziej, D., Bauer, M., Barsan, N., Kvashnina, K., Rossell, M. D., Weimar, U., and Grunwaldt, J. D.: The Structure and Behavior of Platinum in SnO2-Based Sensors under Working Conditions, Angew. Chem. Int. Edit., 50, 2841–2844, 2011.

Loopstra, B. O. and Rietveld, H. M.: Further refinement of the structure of WO$_3$, Acta Crystallogr. B, 25, 1420–1421, 1969.

Vallejos, S., Stoycheva, T., Umek, P., Navio, C., Snyders, R., Bittencourt, C., Llobet, E., Blackman, C., Moniz, S., and Correig, X.: Au nanoparticle-functionalised WO$_3$ nanoneedles and their application in high sensitivity gas sensor devices, Chem. Commun., 47, 565–567, 2011.

Vallejos, S., Umek, P., Stoycheva, T., Annanouch, F., Llobet, E., Correig, X., De Marco, P., Bittencourt, C., and Blackman, C.: Single-Step Deposition of Au-and Pt-Nanoparticle-Functionalized Tungsten Oxide Nanoneedles Synthesized Via Aerosol-Assisted CVD, and Used for Fabrication of Selective Gas Microsensor Arrays, Adv. Funct. Mater., 23, 1313–1322, 2013.

Viscarra Rossel, R. A., McGlynn, R. N., and McBratney, A. B.: Determining the composition of mineral-organic mixes using UV–vis–NIR diffuse reflectance spectroscopy, Geoderma, 137, 70–82, 2006.

Xi, G., Ye, J., Ma, Q., Su, N., Bai, H., and Wang, C.: In situ growth of metal particles on 3D urchin-like WO$_3$ nanostructures, J. Am. Chem. Soc., 134, 6508–6511, 2012.

19

Encapsulation of implantable integrated MEMS pressure sensors using polyimide epoxy composite and atomic layer deposition

P. Gembaczka[1], **M. Görtz**[1], **Y. Celik**[1], **A. Jupe**[1], **M. Stühlmeyer**[1], **A. Goehlich**[1], **H. Vogt**[1], **W. Mokwa**[2], and **M. Kraft**[1]

[1]Fraunhofer Institute for Microelectronic Circuits and Systems, Duisburg, Germany
[2]Institute of Materials in Electrical Engineering, RWTH Aachen University, Aachen, Germany

Correspondence to: P. Gembaczka (pierre.gembaczka@ims.fraunhofer.de)

Abstract. Implantable MEMS sensors are an enabling technology for diagnostic analysis and therapy in medicine. The encapsulation of such miniaturized implants remains a largely unsolved problem. Medically approved encapsulation materials include titanium or ceramics; however, these result in bulky and thick-walled encapsulations which are not suitable for MEMS sensors. In particular, for MEMS pressure sensors the chip surface comprising the pressure membranes must be free of rigid encapsulation material and in direct contact with tissue or body fluids. This work describes a new kind of encapsulation approach for a capacitive pressure sensor module consisting of two integrated circuits. The micromechanical membrane of the pressure sensor may be covered only by very thin layers, to ensure high pressure sensitivity. A suitable passivation method for the high topography of the pressure sensor is atomic layer deposition (ALD) of aluminium oxide (Al_2O_3) and tantalum pentoxide (Ta_2O_5). It provides a hermetic passivation with a high conformity. Prior to ALD coating, a high-temperature resistant polyimide–epoxy composite was evaluated as a die attach material and sealing compound for bond wires and the chip surface. This can sustain the ALD deposition temperature of 275 °C for several hours without any measurable decomposition. Tests indicated that the ALD can be deposited on top of the polyimide–epoxy composite covering the entire sensor module. The encapsulated pressure sensor module was calibrated and tested in an environmental chamber at accelerated aging conditions. An accelerated life test at 60 °C indicated a maximum drift of 5 % full scale after 1482 h. From accelerated life time testing at 120 °C a maximum stable life time of 3.3 years could be extrapolated.

1 Introduction

The development of small MEMS (micro-electro-mechanical systems) sensors with functional surfaces enables the ability to perform diagnostic and therapeutic tasks inside the human body in the form of a long-term implant. Application examples include continuous pressure monitoring (Mokwa, 2007; Gembaczka et al., 2013), nerve and muscle stimulation (Eick et al., 2009) and drug delivery systems (Hang Tng et al., 2012). Usually medically approved micro-electronic implants such as pacemakers (Park and Lakes, 2007), cochlear implants (Loeb, 1990) or brain stimulators (Rezai et al., 2002) are encapsulated by non-flexible materials such as ti-tanium or ceramics with considerable sidewall thickness in the range of 100 µm to 1 mm. Such hermetic encapsulations do not allow individual regions of the sensor to be in direct contact with the tissue or body fluids, as required for pressure sensing. Furthermore, in applications like retinal implants, where bendability of the implant is needed, the limits of miniaturization of thick-walled encapsulation materials are quickly reached. To take full advantage of MEMS sensors a paradigm shift from voluminous and solid encapsulations towards miniaturized and flexible encapsulations is required.

The measurement of pressure in different areas of the human body, such as blood pressure (in large arteries or veins), intraocular pressure or intracranial pressure is important for medical diagnostics and therapy (Mokwa, 2007). The continuous monitoring of the corresponding pressure values with a permanent implant can be beneficial for the treatment of medical conditions such as intracranial and arterial hypertension, heart insufficiency and glaucoma.

The development of a miniaturized encapsulation technology for MEMS pressure sensors is particularly challenging as the pressure outside of the housing must be transferred to a pressure-sensitive membrane. A number of approaches for the encapsulation of implantable pressure sensors have been suggested previously (Mokwa, 2007). Silicone is often used as encapsulation material because it is soft and can transmit the pressure well (Cleven et al., 2012; Bradford et al., 2010; Stangel et al., 2001). However, silicones absorb water and the material properties change over time leading to a drift of the pressure value measured by the sensor. Preconditioning of the material and offset compensation of the sensor can alleviate the problem (Gräfe et al., 2009). Nevertheless, water diffusion remains problematic especially with electrical components inside the human body. Materials like Parylene-C were applied to further reduce the water uptake (Schlierf et al., 2005) or the membrane of the pressure sensor was directly made of parylene (Ha et al., 2012; Chen et al., 2008). However, a hermetically sealed encapsulation is not achievable based on polymer alone, therefore novel solutions must be developed.

The pressure sensor employed in this work is fabricated in CMOS technology allowing the integration of a capacitive pressure sensor with the read-out electronics on one chip. For some applications a second chip is required for signal processing. A hard and thick passivation would change the characteristics of the sensitive membranes of the pressure sensor chip so strongly that no sufficient pressure transmission would be possible. In previous preliminary work, an encapsulation concept for the pressure sensor surface based on atomic layer deposition (ALD) was successfully tested (Betz, 2011). A passivation layer consisting of 50 nm Al_2O_3 and 50 nm Ta_2O_5 was deposited by ALD on the pressure sensor chip only. Figure 1 shows the pressure membranes and part of the readout electronics after an accelerated life time test in phosphate-buffered saline (PBS) at 150 °C for 1 h. Although the test results were encouraging, first indications of corrosion were observed at the aluminium bond pads and the saw edges of the chip. This was attributed to the inhomogeneity of the sawing process (Betz, 2011).

In a novel encapsulation concept, therefore, the electrical contacts of the chip, the bond wires and the dicing-edges have to be protected before ALD passivation. ALD deposition is performed at a temperature of 275 °C for several hours. All components of the system, including the protective material, have to be resistant to this temperature. In this study we used the high-temperature resistant polyimide–epoxy composite *Polytec EP P-690* (Polytec PT GmbH, 2011) as a protective casting material, which serves as a die-attach material, sealing compound for the bond wires and the

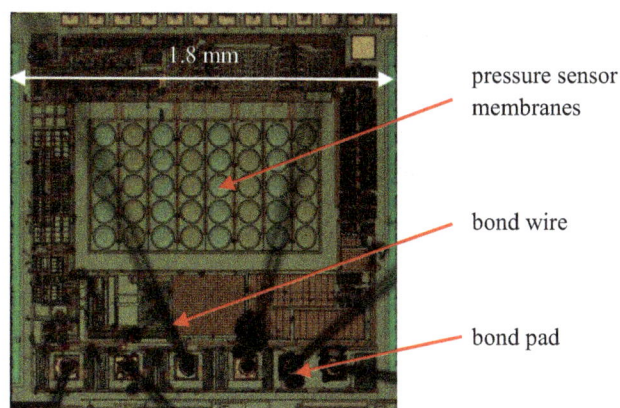

Figure 1. Photo of the pressure sensor chip after 1 h in PBS at 150 °C (Betz, 2011).

chip surface. However, there is no commercially available material optimized for such an application.

Particularly problematic is that polyimide–epoxy significantly shrinks during the curing phase. A process had to be developed to circumvent the shrinkage. For the first time, this work reports on a biocompatible encapsulation method combining ALD and polymers.

The paper is organized as follows: in Sect. 2 an assembly of a two-chip pressure sensor module (pressure sensor chip including read-out electronics and signal processing chip) is introduced to facilitate the development of novel encapsulation and assembly technique. Section 3 describes the high-temperature resistant polyimide–epoxy composite and its deployment. In Sect. 4, the ALD layer is discussed and test results regarding the hermetic seal of the sensor are presented. Section 5 presents pressure sensor module calibration and measurement results at various temperatures, and Sect. 6 accelerated life time testing results. Finally, in Sect. 7 some conclusions are drawn.

2 Components and materials of the pressure sensor module

The pressure sensor module mainly consists of two application-specific integrated circuits (ASICs). The first is the capacitive pressure sensor chip (PS) including the read-out electronics, the second a signal post-processor chip (SPP). Both are mounted on a silicon chip carrier with bond pads and signal tracks. Additionally, some surface mount components (SMD) are required. The cross-section of the pressure sensor module is depicted in Fig. 2. In the first step, the chips were glued on the chip carrier using a polyimide–epoxy composite and electrically connected to the carrier with bond wires. The bond wires and the entire module, except for the pressure sensor membranes, were covered with the same polyimide–epoxy composite (as described in

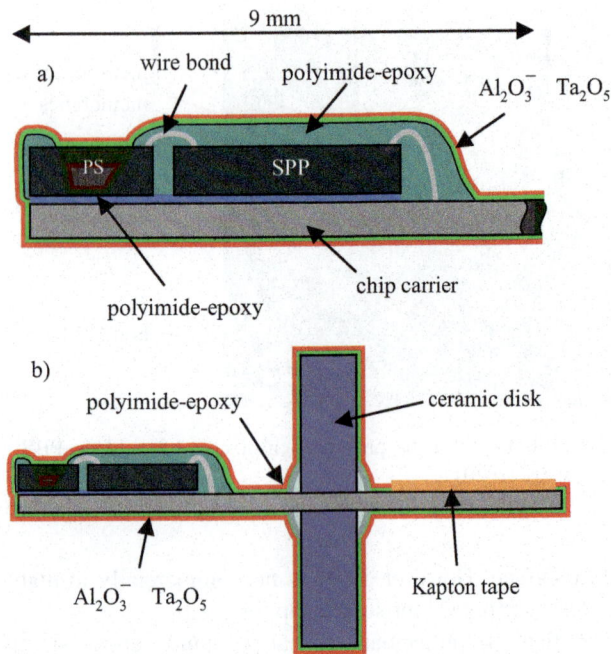

Figure 2. (a) Sketch of the pressure sensor module and encapsulation concept in cross-section. (b) Schematic representation of the entire sensor assembly with ceramic disc separator and taped pad structures.

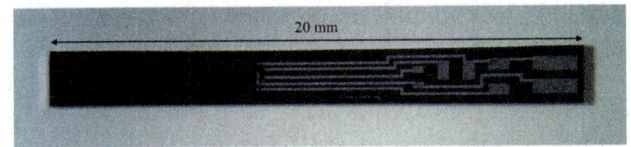

Figure 3. Photo of the silicon chip carrier.

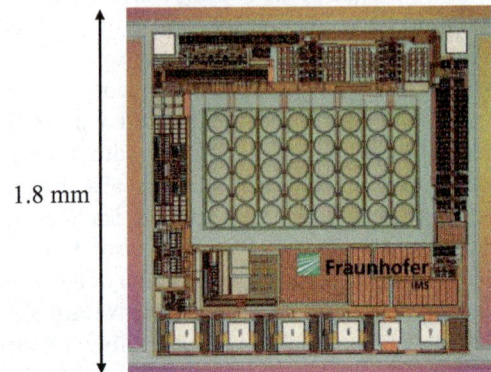

Figure 4. Photograph of the capacitive pressure sensor chip before encapsulation (Betz, 2011).

Sect. 3) used for die attachment to minimize thermal expansion stress.

We choose here the approach to encapsulate the chips by polyimide–epoxy first, and then by an ALD layer to achieve a hermetically sealed passivation. Considering the opposite approach, application of a polymer potting layer on top of the ALD layer is not a viable solution, as the bond wires mechanically deform die to shrinkage of the polyimide–epoxy in the curing process, which, in turn, would lead to miniature fractures of the ALD layer. Another disadvantage is the water uptake of the polyimide–epoxy which leads to changes of the material properties of the polymer over time. The water uptake of polymer is very low but the pressure sensor is highly sensitive to any kind of mechanical stress and, in general, any change of the material properties leads to drift of the measured pressure value. Here, it was therefore chosen to apply first a layer of polyimide–epoxy followed by ALD deposition to create a water impermeable, hermetic encapsulation.

A ceramic disc with an opening for the chip carrier was placed next to the ASICs and fixed with the same polymer. The disc serves as a separation between the ASICs that are exposed to the medium on one side and the discrete circuit components on the other side; this is depicted in Fig. 2b. The overall assembly of the separation disc and the discrete components is not the subject of this paper and therefore shall not be described in further detail. The final step (described in more detail in Sect. 4) is the passivation of the entire as-

sembly in an ALD tool with aluminium oxide (Al_2O_3) and tantalum pentoxide (Ta_2O_5).

Only the hermetically sealed area located left of the ceramic disc is later exposed to the liquid medium.

2.1 Silicon chip carrier

The carrier for the entire sensor module is made of a silicon chip with aluminium tracks (for the discrete components), pads and space for the pressure sensor chip and the SPP, as shown in Fig. 3. The chip carriers are processed at wafer level and diced into separate chips.

Silicon as a carrier material was selected because it has the same thermal expansion coefficient as the ASICs. The thermal expansion of different materials is critical in this passivation method, as thermal stress can lead to fractures.

2.2 Capacitive pressure sensor chip

The pressure sensor chip has been fabricated in a CMOS technology allowing the integration of capacitive pressure sensor membranes with the readout electronics in a surface micromachining process. The chip size is 1.8×1.8 mm^2 and is shown in Fig. 4.

The bottom electrode of the pressure sensor is made of n^+ doped silicon, whereas the upper electrode is made of a flexible polysilicon membrane, as shown in Fig. 5. By applying pressure to this assembly, the distance between the two electrodes is changed resulting in change of the capacity C_{sens}. The pressure sensor has already been described in

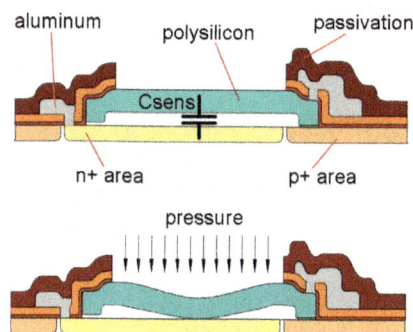

Figure 5. Section through a pressure sensor membrane (Trieu, 2011).

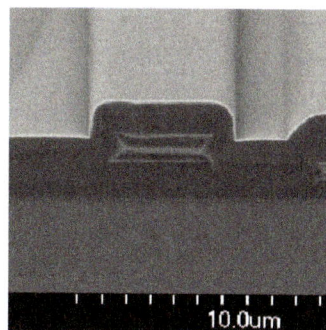

Figure 6. Cross-section SEM image of an area of the pressure sensor illustrating the difficult topography for encapsulation.

more detail in Dudaicevs et al. (1994), Müntjes et al. (2010), Trieu (2011) and Gembaczka et al. (2013).

In order to achieve a sufficiently large change of the capacitance, 20 identical pressure sensor membranes are combined into an array. The pressure sensor chip is not only sensitive to the applied pressure, but due to its micro-mechanical structure, also to mechanical stress induced, for example, by the different thermal expansion coefficients of the chip and the adhesive materials. Consequently, a soft silicone is typically used as adhesive to mechanically decouple the chip. However, due to the ALD deposition temperatures of 275 °C any silicone will exhibit considerable thermal change. Furthermore, the difficult topography of the pressure sensor chip restricts the choice of a suitable passivation material. Figure 6 shows the cross-section of an area of the pressure sensor depicting this issue clearly.

2.3 Signal-post-processing chip

To get a functional pressure sensor module a second ASIC is required to digitize the analogue pressure sensor signals. The SPP chip shown in Fig. 7 was fabricated in a $0.35\,\mu m$ CMOS technology with a size of approximately $1.8 \times 5\,mm^2$.

The width of the ASIC is identical to the width of the pressure sensor chip. The ASIC has a temperature sensor integrated based on a bipolar transistor with the necessary read-out circuits. The temperature measurement is used to compensate the temperature dependence of the pressure sensor. In addition to an analogue-to-digital converter the SPP includes a voltage regulator, a local oscillator and an EEPROM for permanent storage of calibration and identification data (Gembaczka et al., 2013). With a two-wire connection cable, the digitized sensor data is transmitted to a data reader. The connection is used both for data transmission and power supply.

3 Encapsulation with high-temperature polyimide–epoxy composite

3.1 Choice of material

For the first encapsulation layer of the pressure sensor module a polymer had to be found satisfying all requirements. The same material should be used for the die-attach and the potting of the pressure sensor module including the bond wires. This has the advantage that thermal expansion coefficients are identical avoiding tension and thus cracks in the assembly.

Ideally, the material should satisfy the following requirements:

- long-term temperature stability up to 275 °c
- high glass transition temperature
- low thermal expansion
- thermally conductive
- electrically insulating
- good adhesive properties
- smooth surface without pores
- dispensable
- thixotropic
- CMOS compatible.

The desired properties, in particular the temperature stability, drastically reduce the list of candidates of possible polymers. With regard to the processability and adhesion, an epoxy is a good choice for potting. Epoxy resins with different modifiers are available to improve material properties, for example rubber additives for flexibility (Boyle et al., 2001). One disadvantage is the decomposition and deformation of polymers (Beyler and Hirschler, 2001; Cheng et al., 2009). Some electrically insulating high-temperature epoxy

Figure 7. Photograph of the SPP-chip.

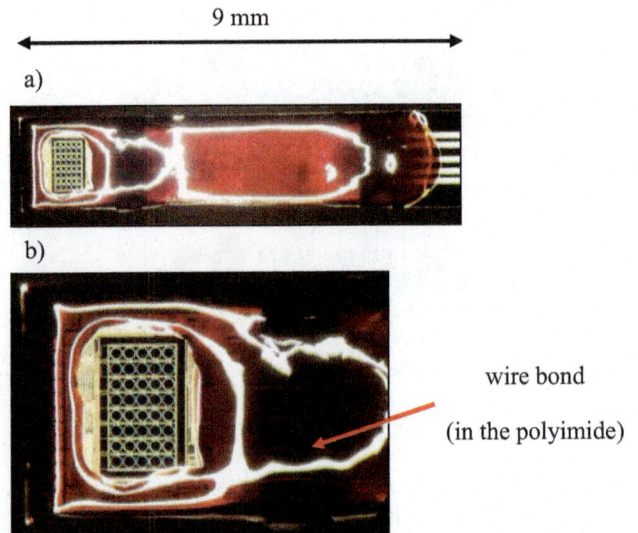

Figure 8. Casting of the entire sensor module before the ALD passivation. (**b**) Detailed image of the pressure sensor chip after the casting.

composites already exist that allow operating temperatures up to about 260 °C, for example the *Duralco 4461* (Polytec PT GmbH, 2010) has a maximal operating temperature of 260 °C but the deformation temperature is about 210 °C hence it is not suitable for the envisaged application. To obtain a homogeneous and defect-free ALD passivation layer the polymer must not deform during heating or cooling processes.

Alternatively, polyimides are materials satisfying nearly all requirements and are well established for applications in microsystem technology (Walewyns et al., 2013). However, processing of polyimide is challenging due to the evaporation of the solvent (for example N,N′-Dimethylformamid, N,N-Dimethylacetamid or N-Methylpyrrolidon) (Calderón, 2005) and water during the curing process (or imidization) leading to a significant shrinkage of the material which is especially problematic for the encapsulation of the bond wires. Another critical point is the adhesion of pure polyimide which is typically rather poor (Gaw and Kakimoto, 1999). The bonding of two objects is difficult because of the evaporation of the solvent creating pore-like structures during the curing process. The pores are also problematic for a hermetic ALD passivation layer; therefore pure polyimides are also not suitable for the pressure sensor module encapsulation. In particular synthesized polyimides with additives allow improving some material properties; examples include *U-Varnish-S* (UBE) or *PI2611* (HD Microsystems) which exhibit lower water absorption than basic polyimides (Rubehn and Stieglitz, 2010) and were used for neural implants (Hassler et al., 2011).

In particular, polyimide–epoxy composite was identified as a suitable material; here a small amount of epoxy (typically 10–15 %) is added to the polyimide partially replacing the solvent (Gaw and Kakimoto, 1999). This increases adhesive strength, shape stability and reduces the formation of pores.

For the encapsulation of the pressure sensor module we used *Polytec EP P-690* (Polytec PT GmbH, 2011). The shrinkage during curing was still significant as well as the formation of pore-like structures, but less than in pure polyimides. To address the remaining problems, we developed a curing process to minimize these issues

3.2   Polyimide–epoxy composite deposition

The potting of the chips and the bond wires with a polyimide–epoxy composite is cumbersome due to the lack of dimensional stability during the application process. The polyimide–epoxy composite flows over the chip surface depending on the material and topography making it difficult to control and predict area coverage. The pressure sensor membranes need to remain free of the polyimide–epoxy composite. Pre-heating of the chip carrier allows application of the polyimide–epoxy composite in certain areas only. The polyimide–epoxy composite was applied manually using a dispenser. For this reason the edge of the polyimide–epoxy composite around the area of the membranes is irregular.

A further critical point is the shrinkage of the material due to the evaporation of the solvent during curing. Preliminary tests showed that the 25 µm AlSi bond wires were deformed by volume reduction of the polyimide–epoxy composite. In some samples, the deformation was so strong that the bond wires were pressed onto the chip edge resulting in electrical short circuits.

To avoid this problem, a high degree of imidization was achieved by heating the pressure sensor module up to 290 °C for 1 h. This temperature was selected to prepare the setup for the ALD deposition at a maximum temperature of 275 °C. In this way, no measurable shrinkage of the polyimide–epoxy was observed during the ALD deposition. The result of the polyimide–epoxy application after curing is shown in Fig. 8.

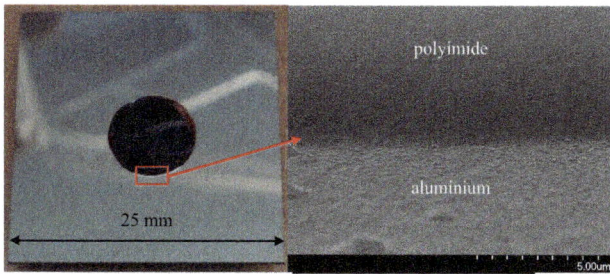

Figure 9. Photograph of a test chip (left) and SEM image of the transition region at a 45° angle (right).

4 ALD passivation

4.1 Material selection and layer thickness

The requirements for the ALD passivation material are as follows:

- good adhesion to all materials (silicon, silicon nitride, polyimide–epoxy composite)

- must not alter the mechanical characteristic and properties of the pressure membranes

- conformal deposition of high surface topography

- biocompatible

- electrically insulating

- impermeable to water vapour

- corrosion resistant

- deposition at moderate temperatures to preserve the underlying potting material and CMOS circuitry.

Based on a previous study (Betz, 2011) aluminium oxide (Al_2O_3) and tantalum pentoxide (Ta_2O_5) were chosen as ALD layers. Al_2O_3 ensures good adhesion and conformity whereas the final Ta_2O_5 layer ensures excellent chemical resistance (Härkönen et al., 2011). It has also been demonstrated that NaCl solution has no measurable influence on Ta_2O_5 (Betz, 2011), ensuring biocompatibility.

It should be noted that the pressure sensor membrane deflection with 100 nm Ta_2O_5 has been already examined by Betz (2011) with an optical interferometer. It was observed that the pressure-dependent deflection of a membrane was decreased by 19 % compared to the unpassivated membrane. A reduction in pressure sensitivity of this magnitude can be easily compensated by the gain setting of the integrated readout circuit of the pressure sensor. Furthermore, the pressure sensor is designed so that the membrane touches the underlying substrate at a pressure of approximately 1.6 bar; therefore the maximum pressure survivability is rather high (~ 5 bar), definitely much higher than the maximum pressure experienced in an implant.

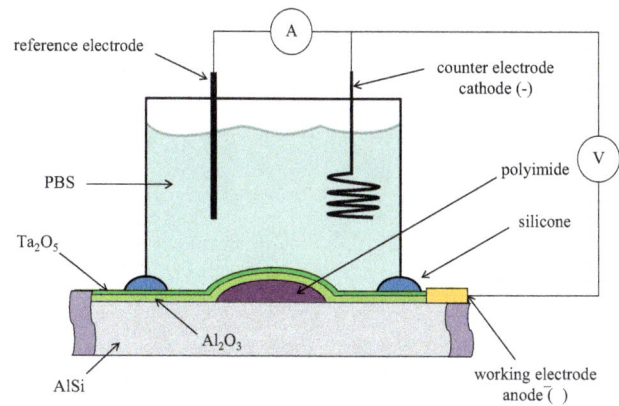

Figure 10. Schematic representation of the test setup for the amperometric defect investigation.

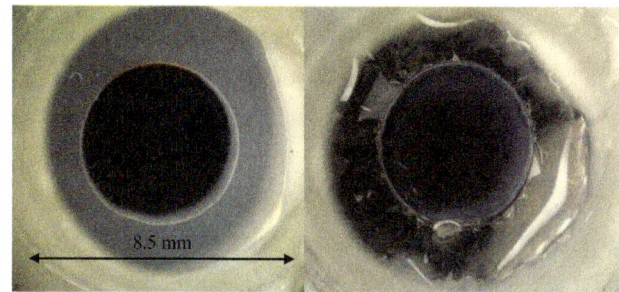

Figure 11. Intact probe after an amperometric defect test (left) and a sample having a defect in the ALD passivation (right). The sacrificial AlSi layer is completely decomposed.

Table 1 summarizes the process parameters for the deposition of the ALD layers.

4.2 Investigation of the ALD passivation

4.2.1 Amperometric defect investigation of the ALD passivation layer on a test structure

The interface between the chip surface and the polyimide is a critical area which had to be examined in more detail since fractures due to stress caused by the cooling process may damage the ALD passivation layers. Another failure mechanism is particle contamination. In order to locate defects in the passivation a destructive amperometric measurement test was carried out. For this purpose a silicon test wafer was fabricated with the following layers: 10 nm Ti, 40 nm TiN, 900 nm AlSi deposited by sputtering.

The wafer was diced with a wafer-saw into individual chips with a size of 25×25 mm^2. Then, in a circular area of approximately 5 mm diameter polyimide–epoxy composite was applied manually with a dispenser and cured with the same parameters used for the pressure sensor module. This entire structure was coated with the ALD passivation layers. A first inspection by SEM, as shown in Fig. 9, revealed that

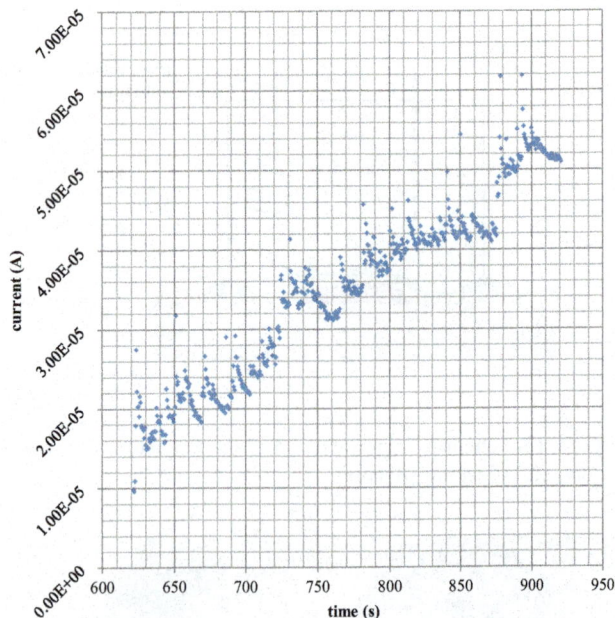

Figure 12. Current flow at a defective passivation at 3 V.

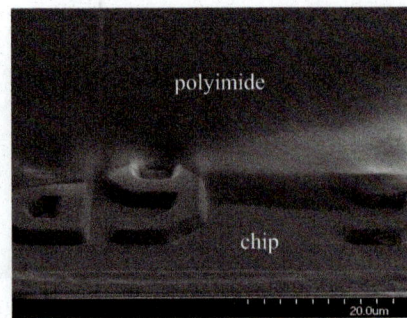

Figure 13. SEM image of boundary between polyimide–epoxy composite and chip surface with an ALD layer (not visible) over the entire area.

Table 1. Process parameters for the deposition of the passivation layers.

Material	Thickness [nm]	Temperature [°C]	Cycles	Growth rate per cycles [Å]
Al_2O_3	50	270	500	1
Ta_2O_5	50	275	1300	0.4

in the transition region between the silicon sample and the polyimide no cracks were visible.

For the amperometric defect investigation, the test chips were prepared by gluing a plastic tube with silicone in the middle of the encapsulated region thus creating a container for PBS. The AlSi layer directly underneath the passivation test layers was used as the anode. The cathode and the reference electrodes were placed in the PBS. Figure 10 shows the measurement setup.

For the measurements an impedance analyser (*Autolab μAUTOLABIII/FRA2*) was used. An applied voltage between the electrodes resulted in no measurable current flow in the case of intact passivation. Contrarily, in the case of a defect in the ALD passivation layer, the AlSi layer began to decompose due to the contact with the electrolyte resulting in a measurable current flow of about 1 μA which slowly increased. The AlSi layer directly below the passivation thus formed a sacrificial layer. To get an electrical contact to the AlSi layer the passivation was destroyed with a needle in the edge region. Figure 11 shows an optically intact sample (left) and a sample with a severe defect (right) after the amperometric test.

With the test samples the following amperometric measurements were carried out at room temperature: increasing the voltage from 1 to 5 V in 1 V steps for 5 minutes each and continuous measurement for 25 h at 5 V. A voltage of 5 V was selected since many medical implants operate at this voltage.

The graph in Fig. 12 shows the current versus time for a sample with a defect. For the measurement using an increasing voltage the sample showed a first measurable current flow at a voltage of 3 V which continuously increased with time.

A sample without defect showed no measurable current flow, even after 25 h at 5 V.

The preliminary tests showed that the adhesion of the polyimide–epoxy composites was sufficient despite the high-temperature stress it was exposed to. Dust particles were identified as the main source for defects; this is attributed to the fact that the test chips were not coated at wafer level. We tested four samples which were cleaned only with a stream of nitrogen before ALD. All of these samples showed defects. Four further samples were cleaned first with acetone, isopropanol and finally with DI-water; these showed no defects. The experiments indicated that the cleaning of the surface and full processing in clean-room conditions are important for a defect-free coating.

4.2.2 Optical investigation of the sensor module

The amperometric defect investigation showed that the transition region between silicon and polyimide–epoxy composite exhibited sufficient adhesion and that ALD passivation of polyimide–epoxy composite without fractures is in principle possible. Therefore, the pressure sensor modules were coated in the ALD tool. After the coating, a sample was analysed in the SEM. Figure 13 shows a close-up of the transition region between the chip surface and the polyimide–epoxy composite at an angle of 45°. It confirms that the transitional area could be coated despite the high topography of the ASICs without cracking or fractures.

Figure 14 shows the encapsulation of the entire pressure sensor chip as a SEM picture. The brightness is due to charg-

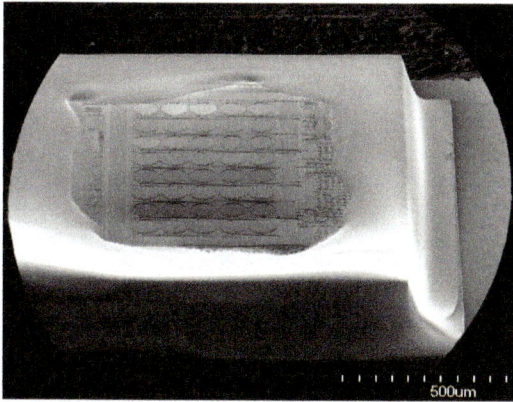

Figure 14. SEM image of a defect-free passivation.

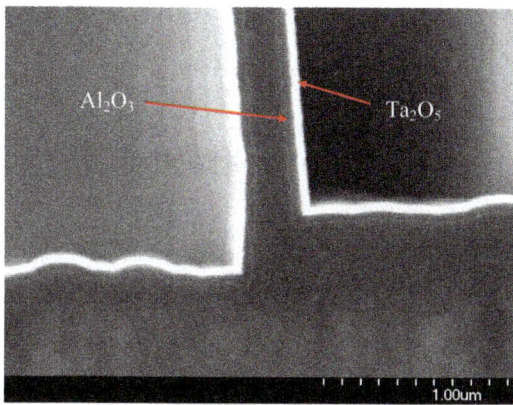

Figure 15. Cross-section SEM image of the ALD passivation layers.

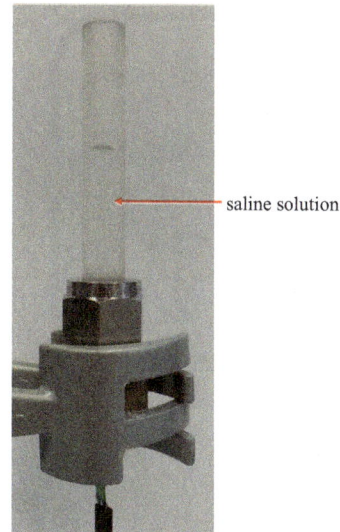

Figure 16. Photo of the housing with the transparent tube for holding the test fluid and creating a well-defined pressure.

ing of the tantalum pentoxide layer in the SEM, which has the benefit that defects can be detected easily.

Additionally, wafer pieces of pressure sensor chips were coated in the same ALD process step to control the layer thickness and conformity. Figure 15 shows an area of the pressure sensor after the ALD coating. The bright tantalum pentoxide layer on the surface is clearly visible and indicates excellent conformal coverage.

5 Test and characterization of the sensor module after encapsulation

Test measurements to evaluate and characterize the pressure sensor module after encapsulation were carried out. Prior to the measurement the sensor modules were calibrated allowing the temperature dependency of the pressure sensor to be compensated. The calibration took place in a pressure range from 800 to 1400 hPa and a temperature range of 24 to 40 °C. The calibration parameters were controlled with a pressure calibrator *DPI-520* and a PT-100 temperature sensor.

5.1 Housing for the sensor module

To apply a defined pressure in a test fluid the pressure sensor module needed a suitable housing. The housing for the sensor module was made of polyether ether ketone (PEEK) which offers excellent chemical resistance and good temperature stability. The housing is designed in a modular way to facilitate different types of measurements. Figure 16 shows the calibration setup consisting of the PEEK housing and a transparent tube screwed onto the housing. Test liquid could then be dispensed into the tube creating a well-defined pressure.

The resulting liquid column provided a sufficient fluid reservoir. It also allowed connecting an air hose to apply a defined external pressure required for the calibration. The entire setup is placed in a climatic chamber to set the desired temperature.

5.2 Measurement setup

With the presented measuring system it is possible to measure not only pressure but also temperature. The measuring system consisted of the housed pressure sensor module placed in a climate chamber and connected with a hose to the pressure calibrator. The pressure sensor module transmits the measured values through a two-wire cable to a data reader device outside of the climatic chamber. The data reader contains the interface and energy controller (IEC) and a microcontroller. The IEC represents the link between the pressure sensor module and the microcontroller to control the communication and power supply. The data reader transmits the data via USB to a PC running a purpose programmed *Lab-View* software. The software visualizes and saves the measured values. The software also controls the pressure calibra-

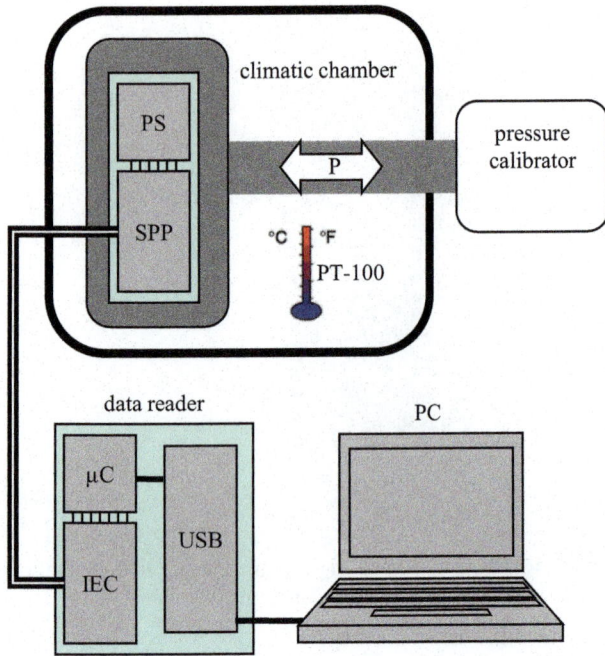

Figure 17. Overview of the pressure measurement system.

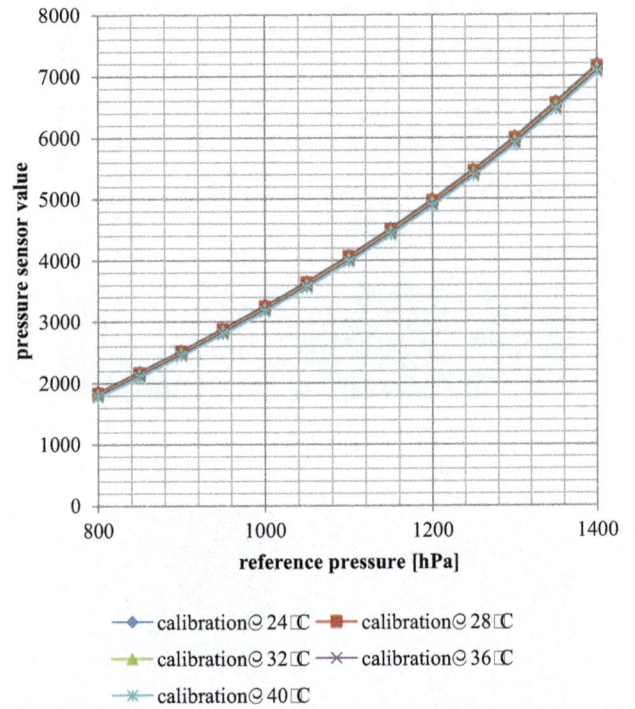

Figure 18. Pressure value of an encapsulated sensor module at various temperatures in saline solution.

tor, measures the temperature in the climatic chamber with a PT-100 and programmes the climatic chamber to adjust the temperature. Figure 17 shows the main components of the measurement setup.

5.3   Adjustment and calibration

The digital output of the pressure sensor module is in a form of values between 0 and 8191. The calibration of the pressure sensor module associates the digital values with the real pressure measured by a reference pressure sensor (*DPI-520*). To create a pressure calibration curve equidistant measurement points were selected. With two parameters it was possible to set the offset and the gain of the pressure sensor module and optimize the curve within certain limits. These parameters were introduced for compensating manufacturing tolerances and encapsulation influences.

The pressure curves used for calibration were recorded at various temperatures in 0.9 % saline solution. Figure 18 shows an example of a calibration measurement against the reference pressure sensor. The pressure curves were recorded in a range of 800 to 1400 hPa in a temperature range between 24 and 40 °C. In this resolution, the pressure curves almost overlap, indicating that the temperature coefficient of the pressure sensor module is very low.

With the measurement output values for the pressure (z_P) and the temperature (z_T) the real pressure is approximated

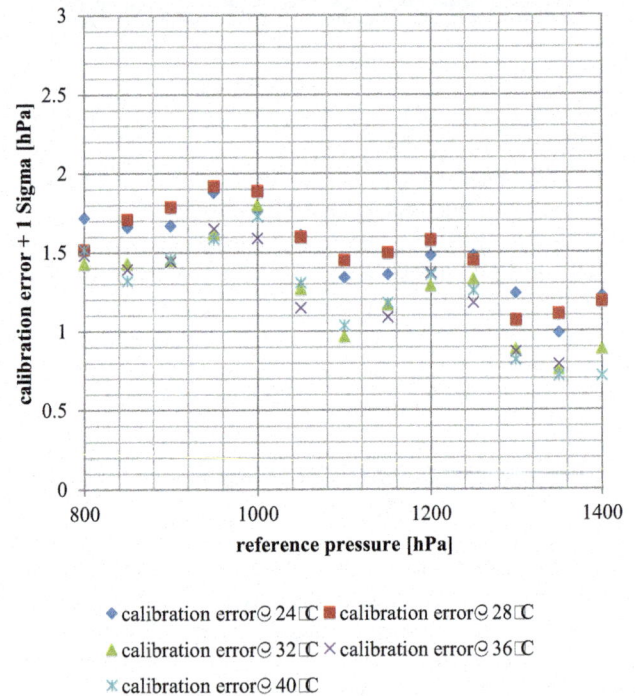

Figure 19. Pressure–calibration error (including 1-sigma noise) at different temperatures.

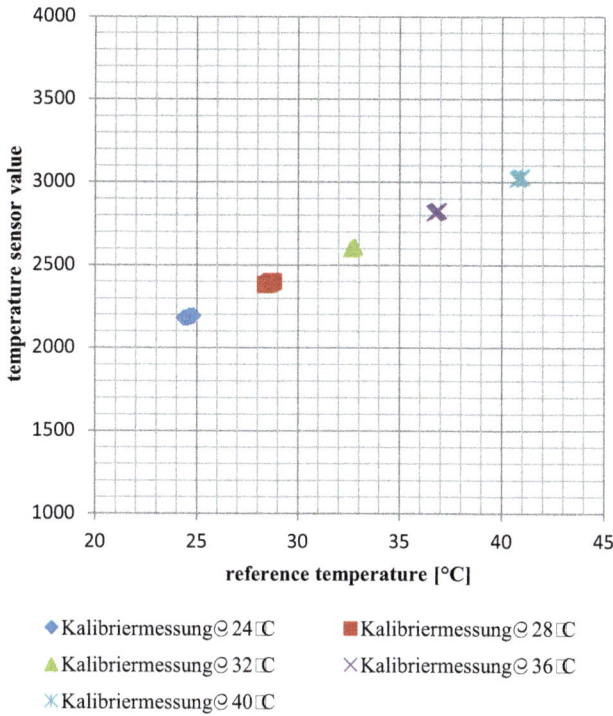

Figure 20. Temperature value of an encapsulated sensor module at various temperatures in saline solution.

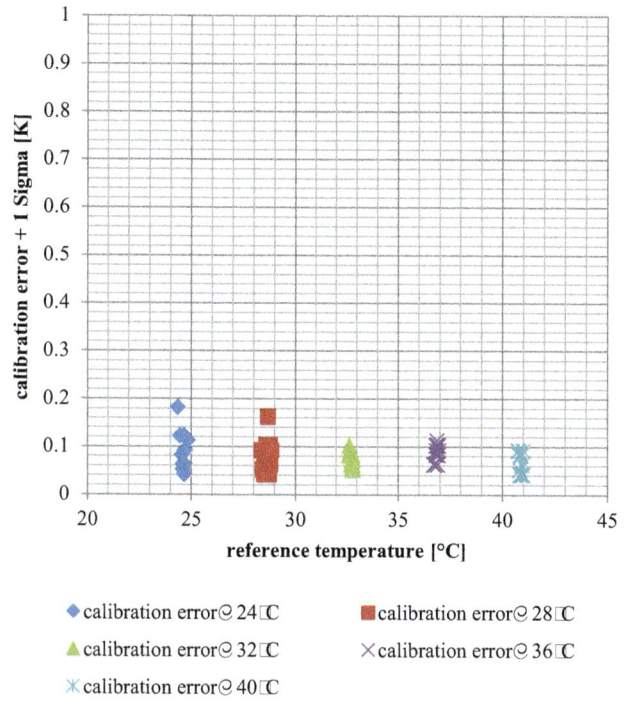

Figure 21. Temperature–calibration error (including 1-sigma noise) at different temperatures.

with a multivariable regression polynomial P_K:

$$P_K = \sum_{i=0}^{4} \left[(z_P)^i \sum_{j=0}^{2} a_{i,j}(z_T)^j \right]. \tag{1}$$

The variable $(a_{i,j})$ are the calibration coefficients of the polynomial witch are determined by the method of least squares.

For low temperature coefficients of the pressure sensor module (as for the measurement result shown in Fig. 18) calibration with the polynomial is possible. In Fig. 19 the calibration error of the pressure sensor module is shown, which is below 2 hPa (including 1-sigma noise). Based on the measurement range this corresponds to a total error of about 0.34 % full scale span.

Figure 20 shows the temperature sensor values at different pressures between 24 and 40 °C. The data points for the different pressures are very close together which demonstrates the good stability and reproducibility. As temperature reference a PT-100 was used. The calibration error of the temperature sensor shown in Fig. 21 is less than 0.2 °C (including the 1-sigma noise)

6 Accelerated life testing

To estimate the long-term stability accelerated life tests of the pressure sensor modules were carried out. Several calibrated pressure sensor modules were placed in a sealed housing filled with 0.9 % saline solution. The housed pressure

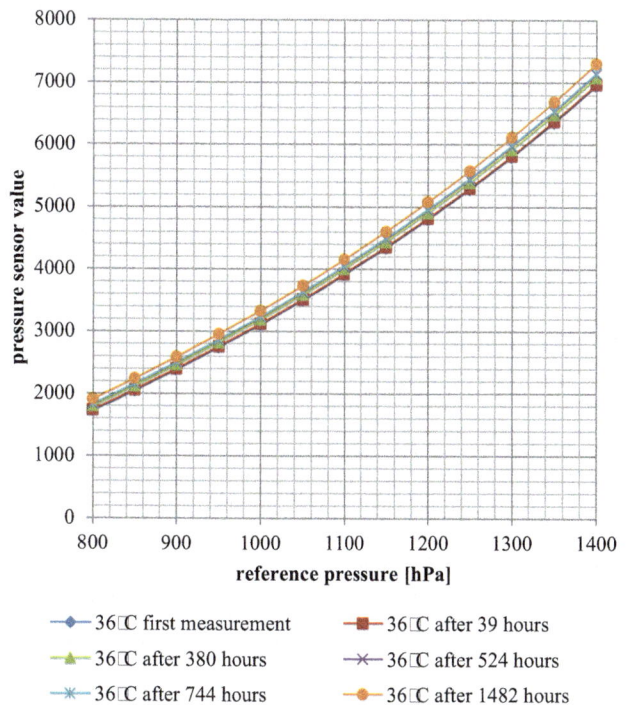

Figure 22. Pressure curves of different calibrations at 36 °C after different periods of the accelerated life time test at 60 °C in saline solution.

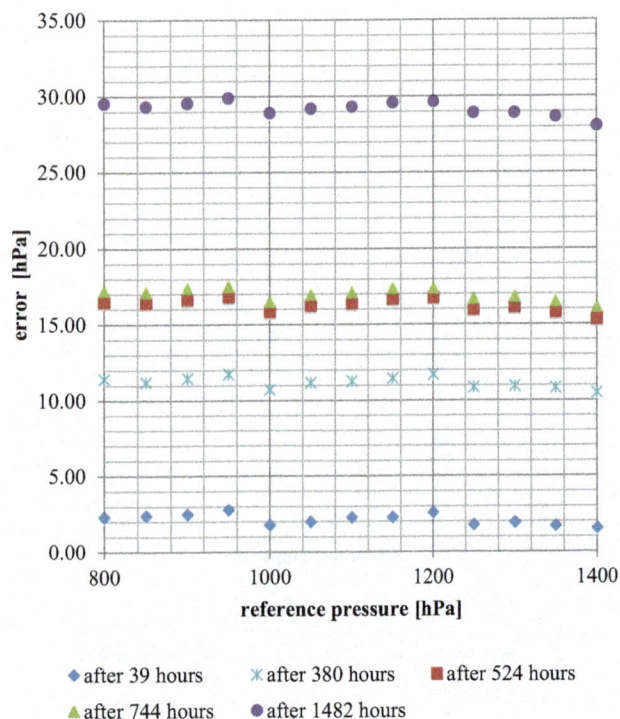

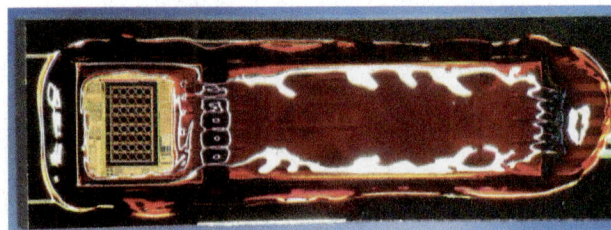

Figure 24. Photo of the sensor module with the wider silicon chip carrier.

Figure 23 chart region:

y-axis: error [hPa], ranging 0.00 to 35.00
x-axis: reference pressure [hPa], ranging 800 to 1400

Legend:
- ◆ after 39 hours   × after 380 hours   ■ after 524 hours
- ▲ after 744 hours   ● after 1482 hours

Figure 23. Error of the pressure curves after a different accelerated life test time at 60 °C in saline solution (the measurements were recorded at 36 °C).

sensor modules were stored in an oven at a temperature of 60 °C. This temperature was selected because the communication with the housed pressure sensor module was still functional at this temperature. The pressure sensor module was continuously electrically powered-up and read-out during the test. Higher temperatures would disturb the communication. To detect drift caused by environmental effects, the aging test was interrupted after certain time intervals and a control measurement was done to compare the results with the measurement before the start of the aging test. Control measurements were carried out at different temperatures and pressures. To illustrate the drift of the sensor module, pressure curves at 36 °C of the control measurements at different time intervals of the aging test are shown in Fig. 22. After the control measurements, the sensor modules were placed back in the environmental chamber.

An upward shift of the pressure curves can be observed with increasing aging time. The calibration polynomial of the first measurement before the aging period was used to calculate the pressure value error. The measurement total calibration error in hPa is shown in Fig. 23. It should be noted that the housed pressure sensor module had to be removed from the oven and the housing opened before every control measurement to allow mounting the transparent tube for the pressure calibrator. This process is likely to cause mechanical stress on the housing and thus influence the sensor output. The error after 524 h and after 744 h of aging was about

17 hPa. This corresponds to a total error of about 2.8 % full scale. Between the measurement curves at 744 and 1482 h the complete pressure sensor module housing was opened for a visual investigation. The last measurement curve after 1482 h indicates that the system changed after the inspection as the error increased up to 30 hPa.

The sensor modules failed in continuous operation after a total of 1500 h at 60 °C in saline solution. The reason for the failure is probably the saw line of the chip carrier which was not completely potted with the polyimide–epoxy composite in this design. Figure 24 shows a new design of a wider chip carrier, so that the polyimide does not extend to the saw lines of the chip carrier; this is likely to solve the problem.

The interpretation of the measurement data with respect to a prediction of long-term stability is not straightforward. Further accelerated aging tests at 100 % humidity and a temperature of 120 °C were carried out. The longest aging time here was 116 h. We used an Arrhenius approach to calculate the operating life time at 37 °C (Betz, 2011). We choose the activation-energy of 0.7 eV to calculate operating life time as it is used for the aging of semiconductors in CMOS technology (White et al., 2013; Bayle and Mettas, 2010). Here, the influence of the liquid is not taken into account. The calculated life time at 37 °C based on these tests was 3.3 years. An incorrect assumption on the value of the activation energy can lead to an incorrect deduction, which is the weakness of the acceleration model (Bayle and Mettas, 2010). If we use an activation-energy of 0.6 eV, the calculated life time at 37 °C is 1.5 years and with an activation-energy of 0.8 eV the calculated life time at 37 °C is 7.5 years. For an unambiguous prediction of the life time a greater number of sensor modules have to be tested.

7   Conclusion

An encapsulation process has been developed to use a high-temperature resistant polyimide–epoxy composite as a die-attach material and sealing compound for the bond wire and parts of the chip surface excluding the MEMS pressure membranes. The process prepares the pressure-sensor module for ALD passivation. It could be shown that a conformal ALD of Al_2O_3 and Ta_2O_5 on a polyimide–epoxy composite is possible, including the transition regions to the chip surface.

Moreover, it has been shown that a complete pressure sensor assembly can be encapsulated and hermetically sealed. The passivated pressure sensor was calibrated and tested in a 0.9 % saline solution showing excellent results.

An accelerated life test at 60 °C with permanent electric read-out indicated a maximum error of 5 % full scale after 1482 h. In a second accelerated life test at 120 °C the passivated pressure sensor system reached an operation time of 116 hours until it fails. Using an Arrhenius approach with activation energy of 0.7 eV the calculated life time at 37 °C is 3.3 years. In order to get a better assessment of the life time, a greater number of sensor modules have to be tested.

Further optimization and automation of the fabrication process may allow a new kind of hermetic encapsulation for human implants based on MEMS sensors. As a next step, it is intended to test the pressure sensor module in blood and in an animal model in the near future. However, similar results are expected at least for the Ta_2O_5 layer, as has already been found to be more corrosion resistant in blood compared to bare Ti (Sun et al., 2013).

Edited by: R. Kirchner
Reviewed by: two anonymous referees

References

Bayle, F. and Mettas, A.: Temperature Acceleration Models in Reliability Predictions: Justification & Improvements, Proceedings of Reliability and Maintainability Symposium (RAMS), Vol. 1, 25–28, 2010.

Betz, W.: Flexible mikroelektromechanische Implantate für den chronischen Einsatz: Verkapselungskonzepte und Testverfahren für die Materialcharakterisierung, PhD-Thesis, Faculty of Engineering, University Duisburg-Essen, 2011.

Beyler, C. L. and Hirschler, M. M.: Thermal Decomposition of Polymers, Chapter in SFPE Handbook of Fire Protection Engineering, 3rd Edn., 2001.

Boyle, M. A., Martin, C. J., and Neuner, J. D.: Epoxy Resins, ASM Handbook, Composites, Vol. 21, 78–89, 2001.

Bradford, B., Krautschneider, W., and Schröder, D.: Wireless Power and Data Transmission for a Pressure Sensing Medical Implant, Proceedings of BMT 2010, Rostock-Warnemünde, Germany, 2010.

Calderón, J. B.: Oberflächenmodifizierung und -analytik von Polyimid, Dissertation, Faculty of Mathematics, Informatics and Natural Sciences, RWTH Aachen University, 2005.

Chen, P.-J., Rodger, D. C., Saati, S., Humayun, M. S., and Tai, Y.-C.: Microfabricated Implantable Parylene-Based Wireless Passive Intraocular Pressure Sensors, J. Microelectromech. S., 17, 1342–1351, 2008.

Cheng, J., Li, J., and Zhang, J. Y.: Curing Behavior and Thermal Properties of Trifunctional Epoxy Resin Cured by 4, 4′-Diaminodiphenyl Sulfone, Express Polym. Lett., 3, 501–509, 2009.

Cleven, N. J., Müntjes, J. A., Fassbender, H., Urban, U., Görtz, M., Vogt, H., Gräfe, M., Göttsche, T., Penzkofer, T., Schmitz-Rode, T., and Mokwa, W.: A Novel Fully Implantable Wireless Sensor System for Monitoring Hypertension Patients, IEEE T. Bio-Med. Eng., 59, 3124–3130, 2012.

Dudaicevs, H., Kandler, M., Manoli, Y., Mokwa, W., and Spiegel, E.: Surface Micromachined Pressure Sensors with Integrated CMOS Read-Out Electronics, Sensor Actuat. A-Phys., 43, 157–163, 1994.

Eick, S., Wallys, J., Hofmann, B., van Ooyen, A., Schnakenberg, U., Ingebrandt, S., and Offenhäusser, A.: Iridium Oxide Microelectrode Arrays for in Vitro Stimulation of Individual Rat Neurons from Dissociated Cultures, Frontiers in Neuroengineering, Vol. 2, doi:10.3389/neuro.16.016.2009, 2009.

Gaw, K. O. and Kakimoto, M.: Polyimide-Epoxy Composites, Progress in Polyimide Chemistry I, Advances in Polymer Science, 140, 107–136, 1999.

Gembaczka, P., Görtz, M., Kordas, N., Lerch, R., Müntjes, J., Kraft, M., and Mokwa, W.: Integrated Capacitive Pressure Sensor for an Implantable Wireless System for Measuring Pressure in the Pulmonary Artery, Proceedings of Mikrosystemtechnik 2013 – Von Bauelementen zu Systemen, Aachen, Germany, 2013.

Gräfe, M., Göttsche, T., Osypka, P., Görtz, M., Trieu, H. K., Fassbender, H., Mokwa, W., Urban, U., Schmitz-Rode, T., Hilbel, T., Becker, R., Bender, B., Coenen, W., Fähnle, M., and Glocker, R.: HYPER-IMS: A Fully Implantable Blood Pressure Sensor for Hypertensive Patients, Proceedings of Sensor + Test Conference, 2009.

Ha, D., de Vries, W. N., John, S. W. M., Irazoqui, P. P., and Chappell, W. J.: Polymer-Based Miniature Flexible Capacitive Pressure Sensor for Intraocular Pressure (IOP) Monitoring Inside a Mouse Eye, Biomedical Microdevices, 14, 207–215, 2012.

Hang Tng, D. J., Hu, R., Song, P., Roy, I., and Yong, K.-T.: Approaches and Challenges of Engineering Implantable Microelectromechanical Systems (MEMS) Drug Delivery Systems for in Vitro and in Vivo Applications, Micromachines, 3, 615–631, 2012.

Härkönen, E., Díaz, B., Światowska, J., Maurice, V., Seyeux, A., Vehkamäki, M., Sajavaara, T., Fenker, M., Marcus, P., and Ritala, M.: Corrosion Protection of Steel with Oxide Nanolaminates Grown by Atomic Layer Deposition, J. Electrochem. Soc., 158, 369–378, 2011.

Hassler, C., Boretius, T., and Stieglitz, T.: Polymers for Neural Implants, J. Polym. Sci. Pol. Phys., 49, 18–33, 2011.

Loeb, G. E.: Cochlear Prosthetics, Annu. Rev. Neurosci., 13, 357–371, 1990.

Mokwa, W.: Medical implants based on microsystems, Measurement, Sci. Technol., 18, R47, doi:10.1088/0957-0233/18/5/R01, 2007.

Müntjes, J., Meine, S., Flach, E., Görtz, M., Hartmann, R., Schmitz-Rode, T., Trieu, H. K., and Mokwa, W.: Assembly of a Pulmonary Artery Pressure Sensor System, Acta Polytechnica, 50, 56–59, 2010.

Park, J. and Lakes, R. S.: Biomaterials: An Introduction, 3rd Edn., Springer, 2007.

Polytec PT GmbH: Duralco 4461, Data Sheet, 2010.

Polytec PT GmbH: Polytec EP P-690, Data Sheet, 2011.

Rezai, A. R., Finelli, D., Nyenhuis, J. A., Hrdlicka, G., Tkach, J., Sharan, A., Rugieri, P., Stypulkowski, P. H., and Shellock, F. G.: Neurostimulation Systems for Deep Brain Stimulation: In Vitro Evaluation of Magnetic Resonance Imaging–Related Heating at 1.5 Tesla, JMRI-J. Magn. Reson. Im., 15, 241–250, 2002.

Rubehn, B. and Stieglitz, T.: In Vitro Evaluation of the Long-Term Stability of Polyimide as a Material for Neural Implants, Biomaterials, 31, 3449–3458, 2010.

Schlierf, R., Görtz, M., Schmitz-Rode, T., Mokwa, W., Schnakenberg, U., and Trieu, H. K.: Pressure Sensor Capsule to Control the Treatment of Abdominal Aorta Aneurisms, Proceedings of the 13th International Conference on Solid-State Sensors, Actuators and Microsystems, 2, 1656–1659, Seoul, South Korea, 2005.

Stangel, K., Kolnsberg, S., Hammerschmidt, D., Hosticka, B. J., Trieu, H. K., and Mokwa, W.: A Programmable Intraocular CMOS Pressure Sensor System Implant, IEEE J. Solid-St. Circ., 36, 1094–1100, 2001.

Sun, Y.-S., Chang, J.-H., and Huang, H.-H.: Corrosion Resistance and Biocompatibility of Titanium Surface Coated with Amorphous Tantalum Pentoxide, Thin Solid Films, 528, 130–135, 2013.

Trieu, H. K.: Surface Micromachined Pressure Sensors for Medical Applications, Proceedings of Sensor + Test Conferences, Nuremberg, Germany, 2011.

Walewyns, T., Reckinger, N., Ryelandt, S., Pardoen, T., Raskin, J. P., and Francis, L. A.: Polyimide as a Versatile Enabling Material for Microsystems Fabrication: Surface Micromachining and Electrodeposited Nanowires Integration, J. Micromech. Microeng., 23, 095021, doi:10.1088/0960-1317/23/9/095021, 2013.

White, M., Cooper, M., Yuan Chen, and Bernstein, J.: Impact of Junction Temperature on Microelectronic Device Reliability and Considerations for Space Applications, Proceedings of Integrated Reliability Workshop Final Report, IEEE International, 133–136, 2003.

20

Characterization of ash particles with a microheater and gas-sensitive SiC field-effect transistors

C. Bur[1,2], **M. Bastuck**[1], **A. Schütze**[1], **J. Juuti**[3], **A. Lloyd Spetz**[2,3], **and M. Andersson**[2,3]

[1]Lab for Measurement Technology, Saarland University, Saarbrücken, Germany
[2]Div. of Applied Sensor Science, Linköping University, Linköping, Sweden
[3]Microelectronics and Material Physics Laboratories, University of Oulu, Oulu, Finland

Correspondence to: C. Bur (c.bur@lmt.uni-saarland.de, chrbu@ifm.liu.se)

Abstract. Particle emission from traffic, power plants or, increasingly, stoves and fireplaces poses a serious risk for human health. The harmfulness of the particles depends not only on their size and shape but also on adsorbates. Particle detectors for size and concentration are available on the market; however, determining content and adsorbents is still a challenge.

In this work, a measurement setup for the characterization of dust and ash particle content with regard to their adsorbates is presented. For the proof of concept, ammonia-contaminated fly ash samples from a coal-fired power plant equipped with a selective non-catalytic reduction (SNCR) system were used. The fly ash sample was placed on top of a heater substrate situated in a test chamber and heated up to several hundred degrees. A silicon carbide field-effect transistor (SiC-FET) gas sensor was used to detect desorbing species by transporting the headspace above the heater to the gas sensor with a small gas flow. Accumulation of desorbing species in the heater chamber followed by transfer to the gas sensor is also possible.

A mass spectrometer was placed downstream of the sensor as a reference. A clear correlation between the SiC-FET response and the ammonia spectra of the mass spectrometer was observed. In addition, different levels of contamination can be distinguished. Thus, with the presented setup, chemical characterization of particles, especially of adsorbates which contribute significantly to the harmfulness of the particles, is possible.

1 Introduction

Particle emission from traffic or huge power plants poses a serious risk for human health. The harmfulness of the particles depends mainly on their size, shape and content (Buzea et al., 2007). In recent years, the amount of nano-sized particles has increased considerably, which increases the risk for human beings. (Buzea et al., 2007; NIOSH, 2013).

Particle detectors for size and concentration are available on the market and are usually based on optical systems, e.g., light scattering (Xu, 2014) or charging of particles (Ntziachristos et al., 2011; Lanki et al., 2011; Amanatidis et al., 2013). Surface acoustic wave resonators (SAWR) have been used to detect submicron-sized particles with a mass below 1 ng (Thomas et al., 2013). For soot detection, sensor systems based on thermophoresis (Bjorklund, 2010) and electri-

cal impedance spectroscopy (EIS) of interdigital electrodes (IDE) (Messerer, 2003; Bartscherer and Moos, 2013) are also presented. Impedance spectroscopy is also being developed to reveal particle size (Osite et al., 2011; Lloyd Spetz et al., 2013). Geiling et al. (2013) presented a hybrid particle detector based on low-temperature cofired ceramics (LTCC) which measures the interaction of single particles with an electrical field. Not only the particle itself but also its composition can be harmful, and, in particular, adsorbed substances raise the potential risk significantly. Particularly in heavy industry work place environments, workers are exposed to high concentrations of ash and dust particles, which can affect their health (Lanki et al., 2011). Identification and quantification of such particles or adsorbates may potentially be used as a method for assessment of their health effects. However, determining the content of particles is still a challenge.

Gas-sensitive field-effect transistors based on silicon carbide as a substrate material (SiC-FET) are suitable sensors to operate in harsh environments. Development of different applications ranging from exhaust monitoring related to vehicles (Larsson et al., 2002) and small- and medium-scale power plants (Andersson et al., 2007), to ammonia detection in selective catalytic reduction (SCR) systems (Andersson et al., 2013), to sulfur dioxide detection in huge power plants (Darmastuti et al., 2014) have been demonstrated in the last years. The outstanding performance of the sensors in withstanding these environments is largely due to the chemical inertness of silicon carbide (SiC). In addition, SiC has a wide band gap (3.2 eV for 4H-SiC), which allows for operating temperatures up to 1000 °C without loss of its semiconducting behavior (Lloyd Spetz et al., 1997, 2003). The SiC field-effect transistor can be made gas sensitive by using a catalytic gate material, like palladium (Pd), platinum (Pt) or iridium (Ir) (Lundström et al., 2007). The sensing properties of SiC-FETs depend mainly on the gate material, its structure (porosity and number of three phase boundaries), the underlying oxide and the operating temperature. Gas molecules arriving at the catalytic surface of the gate can directly adsorb, dissociate and/or react with, for example, adsorbed oxygen. Adsorbed species on the surface of the sensor change the gate to a substrate electric field, which in turn influences the concentration of mobile carriers in the channel of the transistor. This causes a shift in the IV curve of the sensor. A detailed description of the sensing mechanism can be found elsewhere (Andersson et al., 2013).

Performance of SiC-FETs in terms of sensitivity and selectivity can be enhanced by dynamic operation. It has been reported that discrimination of typical exhaust gases (Bur et al., 2012a) as well as quantification of nitrogen oxides (NO_x) (Bur et al., 2012b) is possible when using temperature-cycled operation (TCO). Additionally, gate-bias-cycled operation (GBCO) together with TCO can boost the selectivity of the sensors further (Bur et al., 2014).

In this work, a new method is proposed to study the content of particles, i.e., substances adsorbed on the particles. For that, a large-scale laboratory measurement setup based on a ceramic hotplate and a SiC-FET gas sensor is suggested in order to measure the content of particles. The results presented in this paper can be seen as a proof of concept and are part of the ongoing development of a portable particle detector (Lloyd Spetz et al., 2013).

2 Methodology

Measuring the content of particles with a cost-effective, handheld device is a challenging task. Since not only the content of the particle itself can be harmful to humans but also adsorbed substances, we propose a setup in which either the particles themselves or their adsorbates are transformed into the gas phase in order to be detected by a gas sensor. There-

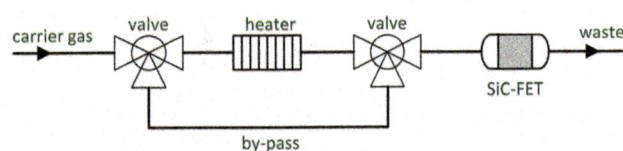

Figure 1. Schematic of the measurement setup.

fore, we suggest placing/collecting the test samples/particles on top of a ceramic heater in order to rapidly heat up the particles. The outgassing substances are then detected by a gas-sensitive field-effect transistor located downstream of the heater. In order to investigate this concept, the measurement setup shown in Fig. 1 is suggested.

The setup consists of a heater chamber, two valves, a bypass to the heater and the sensor chamber. The bypass approach gives rise to several advantages: (1) the gas sensor is always under controlled conditions, and thus no disturbances affect the sensor response, e.g., when placing the particles on top of the heater; (2) using the valves allows well defined exposure of the desorbates to the gas sensor. (3) Asynchronous operation of the valves and heater allows for example accumulation of desorbates before exposing them to the gas sensor.

However, when switching the valves we observed that the sensor response can be affected, which might be due to a change in pressure and/or flow inside the tubing system. Therefore, for preliminary measurements (Sects. 4.1–4.4), we used a simpler setup without the bypass where the sensor is connected directly to the heater chamber. Nevertheless, our suggested bypass approach is in particular interesting for an integrated particle sensor which has to accumulate particles independently on the heater. First results when using the complete setup (cf. Fig. 1) are presented in Sect. 4.5 and can be seen as an extension of the paper.

3 Experimental

3.1 Gas-sensitive field-effect transistor

For all measurements n-channel metal insulator semiconductor field-effect transistors (MISFET) based on silicon carbide (SiC-FET) were used (Fig. 2a). The devices are processed from 4 in. 4H-SiC wafers with mass production technology (SenSiC AB, Kista, Sweden, and ACREO AB, Kista, Sweden). Each sensor chip holds four sensors. As catalytic gate metallizations, 25 nm thick porous platinum and 30 nm thick porous iridium films were used. The gate dimension was 300 µm wide by 10 µm long. A detailed description can be found elsewhere (Andersson et al., 2013). The SiC chip was glued onto a ceramic heater (Heraeus PT-6.8 M 1020, Heraeus Sensor Technology, Kleinostheim, Germany) to allow for precise heating of the sensor. As a reference, a Pt-100 temperature sensor (Heraeus GmbH, Germany) was

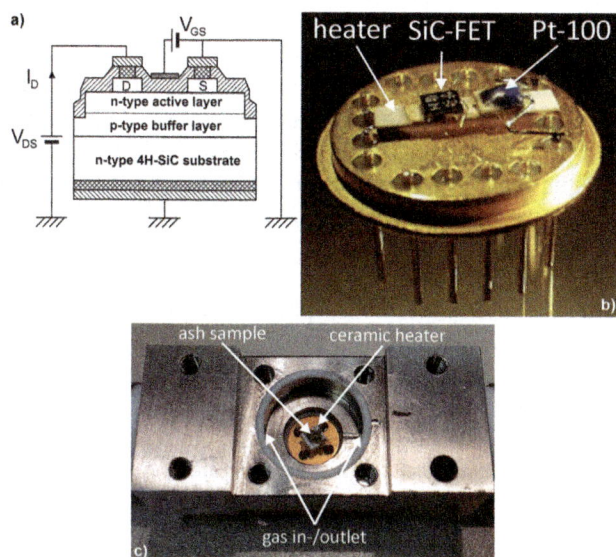

Figure 2. (a) Cross-sectional view of the SiC-FET used. (b) Picture of the SiC-FET sensor mounted on a TO-8 header. (c) Picture of the stainless steel heater chamber holding a ceramic header mounted on a TO-5 header with some ash on top.

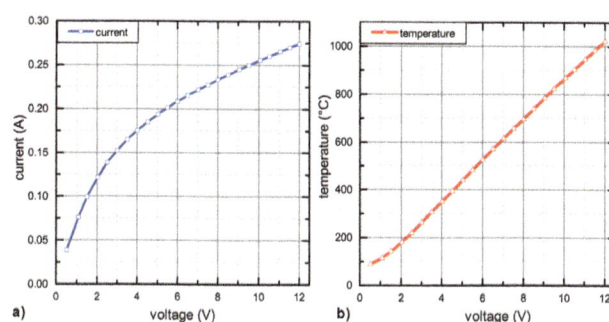

Figure 3. (a) Current–voltage characteristics of the ceramic heater used. (b) Derived temperature–voltage characteristics.

placed next to the SiC chip in order to measure the actual temperature. The SiC-FET chip and the Pt-100 were glued on the surface of the heater using a high-temperature, non-conducting ceramic die. The electrical contacts of the heater substrate and of the Pt-100 were established by spot welding to two pairs of pins of the gold-plated 16-pin TO8 header (cf. Fig. 2b). Electrical contacts to the FET structures on the SiC chip were made via gold wire bonding.

The SiC-FET was operated in a constant drain-current mode while the drain–source voltage was recorded as the sensor signal. Typical values of the current were in the range of 30 to 60 μA. Gate and substrate contacts of the transistor were grounded. In general, the baseline of the transistor can be adjusted by applying a bias to either the substrate or the gate, whereas the gate bias additionally influences the sensing behavior (Bur et al., 2014).

Sensor control and data acquisition was performed using a combined system developed by 3S GmbH, Saarbrücken, Germany. The system controls the sensor temperature with an analog control circuit with a resolution of 1 °C. Data acquisition is performed using a 14 bit analog-to-digital converter (ADC) measuring the drain–source voltage with a theoretical resolution of approximately 0.4 mV. The drain current can be set with an accuracy of less than 1 μA. The acquisition rate for all measurements was 10 Hz.

3.2   Ceramic heater platforms

Sensor substrates from Umwelt Sensor Technik GmbH, Geschwenda, Germany, without a sensing layer were used as ceramic heaters. The substrates consist of a platinum heater

(Pt-10) on an alumina substrate with outer heater dimensions of 3 mm × 3 mm × 0.75 mm (Fig. 2c). In order to set the temperature, a voltage–resistance curve for the heater was recorded by applying a stepwise increasing voltage and simultaneously measuring the current. Based on this, a voltage–temperature characteristic (Fig. 3) was achieved. Temperature of the heater can be calculated accordingly:

$$T = \frac{-A + \sqrt{A^2 - 4 \cdot B \cdot \left(-\frac{R}{R_0} + 1\right)}}{2 \cdot B},$$

where $R = \frac{V}{I}$ is the resistance; R_0 is the resistance at 0 °C, here 10 Ω; and $A = 3.9083 \times 10^{-3}$ °C^{-1} and $B = -5.775 \times 10^{-7}$ °C^{-2} are the parameters of the standard platinum curve.

3.3   Test samples

In order to study the suggested setup for measuring content of particles, ammonia (NH_3)-contaminated fly ash from a coal-fired power plant was used. This power plant uses a selective non-catalytic reduction (SNCR) system to reduce emissions of nitrogen oxides by injecting urea. The amount of ammonia in the fly ash was analytically determined by SGS Institute Fresenius, Germany. Test samples with varyingly high ammonia contaminations, i.e., 34, 64 and 84 mg kg^{-1} (milligram ammonia per kilogram of ash), were used for testing the proposed setup.

3.4   Measurement setup

As described in Sect. 2, the measurement setup consists of a heater chamber, two valves and a sensor chamber. The heater and sensor chamber are made of stainless steel, and three/two-way valves with PEEK housing and Kalrez (FFKM) sealing from Bürkert GmbH, Ingelfingen, Germany (type 6608), were used. The connections between the different parts were made by 1/4 in. stainless steel Swagelok tubing. Dry synthetic air with a flow of 25 mL min^{-1} was used as carrier gas. For preliminary testing and validation, an environmental mass spectrometer (Hiden HPR20 running

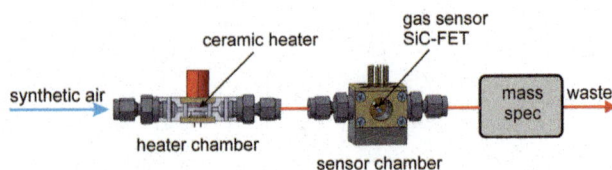

Figure 4. Schematic of the setup used with a mass spectrometer in the downstream.

MasSoft 7 Pro) was additionally placed downstream of the heater, i.e., without using the sensor chamber and later on also downstream of the sensor chamber as shown in Fig. 4. In this setup the valves and the bypass are not used since the mass spectrometer measurements are only applied for validation purposes. In this case, the ash was placed on top of the heater and the heating process started after the baseline of the SiC-FET had been stabilized, i.e., a few minutes after placement of the particles. Besides dry synthetic air, argon was also used as a carrier gas for mass spectrometer measurements. Using argon as carrier gas provides the possibility to follow the carbon monoxide (CO, 28 u) signal, which has the same mass as molecular nitrogen (N_2, 28 u).

For each measurement a small pile of ash, approximately 1 mg, was placed on top of the heater (see Fig. 2c).

4   Results and discussion

In this chapter, results from silicon carbide field-effect transistors (SiC-FET) together with mass spectrometer data are presented. In Sect. 4.1 the desorbates from the ash heated up to several hundred degrees are analyzed by means of an environmental mass spectrometer. In the following section, reference measurements with a gas mixing system and the SiC-FETs are performed in order to allow comparison with the results presented in Sects. 4.3 and 4.4. The last section deals with the suggested bypass approach and can be seen as an extension of the paper.

4.1   Characterization of fly ash

As a first step, the heater chamber was directly connected to a mass spectrometer in order to analyze desorbing substances from the ash. Six different substances of interest, i.e., ammonia (NH_3, 17 u), water vapor (H_2O, 18 u), nitrogen monoxide (NO, 30 u), carbon dioxide (CO_2, 44 u), nitrogen dioxide (NO_2, 46 u) and sulfur dioxide (SO_2, 64 u), were chosen to be monitored during the measurements. Although it is known that the ash samples are ammonia-contaminated, NO_x and SO_x are probably also contained in the ash since it is a byproduct of combustion processes. Figure 5 shows the mass spectra when a small pile ($\sim$ 1 mg) of ammonia-contaminated ash (here: 84 mg kg^{-1}) is heated up to 430 and 860 °C. At 430 °C ammonia and water can be desorbed from the ash (cf. Fig. 5a), whereas there is no signal for the other

substances. The change in mass spectra corresponding to the second heating pulse greatly decreased, which is plausible since most of the contaminations had already desorbed by the first pulse. In addition to ammonia, there is also water vapor adsorbed to the ash particles, which is probably from the lab atmosphere. However, the signal for water is overlapping and similar in shape to the ammonia signal. This can partly be due to measurement errors, since both molecules have almost the same molecular weight (ammonia 17 u and water 18 u). However, water is most probably present and then also influences the sensor response.

When the ash sample was heated up to 860 °C not only NH_3 and H_2O were desorbed but also large amounts of NO_2, SO_2 and CO_2 (cf. Fig. 5b). Similar to ammonia and water, the peak heights decrease for the second and third heating pulse but the signal is still quite high. Whereas CO_2 has almost no influence on the SiC-FET signal, NO_2 and SO_2, in contrast to ammonia, are known to be detected as oxidizing gases. Thus, when heating up the samples to high temperatures, both reducing and oxidizing gases will be desorbed. The effects from oxidizing and reducing gases on the sensor signal may partly cancel out when these gases are simultaneously desorbed from the particles since they give rise to opposing sensor responses. However, since the heater that was used has a time constant of a few seconds, there is a period of temperature increase at the beginning of each pulse. Therefore, ammonia and water are released first, followed by the other substances.

Since there is a large amount of CO_2 outgassing, it is likely that CO, which can be detected by the SiC-FETs, is also present. However, when using synthetic air as a carrier gas, one cannot follow carbon monoxide due to the fact that it has the same mass as molecular nitrogen (N_2, 28 u). A small change due to degassing CO cannot be resolved by the mass spectrometer when using synthetic air. Hence, argon was used as a carrier gas instead. For heating pulses up to 430 °C, neither CO_2 nor CO is released from the ash (Fig. 6a). However, at higher temperatures (e.g., 860 °C) CO appears, which is a reducing gas as well (Fig. 6b). In summary, the heating temperature needs be chosen carefully in order to desorb the correct target gas. However, specific temperatures or a temperature ramp can be used for selective desorption and fingerprint detection of desorbants.

4.2   Reference measurement of ammonia and humidity

Before measuring desorbates from ash samples, reference measurements were performed with a platinum gate SiC-FET. Figure 7a shows the sensor responses to 1, 2.5, 4, 5.5 and 7 ppm ammonia in synthetic air under dry conditions. The sensor response of a Pt-gate SiC-FET at 220 °C is 110 mV for 1 ppm ammonia. The response of an Ir-gate SiC-FET at 280 °C is much lower, i.e., 53 mV for 1 ppm ammonia; however, Ir is more selective over, for example, hydrocarbons as compared to the Pt-gate SiC-FET (Andersson

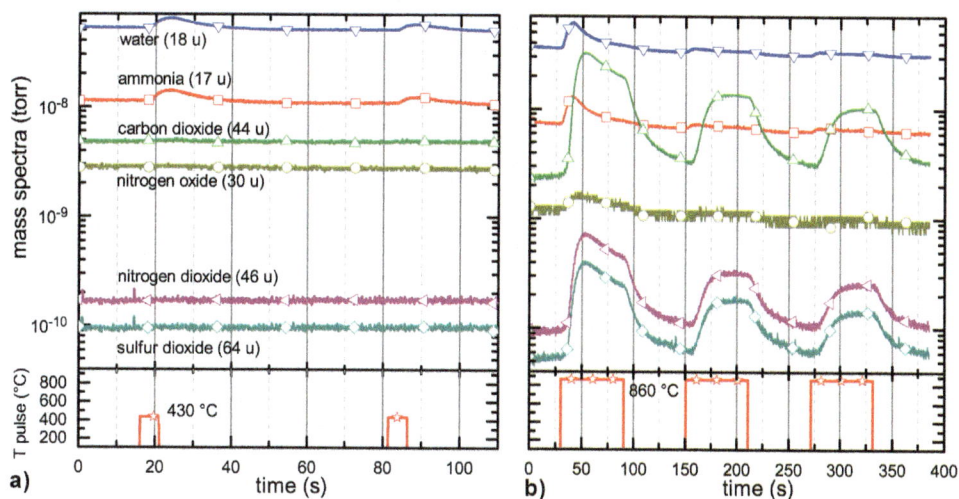

Figure 5. Mass spectra of ammonia-contaminated fly ash ($84\,mg\,kg^{-1}$) when heating up the ash to $430\,°C$ **(a)** and $860\,°C$ **(b)**. Carrier gas is dry synthetic air.

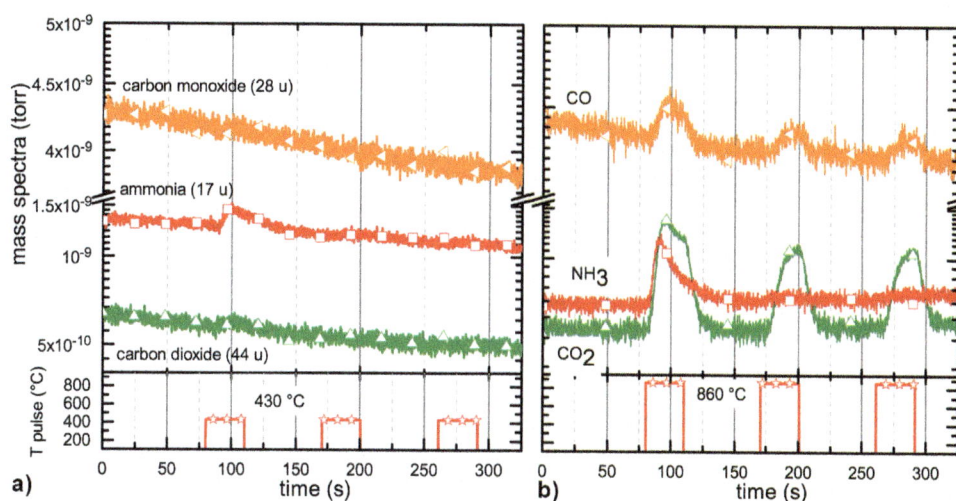

Figure 6. Mass spectra of ammonia-contaminated fly ash ($84\,mg\,kg^{-1}$) when heating up the ash to $430\,°C$ **(a)** and $860\,°C$ **(b)**. Carrier gas is argon.

et al., 2004). Iridium-gate SiC-FETs have been successfully used as ammonia sensors in diesel engine selective catalytic reduction (SCR) systems (Wingbrant et al., 2005).Therefore, iridium-gate SiC-FETs should also be considered in this work.

As shown in Fig. 7b, humidity only has a minor impact on the sensor response (Wingbrant et al., 2005). There is a large difference in baseline between 0 and 10 % relative humidity; however, with increasing humidity the impact becomes smaller. Since there is, besides ammonia, also water vapor desorbing from the ash particles, the corresponding sensor response is to some extent also due to a change in humidity.

Interested readers are referred to Andersson et al. (2004, 2013), in which the sensor response towards, for example, ammonia and carbon monoxide over a wide temperature range is studied. As mentioned earlier, the selectivity of the SiC-FET can be increased by dynamic operation (Bur et al., 2012a, b, 2014).

4.3 SiC-FET response

For preliminary measurements with a gas-sensitive SiC-FET, the setup shown in Fig. 4, where the sensor chamber is directly connected to the heater chamber, was used. In Fig. 8 the sensor response of a Pt-gate SiC-FET at $200\,°C$ is given. In this example, the carrier gas stream has been humidified using a commercial PermaPure tube (Perma Pure, 2014) to the humidity level of the laboratory environment (approximately 30 %). This reduces the influence of degassing water vapor since SiC-FET sensors show almost no sensitivity to

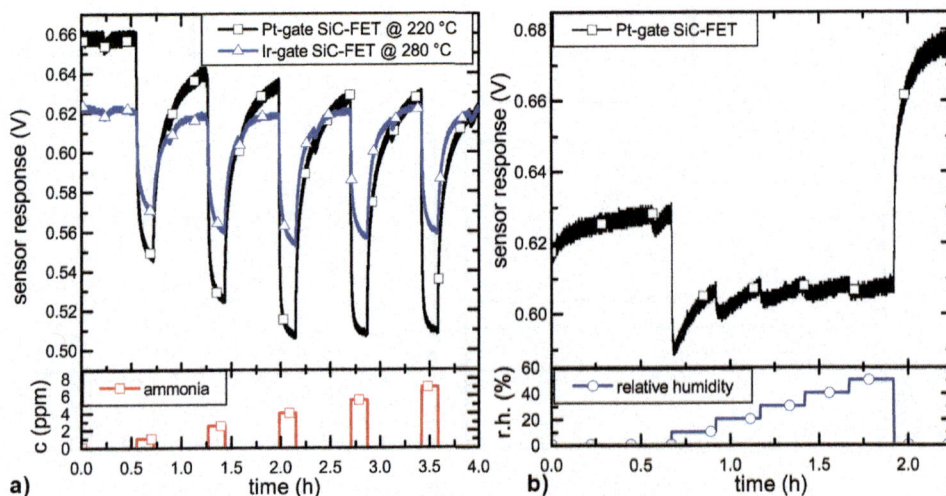

Figure 7. (a) Reference measurements of a Pt-gate SiC-FETs at 220 °C and Ir-gate SiC-FET at 280 °C to different concentrations c (1, 2.5, 4, 5.5 and 7 ppm) of ammonia under dry conditions. **(b)** Influence of relative humidity RH.

variations in humidity for RH higher than 10 % (Wingbrant et al., 2005). Ammonia-contaminated fly ash (84 mg kg^{-1}) was placed on top of the heater which was mounted in the heater chamber. First, two heating pulses up to 430 °C with 30 s each and a break of 120 s in between were applied, followed by two pulses up to 860 °C. The first pulse desorbs most of the ammonia as proven by mass spectrometer measurements. The response of the SiC-FET decreases by approximately 80 mV due to the release of ammonia and water. The two high-temperature pulses lead to desorption of nitrogen dioxide and sulfur dioxide, as well as additional ammonia, water and carbon monoxide (shown in Fig. 6b). Thus, the response of the SiC-FET decreases first (reducing gases) due to the long thermal time constant of the heater (approximately 3–4 s), but shortly thereafter the response increases significantly, which is due to the reaction of oxidizing gases. The response is lowered for the second heating pulse, which suggests that substances are indeed desorbing from ash particles.

In order to see the influence of the heating pulses, the temperature of the heater was increased stepwise from 180 to 860 °C. As can be seen in Fig. 9, some ammonia can already be desorbed at low temperatures; however, the maximum sensor response is reached around 400 °C. Of course, with each pulse, ammonia is released and the amount of ammonia still adsorbed to the ash decreases. Nevertheless, in comparison with other results, higher temperatures (above 600 °C) cause the release of oxidizing (here unwanted) gases. This is also visible in the SiC-FET response in Fig. 9. The decreasing peaks in the response corresponding to 700, 780 and 860 °C pulses are due to reducing gases, probably carbon monoxide, as shown in Fig. 6b. However, the influence of the oxidizing gases is still visible as indicated in Fig. 9.

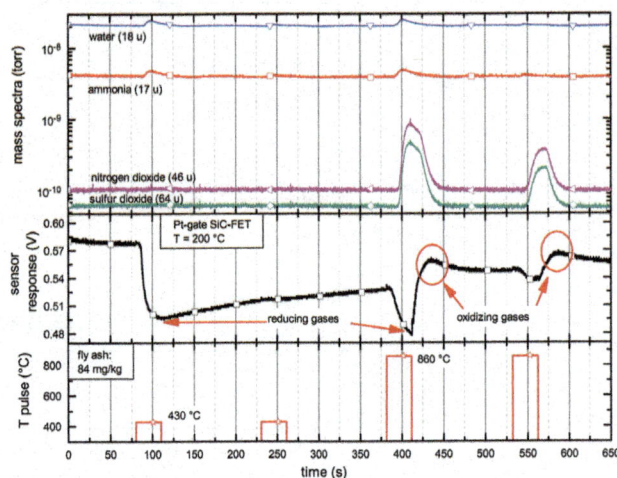

Figure 8. Top: mass spectra; middle: sensor response of a Pt-gate SiC-FET at 200 °C; bottom: heating pulses to 430 and 860 °C of ammonia-contaminated fly ash (84 mg kg^{-1}). With PermaPure tube for humidifying the carrier gas (synthetic air).

In further measurements, SiC-FETs with platinum as well as with iridium were used as the gate metallization. Results shown in Fig. 10 (without a PermaPure tube) indicate that iridium has a shorter recovery time after exposure to ammonia and is slightly more sensitive since the sensor also detects the release of ammonia from the second and third pulse. The operating temperature of the sensor was 280 °C and therefore higher than for platinum, which is due to the fact that iridium is less catalytically active than platinum, making a higher temperature necessary (Andersson et al., 2004).

When applying 650 °C pulses to the ash sample, the Ir-gate FET only shows a decreasing response which corresponds to reducing gases (cf. Fig. 11), i.e., ammonia and

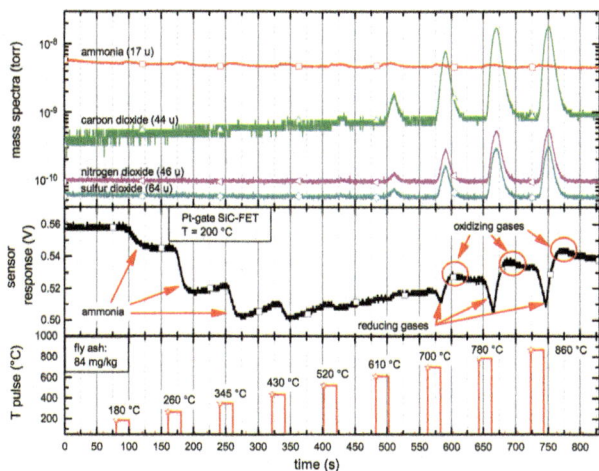

Figure 9. Top: mass spectra; middle: sensor response of a Pt-gate SiC-FET at 200 °C; bottom: heating pulses from 180 to 860 °C of ammonia-contaminated fly ash (84 mg kg^{-1}). With PermaPure tube for humidifying the carrier gas (synthetic air).

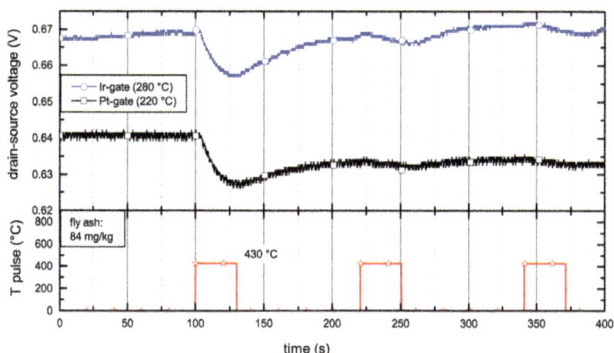

Figure 10. Top: sensor responses of Pt- and Ir-gate SiC-FETs at 220 and 280 °C respectively. Bottom: heating pulses up to 430 °C.

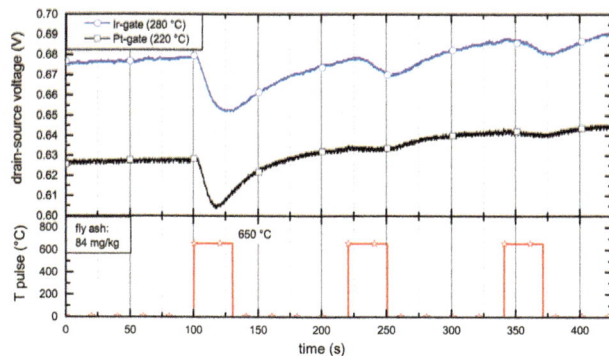

Figure 11. Top: sensor responses of Pt- and Ir-gate SiC-FETs at 220 and 280 °C, respectively. Bottom: heating pulses up to 650 °C.

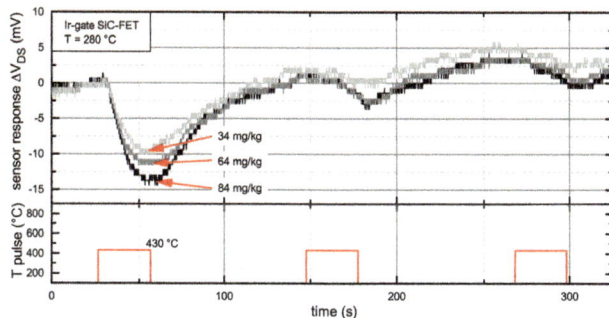

Figure 12. Top: difference signal of an Ir-gate FET at 280 °C to different concentration of ammonia (34, 64, 84 mg kg^{-1}). Bottom: heating pulses up to 430 °C.

to be synchronized with the heating and as appropriate with switching of the valves. Future work will address TCO.

4.4 Different concentrations

Figure 12 shows the sensor response to differently contaminated ash particles when they are heated up to 430 °C. As can be seen, the absolute value of the sensor response changes with the ammonia concentration. However, not only the absolute response but also the shape of the response depends on the ammonia concentration. Shape-describing features in particular can be used in pattern recognition for discrimination purposes (Marco and Gutiérrez-Gálvez, 2012; Bur et al., 2012a, b, 2014).

For lab measurements performed here, the amount which is placed manually on top of the heater is quite well defined. However, in general the amount of ash put on the heater is crucial and strongly affects the height of the sensor response. This issue will have to be solved for a miniaturized portable particle analyzer.

carbon monoxide. For the Pt-gate SiC-FET there is almost no response for the second and third pulse. In general the response for the second and third pulse decreases greatly as compared to the first pulse, which is due to the fact the first pulse already releases most of the adsorbates. However, the FET sensors detect oxidizing gases, but only if the sample is heated up to more than 700 °C (cf. Figs. 8 and 9).

The results in Figs. 10 and 11 suggest increasing the operating temperature of the sensor in order to reduce the recovery time, i.e., to clean the surface. However, at higher temperatures the sensitivity to ammonia is lower. Thus, temperature-cycled operation (TCO) could be used with a temperature cycle consisting of two phases: a low-temperature phase, i.e., 220 °C for platinum and 280 °C for iridium-gate SiC-FETs, at which the sensing takes place, and a high-temperature phase just for cleaning the sensor surface. When using TCO, the temperature cycling of the FET needs

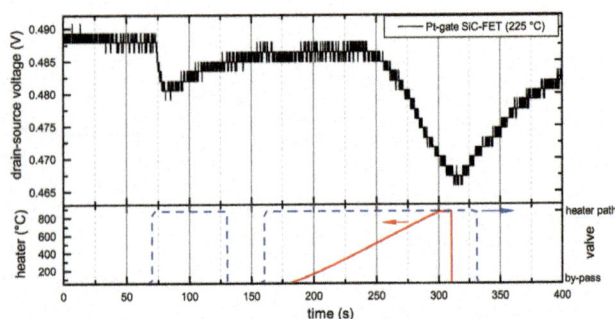

Figure 13. Sensor response of a Pt-gate SiC-FET at 225 °C. Bottom: heating ramp and valve position.

4.5   Measurements with valves

As mentioned in the beginning, our first idea was to use a bypass approach (shown in Fig. 1) in order to always have a well-defined and controlled flow over the gas sensor for baseline stability purposes. This section will briefly show the possibilities the proposed setup shown in Fig. 1, and thus the section should be seen as an outlook. While the ash sample is placed on the heater, the carrier gas is flowing through the bypass to the gas sensor. During that time the baseline of the sensor stabilizes. Before the heating starts, the valves need to switch once in order to exchange the air in the heater chamber since there is a huge difference in baseline if lab air or synthetic air is used. Figure 13 shows the SiC-FET sensor response for a measurement where the bypass is used. As can be seen, the sensor signal decreases during the first switching of the valves (not heating at this point). This is mainly the influence of the trapped lab air in the heater chamber. The actual measurement with fly ash can first begin when the baseline has stabilized again. After that, there are several options how to perform the test, which will be the aim of future work: heating and switching of the valves can be done asynchronously, which means that ammonia can first be accumulated and then be transferred to the gas sensor. Another option is to open the valves right before starting the heating process (shown in Fig. 13). Not only the measurement procedure itself but also the heating process can be changed. In Fig. 13 a heating ramp was used instead of the pulses used before. However, from our experience, a quick raise of temperature is more adequate to desorb substances from the particles. With a temperature ramp the ammonia is desorbed slowly, whereas a quick heating-up leads to a rather sharp release. As mentioned earlier, the different ammonia concentrations can be distinguished not only by the absolute value of the SiC-FET response but to some extent also by the slope of the response. For the later one, a temperature ramp would not be possible, but a heating cycle with a few defined temperatures may still be useful, especially together with smart data evaluation, in order to detect different gas molecules.

5   Conclusions and outlook

In this work a measurement methodology for investigating content of particles has been presented. It was shown that attached substances were desorbed by heating up particles (here: fly ash) with a small ceramic heater to temperatures up to 860 °C. Mass spectrometer measurements proved that ammonia and water are released from these particular particles mainly at lower temperatures, whereas temperatures above 600 °C lead to the formation and/or release of carbon monoxide, nitrogen- and sulfur dioxides, as well as carbon dioxide. The released gases were detected by silicon carbide field-effect transistors (SiC-FETs) with platinum or iridium as gate materials.

Future work will address the optimization of the gas sensor, e.g., operating temperature and operating mode. For discrimination and quantification, temperature- and/or bias modulation of the gas sensor is an appropriate method. However, since the release of ammonia is quite short, the corresponding temperature or bias cycle of the gas sensor needs to be short as well. Another benefit of the dynamic operation is that a high-temperature phase in the cycle can be used to clean the surface of the sensor and thus shorten the recovery time. Alternatively, the gas sensor can be operated at constant temperature and discrimination/quantification of outgassing substances can tentatively be achieved by using features describing the shape of the response similar to temperature-cycled operation (Bur et al., 2012a, b). The presented method is currently under development for integration into a low-temperature cofired ceramics (LTCC)-based package (Sobocinski et al., 2014) by our colleagues at the University of Oulu, Finland.

Acknowledgements. The authors would like to thank SenSiC AB, Kista, Sweden, for providing the sensors and 3S – Sensors, Signal Processing, Systems GmbH, Saarbrücken, Germany, for providing the hardware for sensor operation and read-out.

C. Bur acknowledges support through the "European Network on New Sensing Technologies for Air-Pollution Control and Environmental Sustainability, " (EuNetAir) for a "short-term scientific mission" (STSM) at the University of Oulu, Finland.

A. Lloyd Spetz and M. Andersson acknowledge a grant from the Funding Agency for Innovations – TEKES, Finland, (project CHEMPACK, no. 1427/31/2010.)

Edited by: M. Meyyappan
Reviewed by: two anonymous referees

References

Amanatidis, S., Ntziachristos, L., and Samaras, Z.: Applicability of the Pegasor Particle Sensor to Measure Particle Number, Mass and PM Emissions, SAE Technical Paper, 2013-24-0167, doi:10.4271/2013-24-0167, 2013.

Andersson, M., Ljung, P., Mattson, M., Löfdahl, M., and Lloyd Spetz, A.: Investigations on the possibilities of a MISiCFET sensor system for OBD and combustion control utilizing different catalytic gate materials, Top. Catal., 30/31, 365–368, 2004.

Andersson, M., Everbrand, L., Lloyd Spetz, A., Nyström, T., Nilsson, M., Gauffin, C., and Svensson, H.: A MISiCFET based gas sensor system for combustion control in small-scale wood fired boilers, Proc. IEEE Sensors, Atlanta, USA 28–31 October, 962–965, 2007.

Andersson, A., Pearce, R., and Lloyd Spetz, A.: New generation SiC based field effect transistor gas sensors, Sensor. Actuat. B, 179, 95–106, 2013.

Bartscherer, P. and Moos, R.: Improvement of the sensitivity of a conductometric soot sensor by adding a conductive cover layer, J. Sens. Sens. Syst., 2, 95–102, doi:10.5194/jsss-2-95-2013, 2013.

Bjorklund, R., Grant, A., Jozsa, P., Johansson, M., Fägerman, P. E., Paaso, J., Andersson, M., Hammarlund, L., Larsson, A., Popovici, E., Lutic, D., Pagels, J., Sanati, M., and Lloyd Spetz, A.: Soot sensor based on thermophoresis for high sensitive soot detection in diesel exhausts, in: Proc. IMCS13, Perth, Australia, 2010.

Bur, C., Reimann, P., Andersson, M., Schütze, A., and Lloyd Spetz, A.: Increasing the selectivity of Pt-Gate SiC field effect gas sensors by dynamic temperature modulation, IEEE Sens. J., 12, 1906–1913, 2012a.

Bur, C., Reimann, P., Andersson, M., Lloyd Spetz, A., and Schütze, A.: New method for selectivity enhancement of SiC field effect gas sensors for quantification of NO_x, Microsyst. Technol., 18, 1015–1025, 2012b.

Bur, C., Bastuck, M., Lloyd Spetz, A., Andersson, M., and Schütze, A.: Selectivity enhancement of SiC-FET gas sensors by combining temperature and gate bias cycled operation using multivariate statistics, Sensor. Actuat. B-Chem., 193, 931–940, 2014.

Buzea, C., Pacheco, I. I., and Robbie, K.: Nanomaterials and nanoparticles: sources and toxicity, Biointerphases, 2, MR17–MR71, 2007.

Darmastuti, Z., Bur, C., Möller, P., Rahlin, R., Lindqvist, N., Andersson, M., Schütze, A., and Lloyd Spetz, A.: SiC–FET based SO_2 sensor for power plant emission applications, Sens. Actuat. B-Chem., 194, 511–520, 2014.

Geiling, T., Dressler, L., Welker, T., and Hoffmann, M.: Fine dust measurement with electrical fields – concept of a hybrid particle detector, 9th IMAPS/ACerS Intern. Conf. and Exhibition on Ceramic Interconnect and Ceramic Microsystems Technologies, CICMT 2013, Orlando, Florida, USA, 23–25 April, 2013.

Lanki, T., Tikkanen, J., Kauko, J., Taimisto, P., and Lehtimäki, M.: An electrical sensor for long-term monitoring of ultrafine particles in workplaces, J. Phys. Conf. Ser., 304, 012013, doi:10.1088/1742-6596/304/1/012013, 2011.

Larsson, O., Göras, A., Nytomt, J., Carlsson, C., Lloyd Spetz, A., Artursson, T., Holmberg, M., Lundström, I., Ekedahl, L.-G., and Tobias, P.: Estimation of air fuel ratio of individual cylinders in SI engines by means of MISiC sensor signals in a linear regression model, AE Technical Paper 2002-01-0847, doi:10.4271/2002-01-0847, 2002.

Lloyd Spetz, A., Baranzahi, A., Tobias, P., and Lundström, I.: High temperature sensors based on metal insulator silicon carbide devices, Phys. Status Solidi A, 162, 493–511, 1997.

Lloyd Spetz, A. and Svage, S.: Silicon Carbide – Recent Major Advances, edited by: Choyke, W. J., Matsunami, H., and Pensl, G., Springer, Berlin, Heidelberg, ISBN 978-3-642-18870-1, 2003.

Lloyd Spetz, A., Huotari, J., Bur, C., Bjorklund, R., Lappalainen, J., Jantunen, J., Schütze, A., and Andersson, M.: Chemical sensor systems for emission control from combustions, Sensor Actuat. B-Chem., 187, 184–190, 2013.

Lundström, I., Sundgren, H., Winquist, F., Eriksson, M., Krantz-Rülcker, C., and Lloyd Spetz, A.: "Twenty-five years of field effect gas sensor research in Linköping", Sens. Actuat. B-Chem., 121, 247–262, 2007.

Marco, S. and Gutiérrez-Gálvez, A.: Signal and data processing for machine olfaction and chemical sensing: a review, IEEE Sens. J., 12, 3189–3214, 2012.

Messerer, A., Niessner, R., and Pöschl, U.: Thermophoretic deposition of soot aerosol particles under experimental conditions relevant for modern diesel engine exhaust gas systems, J. Aerosol Sci., 34, 1009–1021, 2003.

NIOSH – National Institute for Occupational Safety and Health (NIOSH): Current Intelligence Bulletin 65: Occupational Exposure to Carbon Nanotubes and Nanofibers, DHHS (NIOSH) Publication No. 2013–145, 2013.

Ntziachristos, L., Fragkiadoulakis, P., Samaras, Z., Janka, K., and Tikkanen, J.: Exhaust Particle Sensor for OBD Application, SAE Technical Paper 2011-01-0626, doi:10.4271/2011-01-0626, 2011.

Osite, A., Katkevich, J., Viksna, A., and Vaivars, G.: Electrochemical impedance spectra of particulate matter and smoke, IOP Conference Series: Materials Science and Engineering, 23, 2011.

Perma Pure: ME Series moisture exchanger, available at: http://www.permapure.com/, last access: September 2014.

Sobocinski, M., Lloyd Spetz, A., Andersson, M., Juuti, J., and Jantunen, H.: Novel method for integration of SiC in LTCC, Abstract book, 14th Electroceramics conference, Bucharest, Romania, June 16–20, 2014.

Thomas, S., Racz, Z., Cole, M., and Gardner J. W.: Dual high-frequency surface acoustic wave resonator for ultrafine particle sensing, Proc. IEEE Sensors Baltimore, MA, USA, 4–6 November, 1–4, 2013.

Wingbrant, H., Svenningstorp, H., Salomonsson, P., Kubinski, D., Visser, J. H., Löfdahl, M., and Lloyd Spetz, A.: Using a MISiC-FET sensor for detecting NH3 in SCR systems, IEEE Sens. J., 5, 1099–1105, 2005.

Xu, R.: Light scattering: a review of particle characterization applications, Particuology, in press, doi:10.1016/j.partic.2014.05.002, 2014.

21

Carbon monoxide gas sensing properties of Ga-doped ZnO film grown by ion plating with DC arc discharge

S. Kishimoto[1], **S. Akamatsu**[1], **H. Song**[2], **J. Nomoto**[2], **H. Makino**[2], and **T. Yamamoto**[2]

[1]National Institute of Technology, Kochi College, Nankoku-shi, Japan
[2]Kochi University of Technology, Kami-shi, Japan

Correspondence to: S. Kishimoto (kishi@me.kochi-ct.ac.jp)

Abstract. The carbon monoxide (CO) gas sensing properties of low-resistance heavily Ga-doped ZnO thin films were evaluated. The ZnO films with a thickness of 50 nm were deposited at 200 °C by ion plating. The electrical properties of the ZnO films were controlled by varying the oxygen assist gas flow rate during deposition. The CO gas sensitivity of ZnO films with Au electrodes was investigated in nitrogen gas at a temperature of 230 to 330 °C. CO gas concentration was varied in the range of 0.6–2.4 % in nitrogen gas. Upon exposure to CO gas, the current flowing through the film was found to decrease. This response occurred even at the lowest temperature of 230 °C, and is thought to be the result of a mechanism different than the previously reported chemical reaction.

1 Introduction

ZnO is a substance for which various applications such as gas sensors and ultraviolet light sensors are anticipated. In recent years, the properties and applications of ZnO nanostructured films have been also studied (Zhao et al., 2010; Lao et al., 2003).

Carbon monoxide (CO) gas, being both colorless and odorless, is a dangerous gas for which it is hoped that high-sensitivity sensors may be developed. The lethal concentration by CO gas is 1500 ppm. There have been a number of attempts to evaluate the sensitivity of ZnO to CO gas. There have also been reports of carbon nanorods and nanowires reacting with high sensitivity to CO gas (Hassan et al., 2013; Kim et al., 2009). Experiments using catalysts such as Pd have also been reported (Trung et al., 2014). In these cases, the response mechanism is understood to have involved chemical reactions between gas molecules and the ZnO surface. All of these films had a high electrical resistance, and the presence of the gas was indicated by a change in resistance. Several investigations have been conducted on CO gas sensing properties of Ga-doped ZnO nanostructured films (Han et al., 2011; Pearce et al., 2009; Phan and Chung, 2013). Phan and Chung (2013) have reported the effects of Ga-doping on CO sensing properties of Ga-doped ZnO nanorods. Their p–n junctions-based nanorods sensors have a fast response.

We have previously reported transparent low-resistance Ga-doped ZnO (GZO) films produced using ion plating (Shirakata et al., 2003; Yamada et al., 2007a, b, 2010). In the present study, we carried out an evaluation of the sensitivity of these polycrystalline ZnO thin films to CO gas.

2 Experimental

The polycrystalline ZnO films were grown using DC arc discharge ion plating (Yamamoto et al., 2012). The growth temperature was 200 °C and a ZnO tablet incorporating 3 wt % Ga was used as the source material. During growth, flowing oxygen was used as an assist gas, with the flow rate being varied between 5–25 cm^3 in order to control the structure and properties of the films. The film thickness, controlled by means of the growth time, was 50 nm. In the previous report, Ga 3 wt %-doped ZnO film was good polycrystalline with a hexagonal structure (Yamada et al., 2006).

The crystallinity of the films was characterized by high-resolution X-ray diffraction (XRD; ATX-G, RIGAKU). Their electrical properties were evaluated by Hall-effect measurements (HL5500PC, Nonometrics) in the van der Pauw configuration at room temperature.

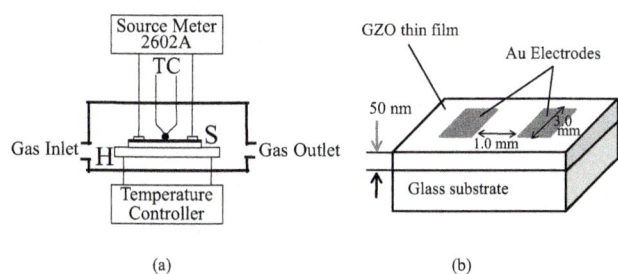

Figure 1. Schematic diagrams of **(a)** measurement setup and **(b)** sample dimension. TC: thermocouple; H: heater; S: Ga-doped ZnO film on glass substrate.

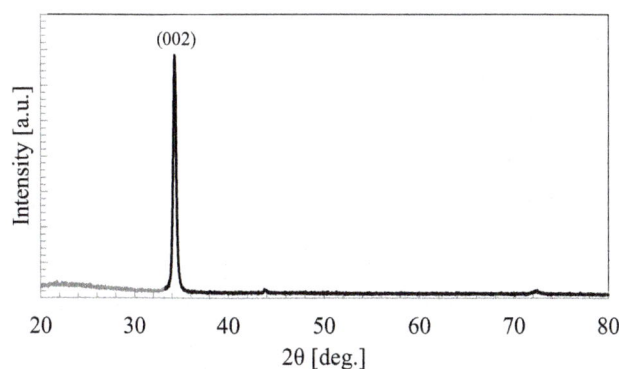

Figure 2. High-resolution XRD pattern result of the ZnO film with 50 nm thickness.

Evaluation of the gas sensitivity was performed using the following method. Figure 1 shows the layout of the sample and electrodes along with the sample chamber. First, Au electrodes (separated by 1 mm) were formed with a thickness of at least 150 nm, and the sample was cut into 5 mm × 5 mm chips, which were placed in a compact chamber for evaluation. The electrodes of the ZnO film were connected to a DC power supply (System SourceMeter 2602A, Keithley), which applied 5 V, and this current was measured for the films. The temperature of the sample films was controlled at between 230 and 330 °C using a ceramic heater set beneath the sample. These temperatures are slightly higher than the desorption temperatures of oxygen from metal oxide (Iwamoto et al., 1978). The sample chamber was flowed continuously with nitrogen gas at a flow rate of 400 sccm. In order to evaluate the intrinsic reaction to CO gas, CO gas sensing properties were investigated in nitrogen gas. Under these conditions, the amount of CO gas necessary to produce the specified concentration was mixed with the nitrogen gas and allowed to flow for 10 s, and the change in the sample current was measured. In order to maintain the reaction satisfactorily, the evaluation was conducted at an extremely high CO gas concentration of 2.5 (0.6 %) to 10 (2.4 %) sccm. The gas was evacuated and released at atmospheric pressure. The gas inlet was placed in such a way that the gas entering the sample

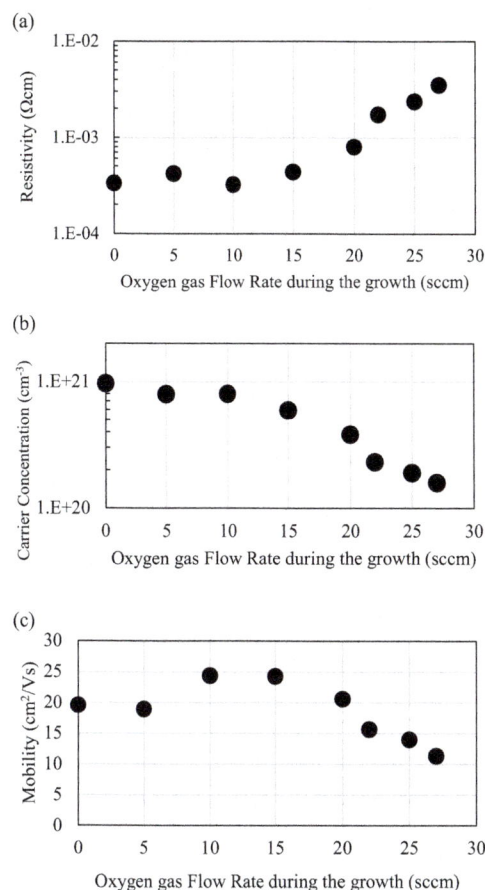

Figure 3. Resistivity as a function of oxygen gas flow rate for GZO films **(a)**. Carrier concentration as a function of oxygen flow rate for GZO films **(b)**. Hall mobility as a function of oxygen flow rate for GZO films **(c)**.

chamber did not directly impinge on the sample. The volume of the cylindrical sample chamber was approximately $200 \, \mathrm{cm^3}$.

3 Results and discussions

Figure 2 shows the XRD results for the deposited ZnO films. It can be seen that the main diffraction peak is associated with (002) planes, so that the film is c axis oriented. Under the same growth condition, these Ga-doped ZnO films were polycrystalline with a wurtzite-type hexagonal structure from a cross-section transmission electron microscopy (TEM) and in-plane XRD measurements (Yamada et al., 2007b).

Figure 3 shows the results of Hall-effect measurements for the samples. Due to the Ga doping, the resistivity was extremely low. The carrier concentration was found to decrease with increasing oxygen flow rate during growth. In all cases, the carrier concentration was $\geq 1.0 \times 10^{20} \, \mathrm{cm^{-3}}$ and, from the temperature dependence of the Hall-effect measurement for samples deposited under same conditions, it was found

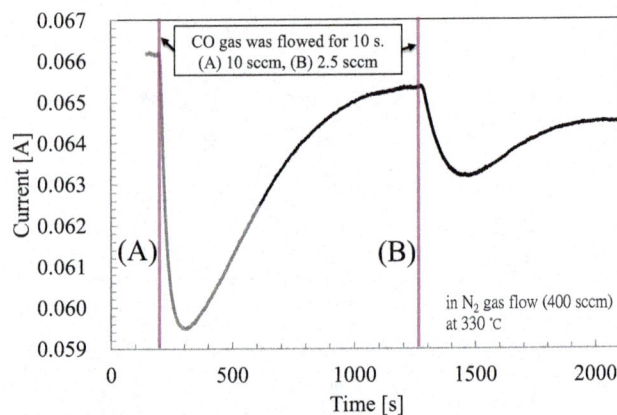

Figure 4. Changes in current in ZnO films at 330 °C due to CO gas. A constant nitrogen gas flow of 400 sccm was used and CO gas was allowed to flow for 10 s.

Figure 5. Dependence of current change in ZnO films on CO gas concentration. The carrier concentration was 1.8×10^{20} cm^{-3}. The sample temperatures were 230, 250 and 330 °C.

that the samples were degenerated. From studies of samples deposited under the same conditions (Yamamoto et al., 2012), both intra-grain scattering and grain boundary scattering are the mechanisms limiting carrier transport.

Figure 4 shows the change in the film current when a sample (oxygen gas flow rate during the growth was 25 sccm) was exposed to a flow of CO gas. In addition to the 400 sccm flow of nitrogen gas, 10 and then 2.5 sccm of CO were sequentially flowed for 10 s, and the change in sample current was measured. The current prior to exposure is determined by the original film resistance. The current shows a decrease upon exposure to the gas, with a minimum occurring at around 100 s following exposure, before showing a tendency to return to its original value. Previously, it was reported that exposure to CO caused a decrease in the resistance of ZnO film due to the chemical reaction (Tanaka et al., 1976) occurring on the surface given in

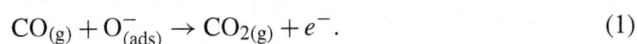

$$CO_{(g)} + O^-_{(ads)} \rightarrow CO_{2(g)} + e^-. \qquad (1)$$

Since the results shown in Fig. 3 indicate the opposite behavior, the mechanism must be different than that previously reported. In other words, it does not involve a chemical reaction with CO molecules on the film surface. It may be that the absorption to the grain boundary of oxygen gas molecules plays the role of a physical barrier for carriers.

In fact, since even at 330 °C there is a large response of several milliamps, it is thought that the reaction involves the entire film rather than just the surface. Since it is difficult to imagine the reaction occurring within the grains in the polycrystalline film, it is possible that it is actually taking place at the grain boundaries. For samples with a relatively low carrier concentration and Hall mobility, the current showed a large decrease of about 5.0 mA at a temperature of 330 °C in response to CO gas. The large current response observed in the present study clearly indicates the practical potential of a ZnO-based CO detector.

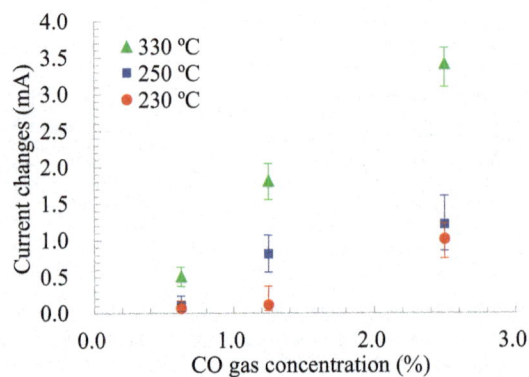

Figure 5 shows the change in the current response as a function of CO gas concentration. The current change increased with increasing the CO gas concentration. Furthermore, even at a low temperature of 230 °C, a large decrease in current was found in response to CO gas. For the elucidation of the reaction mechanism, it requires further experiments.

4   Conclusions

The reaction characteristics of heavily Ga-doped low-resistance ZnO films in response to CO gas were investigated. The ZnO thin films were polycrystalline with a columnar structure and were highly oriented along the c axis. In response to an inflow of CO gas, the current flowing through the film was found to decrease, which is the opposite to the previously reported effect associated with a chemical reaction on the film surface.

In the present study, a CO gas concentration of 6.21 % was chosen in order to observe the reaction clearly, and a large reaction current of 2 mA was obtained. This may have been the result of the heavy Ga doping level.

Acknowledgements. This work has been supported by JSPS KAKENHI grant no. 30320120. 2012–2014.

Edited by: M. Meyyappan
Reviewed by: two anonymous referees

References

Han, N., Liu, H., Wu, X., Li, D., Chai, L., and Chen, Y.: Pure and Sn-, Ga- and Mn-doped ZnO gas sensors working at defferent temperature for formaldehyde, humidity, NH$_3$, toluene and CO, Appl. Phys. A-Mater., 104, 627–633, 2011.

Hassan, J. J., Mahdi, M. A., Chin, C. W., Abu-Hassan, and Hassan, H. Z.: A high-sensitivity room-temperature hydrogen gas sensor based on oblique and vertical ZoN nanorod arrays, Sensors Actuat. B-Chem., 176, 360–367, 2013.

Iwamoto, M., Yoda Y., Yamazoe, N., and Seiyama, T.: Study of Metal Oxide Catalysts by Temprerature Programmed Desorption, J. Phys. Chem., 82, 2564–2570, 1978.

Kim, K., Song, Y., Chang, S., Kim, I., Kim, S., and Lee, S. Y.: Fabrication and characterization of Ga-doped ZnO nanowire gas sensor for the detection of CO, Thin Solid Films, 518, 1190–1193, 2009.

Lao, J. Y., Huang, J. Y., Wang, D. Z., and Ren, Z. F.: ZnO Nanobridges and Nanonails, Nano. Lett., 3, 235, 2003.

Pearce, R., Soderlind, F., Hagelin, A., Kall, P., Yakimova, R., Spetz, E., Becker, E., and Skoglundh, M.: Effect of Water vapour on Gallium doped Zinc Oxide nanoparticle sensor gas response, Sensors IEEE, 2009.

Phan, D. and Chung, G.: Effects of defects in Ga-doped ZnO nanorods formed by a hydrothermal method on CO sensing properties, Sensors Actuat. B-Chem., 187, 191–197, 2013.

Shirakata, S., Sakemi, T., Awai, K., and Yamamoto, T.: Optical and electrical properties of ZNO films prepared by URT-IP method, Thin Solid Films, 445, 278–283, 2003.

Tanaka, M., Tsubone, D., and Yanagida, H.: Dependence of Electrical Conductivity of ZnO on Degree of Sintering, J. Am. Ceram. Soc., 59, 4–8, 1976.

Trung, D. D., Hoa, N. D., Tong, P. V., Duy, N. V., Dao, T. D., Chung, H. V., Nagao, T., and Hieu, N. V.: Effective decoration of Pd nanoparticles on the surface of SnO$_2$ nanowires for enhancement of CO gas-sensing performance, J. Hazard. Mater., 265, 124–312, 2014.

Yamada, T., Ikeda, K., Kishimoto, S., Makino, H., and Yamamoto, T.: Effects of oxygen partial pressure on doping properties of Ga-doped ZnO films prepared by ion-plating with traveling substrate, Surface and Coatings Technology, 201, 4004–4007, 2006.

Yamada, T., Nebiki, T., Kishimoto, S., Makino, H., Awai, K., Narusawa, T., and Yamamoto, T.: Dependences of structural and electrical properties on thickness of polycrystalline Ga-doped ZnO thin films prepared by reactive plasma deposition, Superlattice Microst., 42, 68, 2007a.

Yamada, T., Miyake, A., Kishimoto, S., Makino, H., Yamamoto, N., and Yamamoto, T.: Low resistivity Ga-doped ZnO thin films of less than 100 nm thickness prepared by ion plating with direct current arc discharge, Appl. Phys. Lett., 91, 051915, 2007b.

Yamada, T., Makino, H., Yamamoto, N., and Yamamoto, T.: In-grain and grain boundary scattering effects on electron mobility of transparent conducting polycrystalline Ga-doped ZnO films, J. Appl. Phys., 107, 123534, 2010.

Yamamoto, T., Song, H., Makino, H., and Yamamoto, T. N.: ECS Transactions, edited by: Misra, D., Bauza, D., Chen, Z., Chikyo, T., Iwai, H., Obeng, Y., and Datta, S., 45, 401–410, 2012.

Zhao, S. H., Wang, L. L., Wang, L., and Wang, Z. Y.: Synthesis and luminescence properties of ZnO:Tb3+ nanotube arrays via electrodeposited method, Physica B, 405, 3200, 2010.

22

Room temperature carbon nanotube based sensor for carbon monoxide detection

A. Hannon[1,2], Y. Lu[1,3], J. Li[1], and M. Meyyappan[1]

[1]NASA Ames Research Center, Moffett Field, CA 94035, USA
[2]ERC at NASA Ames Research Center, Moffett Field, CA 94035, USA
[3]ELORET Corporation at NASA Ames Research Center, Moffett Field, CA 94035, USA

Correspondence to: M. Meyyappan (m.meyyappan@nasa.gov)

Abstract. Sulfonated single-walled carbon nanotubes have been used in an integrated electrode structure for the detection of carbon monoxide. The sensor responds to 0.5 ppm of CO in air at room temperature. All eight sensors with this material in a 32-sensor array showed good repeatability and reproducibility, with response and recovery times of about 10 s. Pristine nanotubes generally do not respond to carbon monoxide and the results here confirm sulfonated nanotubes to be a potential candidate for the construction of an electronic nose that requires at least a few materials for the selective detection of CO.

1 Introduction

Carbon monoxide (CO) is an air pollutant and known to have an effect of on global warming. Incomplete burning of coal and hydrocarbon fuels in a variety of applications ranging from power plants to refineries as well as most forms of transport vehicles are the common sources of CO in the atmosphere. From a safety point of view, it can cause explosion at over $\sim 12\%$ in air. But as a colorless and odorless gas, CO poses health risks as well. Upon gaining entry into the lungs through breathing, CO displaces oxygen leading to suffocation and even death. The recommended exposure limit by the National Institute for Occupational Safety and Health is 50 ppm over an 8 h work shift. This points to the need for inexpensive and sensitive detection of CO in the environment, home and office buildings, and other public places.

CO detection technology has been evolving over a long period and many commercial products are available on the market. Some of the common approaches include electrochemical, catalytic combustion and semiconductor devices. Electrochemical sensors provide selectivity for CO, but their lifetime is limited by the electrolyte. Catalytic combustion based sensors rely on oxidation of CO and measuring the change in resistance of a metal electrode, but this is typically a high temperature reaction. Recently, novel catalysts that can oxidize at temperatures as low as 70 °C (Hosaya et al., 2014)

have been proposed. The semiconductor type sensors also measure the change in resistance upon adsorption of CO or any other gas on the surface, and the most common and commercially used technology involves tin oxide thin film based devices (Mishra and Agarwal, 1998). Besides selectivity being an issue, tin oxide and other metal oxide based sensors operate only at elevated temperatures, typically over 200 °C. The nanowire form of the metal oxides has been considered in gas sensor construction in recent years (Meyyappan and Sunkara, 2010) due to the large surface area to volume ratio and other desirable properties. But low or room temperature operation of metal oxide based sensors is not that common, though recently Pd / SnO_2 sensors have been shown to have a good sensitivity in the range of 6–18 ppm of CO at a temperature of 60 °C (Kim et al., 2013).

There is a strong push currently to incorporate sensors into smartphones and other mobile devices for environmental monitoring. This incentivizes development of sensors that operate at room temperature and consume low power. In this regard, carbon nanotubes (CNTs) have long been a candidate for room temperature sensing of various gases and vapors; both single-walled and multiwalled CNTs (SWCNTs, MWCNTs) have been extensively studied for gas/vapor sensors (Meyyappan, 2004; Kaufman and Star, 2008; Fam et al., 2011). Typical small molecules participate in charge transfer reactions with SWCNTs – either donating an electron to or withdrawing from the nanotubes – which leads to a change in the resistance (or other measureable properties such as capacitance) of the CNTs. SWCNTs have been found to be useful to sense even large molecules (such as nitrotoluene, malathion, etc.) that do not participate in change transfer; in these cases, the molecules can make a conducting bridge between adjacent SWCNTs in an intertube modulation mechanism (Li et al., 2003). CO does not participate in a charge transfer process with CNTs according to theoretical and experimental studies (Santucci et al., 2003; Peng and Cho, 2000). However, SWCNTs have been modified, especially with metal loading, to provide a response in the form of a change in resistance upon exposure to CO. Rh-loaded SWCNTs were shown to respond to 2500 ppm of CO at room temperature (Star et al., 2006). Vertically aligned CNTs decorated with Pt, Ru and Ag clusters responded to 0.1 % CO at 150 °C when exposed to a mixture of CO, CO_2, NH_3, CH_4 and NO_2, with the mixture representing a landfill gas (Penza et al., 2010). Adding CNTs to an otherwise nonresponsive polyaniline allows CO sensing in the range of 100–1000 ppm at room temperature (Wanna et al., 2006). Similarly, adding CNTs to a mixture of cobalt and tin oxides has also been shown to improve sensor response for 20–1000 ppm of CO, attributed to the ability of nanorubes to increase the adsorption ability of the mixed oxides (Wu et al., 2008).

Other types of modification to CNTs appear to be more promising in terms of room temperature detection of CO at low concentrations. For example, decorating SWCNTs with a tin oxide nanoparticle enables a CO detection limit of 1 ppm with a response time of 2 s and sensitivity of 0.27 for 100 ppm. Plasma modification of SWCNTs appears to provide a sensitive response to 5 ppm of CO at room temperature (Zhao et al., 2012). MWCNTs with the aid of nitrogen doping also show a good response to CO in the range of 2–20 ppm at ambient and 150 °C conditions (Adjizian et al., 2014). Here we present CO detection results using sulfonated SWCNTs with a good response down to 0.5 ppm.

2 Experimental work

Single-walled carbon nanotubes were purchased from Helix Material Solutions (Richardson, TX) with a purity of ∼ 90 % as claimed by the manufacturer and the sulfonated SWCNTs were provided by South Dakota School of Mines. The nanotubes were first sonicated in concentrated HNO_3 for 2 h and the suspension was refluxed under magnetic stirring at 120 °C for 2 h. This process helps to remove unreacted catalyst particles while introducing oxygen-containing groups, mainly carboxylic groups, on the SWCNTs (Naseh et al., 2009). The pure SWCNTs were filtered and washed several times, then dried at 120 °C overnight under vacuum. Then, 25 mL of concentrated H_2SO_4 and 0.25 g of SWCNTs were mixed and stirred for 5 h at 300 °C. After cool-

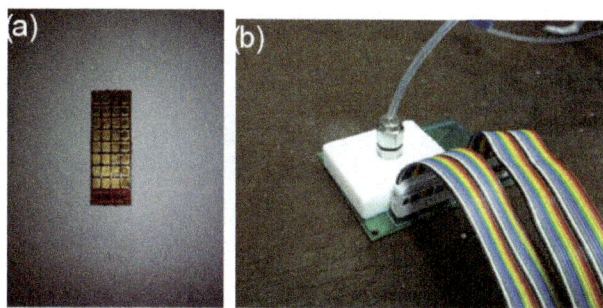

Figure 1. The sensor chip fabricated on a printed circuit board. **(a)** A 32 sensor array chip and **(b)** gas exposure unit.

ing down to room temperature, the suspension was diluted with water and filtered. The solids were washed with distilled water to remove excess acid and dried at 100 °C for 4 h to obtain sulfonated SWNCTs. This process should lead to the opening of the tube caps and formation of sulfonated groups at defect sites along the sidewalls (Yu et al., 2008). To prepare the solution for sensor preparation, the nanotubes were dispersed (0.65 % of nanotubes by weight) in dimethyl formamide (DMF) to form a suspension by sonicating the solution for about 2 h.

Preparation of the interdigitated electrodes on silicon wafers has been described in our previous publications (Lu et al., 2011a, b, c) and a brief account is given below. A printed circuit board (PCB) was used as sensor substrate and a standard photolithography process was used to partially etch away the Au film and define the pattern for the preparation of the sensor chips. The sensor chip consisted of an array of 32 interdigitated electrodes, each with a gap size of 120 μm. A detailed chip configuration schematic is shown in Fig. 1. The sensing materials (0.3 μL solution) were deposited on the interdigitated electrode area manually with the aid of a pipette on the chips. Only eight of the 32 sensors were used here and all were coated with the same sulfonated-SWCNT dispersion. Finally, the sensor chip was heated at 80 °C in air for 1 h and a thin film of nanotubes bridges across the electrodes after the evaporation of the solvent. The base resistance of the sensors was measured to be in the range of 600 ohm– 5 Kohm. The base resistance can be affected by many factors such as the gap size between the interdigitated electrodes, concentration of the nanotubes solution, the amount of solution deposited on the electrode, etc. Indeed, this is the usual range of base resistance that we like to achieve for our sensors, which is normally accomplished by adjusting the solution concentration. Any contribution to base resistance by amorphous carbon can be ruled out, as no amorphous carbon is seen in the SEM image in Fig. 2.

The sensor chip was connected to an interface board through which individual resistances of each sensor channel could be measured. The current-voltage characteristics of each sensor were measured using a Keithley 2700 sys-

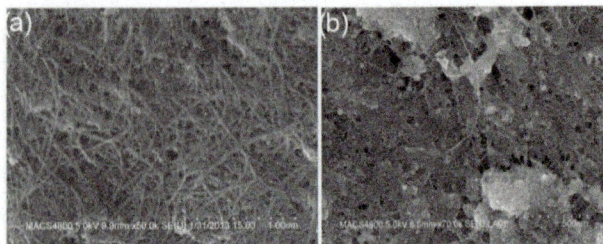

Figure 2. SEM images of **(a)** pristine and **(b)** sulfonated single-walled carbon nanotubes.

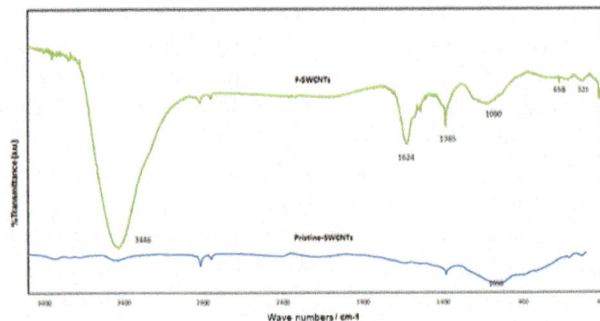

Figure 3. FTIR spectra of pristine and sulfonated SWCNTs.

tem, and a Environics 2000 gas blending and dilution system was used to create different gas concentration streams. A steady total flow of $400\,cm^3\,min^{-1}$ was used during various gas stream exposures. The sensor chip was exposed to the gas stream using a Teflon cover placed right over the chip to prevent gas leaks as shown in Fig. 1. The sensors were first purged with zero air (Airgas Inc.) for 10 min to get a stable baseline, and CO gas exposures were allowed afterwards using a 200 ppm CO balance in air (Airgas Inc.). The sensor tests were performed both in dry air and humid air of 65 % relative humidity (RH).

3 Results and discussion

Field emission scanning electron microscopy (FESEM) (Hitachi S-4800 FESEM) and energy-dispersive X-ray spectroscopy (EDX) (Oxford instruments) were used to study the morphological properties and elemental analysis of the nanostructures. The sensing material was deposited on a silicon substrate for obtaining SEM images instead of directly imaging the sensor chip. A comparison of the SWC-NTs before and after functionalization reveals that their morphology and structure changed after sulfonation. SWCNTs are covered by a layer of a foreign material, which can be groups of the sulfonic acids (Alamdari et al., 2012). The pristine SWCNT film has a uniform morphology, while the sulfonated SWCNTs appear as thickened bundles of nanotubes and a tangled network to make a cluster as shown in Fig. 2. Some changes in the structural integrity of the nanotubes are observed, caused by the strong acid etching of the nanotubes, which leads to tubes of shorter length. The EDS results (data not shown here) show a sulfur content of 2.72 % (by weight) confirming the presence of sulfonic acid groups. But these sulfonic groups may be superficially on the surface and additional characterization is needed to confirm the nature of functionalization.

Fourier transform infrared spectroscopy (FTIR) spectroscopy was used to investigate the nature of these surface groups as shown in Fig. 3. The spectrum for sulfonated SWCNTs has a number of new peaks not present in the spectrum for pristine nanotubes. For example, a newly observed prominent peak at $1624\,cm^{-1}$ could be assigned to the $C = C$

stretching mode of the SWCNT graphitic layer. This band is weak in pristine SWCNTs due to the symmetry of the dipole moment, but it is intensified with defects on the graphic layer (Yu et al., 2008; Alamdari et al., 2012, 2013). The intensified peak in sulfonated SWCNTs indicates the extensive functionalization of SWCNTs. The presence of carboxylic groups $(C = O)$ is indicated by the peak at $1718\,cm^{-1}$. The stretching modes of the sulfate groups in sulfonated nanotubes can be identified with the peaks at 1385 and $1090\,cm^{-1}$ (Peng et al., 2005). The peak at $658\,cm^{-1}$ also indicates the $S = O$ stretching mode of $-SO_3H$ (Yu et al., 2008). The sulfonated nanotubes show a very broad peak in the region of 2990–$3700\,cm^{-1}$, which is assigned to the O–H group. The triplet peaks observed at about $2900\,cm^{-1}$, responsible for the C–H stretching mode, might be the result of hydrocarbon contamination in the spectrometer. In the low-frequency range (see Fig. 3), the peak at $520\,cm^{-1}$ can be assigned to the C–S stretching mode. It is evident from these results that the H_2SO_4 treatment enables covalent sulfonation of SWCNTs.

The covalent functionalization was also verified using Raman spectroscopy and Fig. 4 shows the Raman spectra of SWCNTs before and after the acid treatment. There are two major peaks clearly seen at $\sim 1300\,cm^{-1}$ as the D band and at $\sim 1590\,cm^{-1}$ as the G band. The intensity of the peak at the D band is due to structural defects or impurities. The G band indicates the longitudinal stretching vibrations of the sp2 carbons of semiconducting SWCNTs (Yudianti et al., 2011). The ratio of D and G peak intensities (I_D / I_G) determines the structure of SWCNTs, and a comparison indicates an increase in the D band after sulfonation due to the strong damage to the sidewalls of SWCNTs or the formation of fragments caused by the functionalization (Yu et al., 2008). The ratio of D / G band intensity shows an increase from 12.5 to 33.21 % after the acid treatment. The observed behavior is more due to the increase in I_D rather than the decrease in I_G. The increased I_D / I_G ratio indicates decreasing symmetry in the structure due to the introduction of functionalized groups and the severe damage to the sidewalls of SWCNTs (Yu et al., 2008). The second-order peak of the D band is observed at $2600\,cm^{-1}$, called the G′ peak. Furthermore, the bands are

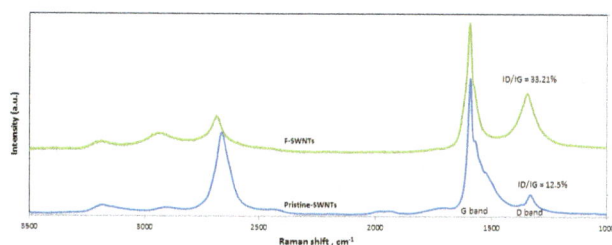

Figure 4. Raman spectra of pristine and sulfonated SWCNTs.

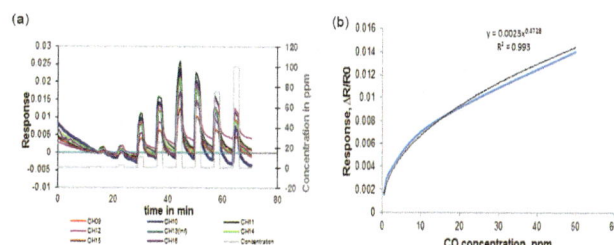

Figure 5. (a) Sensor response to exposure of various dosages of CO and (b) the sensor calibration curve.

shifted by $10\,\mathrm{cm}^{-1}$ due to the chemical charge transfer on oxidized SWCNTs (Yu et al., 2008).

The sensor chip was exposed to CO dosages of 0.5, 2, 10, 25, 50, 60, 75, and 100 ppm at the intervals shown in Fig. 5. The chip was purged with an airflow of $400\,\mathrm{cm}^3\,\mathrm{min}^{-1}$ for the first 10 min before any CO exposure and also in between the CO exposures. Dry air was used to purge the sensor chip and for CO dilution to achieve the required concentrations. Figure 5 shows the response curves for the eight sensors. The response plotted here is normalized resistance $(R - R_0)/R_0$, where R_0 is the baseline resistance before gas exposure and R is the resistance at any time t after the gas exposure. A very stable baseline was observed with this material. Signal processing is done here by looking at the relative change in the slope when gas exposure occurs. The response behaviors of all eight sensors are very close to each other. The sensor-to-sensor variation is due to the manual deposition process used here for adding the nanotubes to the chip, which results in a variation in nanomaterial density in each sensor element. This can be improved with automated ink-jet-type delivery systems in commercial manufacturing of sensor arrays.

The electrical resistance of the sensor film increases upon exposure to CO. The conductivity change of the sensors is concentration dependent and it increases with concentration in the range of 0.5–50 ppm. All eight sensors showed a consistent sensitivity of 0.0014 ± 0.00015 as defined by the slope. The sensitivity is defined as the normalized sensor response/ppm of CO concentration. The overall variation in the sensitivity for the sensors is about $< 5\,\%$, which is similar to common commercial gas/vapor sensors. Pristine SWCNTs did not show any response to CO (data not shown here), while the sulfonated SWCNT sensors showed a clear response to CO between concentrations of 0.5–50 ppm. This behavior is expected since the presence of oxygenated and sulfonated functionalities at the ends of the SWCNTs facilitates electron transfer. The larger response with sulfonated sensors might be the result of the introduction of controlled COOH and sulfonic acid defects, which form low-energy adsorption sites and facilitate charge transfer at defect sites. It is hard to state the exact mechanism of the response, as it requires very thorough investigation. At this stage, based on what is known in the literature, we can state the following possibilities for the sensing mechanism.

COOH-functionalized nanotubes have been widely used as sensing material for sensing NH_3, Cl_2, and CO (Dong et al., 2013; Robinson et al., 2006). As reported by Robinson et al. (2006), the defect sites serve both as low-energy adsorption sites and as nucleation sites for additional condensation of the analyte on the surface of the nanotubes. Sulfonic acids are much stronger acids than the corresponding carboxylic acids. $-SO_3H$ is also an electron withdrawing group similar to the $-COOH$ group and, therefore, it is reasonable to consider that the sulfonic acid groups introduced on the surface cause an enhancement of charge density in the SWCNTs; this can increase the amount of electron transfer between sulfonated SWCNT and CO molecules, which increases the hole current of p-type sulfonated SWCNT (Dong et al., 2013).

The sensors do not show any concentration dependence at concentrations greater than 50 ppm during the systematic increase in concentration shown in Fig. 5. This might be caused by the limited quantity of the functional groups on the nanotubes. As stated earlier, the CO response may be due to reducing gas CO adsorption/interaction at the defect sites; due to the availability of a limited number of defect sites after certain concentration exposure, these sites might be fully occupied. Robinson et al. (2006) reported similar results for acetone and methanol sensing using oxidized nanotubes. The sensors with additional treatment showed better responses due to the introduction of additional defect sites.

The sensor chip was also tested for CO detection under a 65 % RH environment. The chip was first purged with dry air for 5 min and then 65 % humidity was introduced for 25 min. The sensor resistance increased significantly in the presence of humidity, but the base resistance returned to the original level after about 20 min. Next, the chip was exposed to 2, 10, 25, 50, 75, and 100 ppm of CO successively at the time interval shown in Fig. 6. All eight sensor channels showed clear responses to CO and their sensitivity was not significantly reduced. The lack of concentration dependence was seen here as well, but before 50 ppm. This might be due to the residual CO molecules adsorbed on the SWCNT bundles, which are not entirely removed by purging or due to the lack of availability of the active adsorption sites on the nanotube surface. The sensor response and recovery time of about 5–10 s is impressive for room temperature operation.

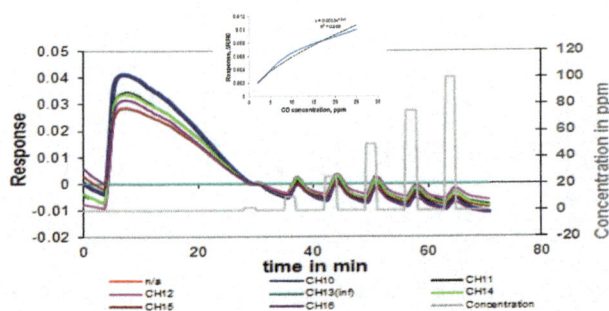

Figure 6. The response of the sensors in Fig. 5 under a 65 % RH environment.

4 Conclusions

Environmental monitoring via smartphones and wearable devices has been gaining popularity, but is in early stages of development. Carbon monoxide is a key pollutant that would be found in any wish list of gases to be monitored. Sensitive detection of CO at room temperature is critical to construct sensors to meet the needs above. Conventional semiconductor sensors based on oxide conductors typically work at elevated temperatures. In this regard, single-walled carbon nanotubes offer an alternative for room temperature gas/vapor sensing, as has been demonstrated in the literature for numerous analytes. But SWCNTs do not respond to CO at any temperature. Construction of an electronic nose consisting of a multisensor array for the detection of any analyte requires at least a few different materials, which respond to that analyte. Thus far, tin oxide loading of SWCNTs (Zhang et al., 2013) and oxygen plasma modified nanotubes (Zhao et al., 2012) have been reported to be effective in CO detection. In the present work, we have shown that sulfonated SWCNTs provide a good response to CO down to 0.5 ppm, and thus can be added to the arsenal of responsive materials for the e-nose construction. Future work will include these different materials and report the operation of our current 32-sensor platform as e-nose. The lifetime of the present CO sensors has not been investigated here, but we have sensors for ammonia, chlorine and NO$_2$, which were made in 2008, still showing sensitivity to these gases. We anticipate a similar lifetime performance for the CO sensors as well, and this will be studied in the future. Future work also should include efforts to improve the detection limit even further while maintaining high selectivity and possibilities to transfer the approach to flexible substrates including paper (Han et al., 2013, 2014).

Acknowledgements. This work was supported by the Nanotechnology Thematic Project in NASA's Game Changing Development Program. The smartphone development was funded by the US Department of Homeland Security, HSARPA Cell-All program via a NASA-DHS interagency agreement (IAA: HSHQDC-08-X-00870). The work conducted by the employees of ERC and ELORET Corporation was supported through subcontracts to the respective organizations. The authors acknowledge H. Hong from South Dakota School of Mines for providing sulfonated SWCNT samples and D. Skiver, M. Shi and A. Nguyen for help with SEM imaging.

Edited by: A. Romano-Rodriguez
Reviewed by: two anonymous referees

References

Adjizian, J. J., Leghrib, R., Koos, A. A., Suarez-Martinez, I., Crossley, A., Wagner, P., Grobert, N., Liobet, E., and Ewels, C. P.: Boron and nitrogen-doped multiwall carbon nanotubes for gas detection, Carbon, 66, 662–673, 2014.

Almdari, R., Golestanzadeh, M., Agend, F., and Zekti, N.: Sulfonic acid functionlized single walled carbon nanotubes: A highly efficient and reusable catalyst for green synthesis of 14-Aryl-14H-dibenzo[a,j] xanthene derivatives under solvent-free conditions, ICNS4, 2012.

Alamdari, R., Golestanzadeh, M., Agend, F., and Zekri, N.: Synthesis, characterization and catalytic activity of sulphonated multi-walled carbon nanotubes as heterogeneous, robust and reusable catalysts for the synthesis of bisphenolic antioxidants under solvent-free conditions, J. Chem. Sci., 125, 1185–1195, 2013.

Dong, K.-Y., Choi, J., Lee, Y. D., Kang, B. H., Yu, Y-Y., Choi, H. H., and Ju, B.-K.: Detection of a CO and NH$_3$ gas mixture using carboxylic acid-functionalized single-walled carbon nanotubes, Nanoscale Res. Lett., 8, 12–18, 2013.

Fam, D. W. H., Palaniappan, A. I., Tok, A. I. R., Liedberg, B., and Moochhala, S. M.: A review on technological aspects influencing commercialization of carbon nanotube sensors, Sens. Actuat. B., 157, 1–7, 2011.

Han, J. W., Kim, B., Li, J., and Meyyappan, M.: A carbon nanotube based ammonia sensor on cotton textile, Appl. Phys. Lett., 102, 193104, doi:10.1063/1.4805025, 2013.

Han, J. W., Kim, B., Li, J., and Meyyappan, M.: A carbon nanotube based ammonia sensor on cellulose paper, RSC Adv., 4, 549–553, 2014.

Hosaya, A., Tamura, S., and Imanaka, N.: A catalytic combustion-type CO gas sensor incorporating aluminum nitride as an intermediate heat transfer layer for accelerated response time, J. Sensors Sensor Syst., 3, 141–144, 2014.

Kaufman, D. R. and Star, A.: Carbon nanotube gas and vapor sensors, Angew. Chem. Int. Edit., 47, 6550–6570, 2008.

Kim, B., Lu, Y., Hannon, A., Meyyappan, M., and Li, J.: Low temperature Pd/SnO$_2$ sensor for carbon monoxide detection, Sens. Actuat. B., 177, 770–775, 2013.

Li, J., Lu, Y., Ye, Q., Cinke, M., Han, J., and Meyyappan, M.: Carbon nanotube sensors for gas and organic vapor detection, Nano Lett., 3, 929–933, 2003.

Lu, Y., Meyyappan, M., and Li, J.: A carbon-nanotube-based sensor Array for formaldehyde detection, Nanotechnology, 22, 055502, doi:10.1088/0957-4484/22/5/055502, 2011a.

Lu, Y., Meyyappan, M., and Li, J.: Fabrication of carbon nanotube-based sensor array and interference study, J. Mater. Res., 26, 2017–2023, 2011b.

Lu, Y., Meyyappan, M., and Li, J.: Trace detection of hydrogen peroxide vapor using carbon-nanotube-based chemical sensor, Small, 7, 1714–1718, 2011c.

Meyyappan, M. (Ed.): Carbon Nanotubes: Science and Applications, CRC Press, 2004.

Meyyappan, M. and Sunkara, M. K.: Inorganic nanowires: applications, properties and characterization, CRC Press, Boca Raton, FL, See Chapter 14, 2010.

Mishra, V. N. and Agarwal, R. P.: Sensitivity, response and recovery time of SnO_2 based think-film sensor array for H_2, CO, CH_4 and LPG, Sens. Actuat. B., 29, 861–874, 1998.

Naseh, M., Khodadadi, A., Mortazavi, Y., Sahraei, O., Pourfayaz, F., and Sedghi, S.: Functionalization of carbon nanotubes using nitric acid oxidation and DBD plasma, World Academy of Science, Eng. Technol., 49, 177–179, 2009

Peng, F., Zhang, L., Wang, H., Lu, P., and Yu, H.: Sulfonated carbon nantotubes as a strong protonic acid catalyst, Carbon, 43, 23978–2429, 2005.

Peng, S. and Cho, K. J.: Chemical control of nanotube electronics, Nanotechnology, 11, 57–60, 2000.

Penza, M., Rossi, R., Aluisi, M., and Serra, E.: Metal-modified and vertically aligned carbon nanotube sensors array for landfill gas monitoring applications, Nanotechnology, 21, 105501, doi:10.1088/0957-4484/21/10/105501, 2010.

Robinson, J. A., Snow, E. S., Badescu, S. C., Reinecke, T. L., and Perkins, F. K.: Role of defects in single-walled carbon nanotube chemical sensors, Nano Lett., 6, 1747–1751, 2006.

Santucci, S., Picozzi, S., Gregorio, F. D., Lozzi, L., Cantalini, C., Valentini, L., Kenney, J. M., and Delley, B.: NO_2 and CO gas adsoprtion on carbon nanotubes: experiment and theory, J. Chem. Phys., 119, 10904–10910, 2003.

Star, A., Joshi, V., Skarupo, S., Thomas, D., and Gabriel, J. C. P.: Gas sensor array based on metal decorated carbon nanotubes, J. Phys. Chem. B., 110, 21014–21020, 2006.

Wanna, Y., Srisukhumbowornchai, N., Tauntranont, A., Wisitsoraat, N., Thavarungkul, P., and Singjai, P.: The effect of carbon nanotube dispersion on CO gas sensing characteristics of polyaniline gas sensor, J. Nanosci. Nanotechno., 6, 3893–3896, 2006.

Wu, R. J., Wu, J. G., Yu, M. R., Tsai, T. K., and Yeh, C. T.: Use of CNT/CO_3O_4-SnO_2 in a carbon monoxide sensor operating at room temperatures, Sensor Lett., 6, 848–851, 2008.

Yu, H., Jin, Y., Li, Z., Peng, F., and Wang, H.: Synthesis and characterization of sulfonated single-walled carbon nanotubes and their performance as solid acid catalyst, J. Solid State Chem., 181, 432–438, 2008.

Yudianti, R., Onggo, H., Surdiram, Satio, Y., Iwata, T., and Azuma, J.: Analysis of functional group sited on multi-wall carbon nanotubes surface, Open Mat. Sci. J., 5, 242–247, 2011.

Zhang, Y., Cui, S., Chang, J., Ocola, L. E., and Chen, J.: Highly sensitive room temperature carbon monoxide detection using SnO_2 nanoparticle-decorated semiconducting single-walled carbon nanotubes, Nanotechnology, 24, 025503, doi:10.1590/1516-1439.235713, 2013.

Zhao, W., Fam, D. W. H., Yin, Z., Sun, T., Tan, H. T., Liu, W., Tok, A. L. Y., Boey, Y. C. F., Zhang, H., Hng, H. H., and Yan, Q.: A carbon monoxide gas sensor using oxygen plasma modified carbon nanotubes, Nanotechnology, 23, 425502, doi:10.1088/0957-4484/23/42/425502, 2012.

23

Objectifying user attention and emotion evoked by relevant perceived product components

R. Schmitt[1], **M. Köhler**[1], **J. V. Durá**[2], and **J. Diaz-Pineda**[2]

[1]Laboratory for Machine Tools and Production Engineering WZL, Chair of Metrology and Quality Management, RWTH Aachen University, Steinbachstr. 19, 52074 Aachen, Germany
[2]Institute of Biomechanics of Valencia (IBV), Universitat Politècnica de València, Camino de Vera s/n, 46022 Valencia, Spain

Correspondence to: M. Köhler (m.koehler@wzl.rwth-aachen.de)

Abstract. A company's aim is to develop products that engage user attention and evoke positive emotions. Customers base their emotional evaluation on product components that are relevant for their perception. This paper presents findings of both identifying relevant product components and measuring emotions evoked by relevant perceived product components. To validate results, the comparison with self-reporting methods identifies similarities and differences between explicit expressed and implicit recorded customer requirements. On the one hand, eye tracking is applied to deduce the attention provoked by perceived product components. In order to link the product strategy with product components, the paper presents results considering the fact that the gaze track is affected by current thoughts. (Köhler et al., 2013, 2014a, b; Köhler and Schmitt, 2012) On the other hand, since self-reporting tools are only useful for obtaining information about the conscious part of customers' emotions, there is a need for measurement methods that measure the changes in physiological signals (bio-signals). Arousal is similar to emotional intensity and is related to the galvanic skin response. Positive or negative emotions are defined by the valence that is measured by facial electromyography. Findings are presented that relate changes in bio-signals on the aesthetical design to the global product impression as well as to emotions and, subsequently, linking changes in physiological signals to the evaluation of semantic concepts and design parameters. The presented approach provides conclusions and valid information about products as well as product components that provoke certain emotions and about product components linked to a certain product concept, which could be part of a product strategy. Consequently, hard facts and special design rules for emotional product design can be deduced.

1 Introduction and motivation

Since the perception of products is linked to emotions and emotions affect the purchasing decision, it is crucial for the success of a company to measure customers' attention and emotions objectively and integrate this awareness into processes of emotional product design. It should be the aim to develop products that engage user attention and evoke positive emotions since customers of the target group base their emotional evaluation on product components that they perceive relevant. This paper presents approaches to systematically survey and objectify that information about the customer's subjective product perception and attention, as well as the emotional evaluation. Perceived quality is one decisive factor for the purchasing decision of almost all types of products. However, human perception is highly complex and cannot be measured efficiently only by customer surveys. Nearly 80 % of all information that is crucial for people's decisions is gathered by the eye (Berghaus, 2005). The developed approach focuses on the visual aspects of perception, since those affect the registration of a bulk of cognitive information. The challenging task is the objectified measurement of attention and related emotions arising while a product is observed.

2 Perceived product quality and emotions

2.1 Perceived quality

The improvement of perceived product quality has become an important determinant in order to stand out from competing companies (Falk et al., 2008). Perceived quality is a subjective and very individual interpretation of product realizations. In order to understand a customer's perception of product quality, the understanding of different levels of perception is of importance (Falk et al., 2008). The customer first perceives an overall impression of the product that is formed by the customer's individual product evaluation. Thereby, the overall impression of the product often corresponds to the system level of the whole product. On this level, the information about, for instance, customer's feelings and product perception is rather subjective. Perception clusters are one level more detailed and formed by different quality attributes that the customers merge on sensory perception such as harmonious aspects in optics. In order to further identify and structure the product, one or several descriptors are defined for each quality attribute. With these descriptors, technical specifications can be described in a very specific way by partial comparisons with terms the customer is familiar with. Finally, the combination of different characteristics of a descriptor with corresponding readings makes it possible to objectively determine relevant technical parameters. The active design of technical parameters again influences each level of perception. Consequently, special metrology is required to gather objective information of specific design parameters and to ensure a harmonious and high-value visual appearance of the product design towards customer's perception. With regard to product development and design, already in early steps of product development, there is a need to create and to ensure products and product attributes that evoke a high quality perception. Therefore, a structured approach to quality perception from the customers' perspective should be systematically applied to survey and to specify product attributes that are relevant to the perception and attention in order to objectify the customer's quality judgments (see Fig. 1 and Falk et al., 2008). Especially during early steps of product development process (PDP), there is a potential to concentrate on objectively measuring the visual impression of products and quality attributes (e.g., by using eye tracking), since it is often more decisive than other impressions (e.g., haptic, acoustic) to evoke emotions (Duchowski, 2007).

2.2 Emotions and purchase decisions

Emotional product development and design – often synonymously referred to a affective engineering (Nagamachi, 2011; Jiao et al., 2006) – are based on the fact that emotions play an essential role in purchasing decisions (Beaujean et al., 2011; Köhler et al., 2013). Those methodologies aim at integrating or typifying emotions and perception in the PDP.

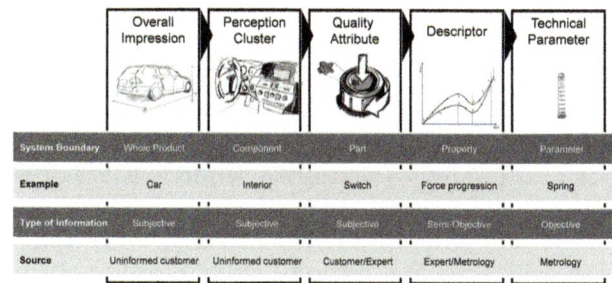

Figure 1. Structured approach towards quality perception referring to Falk et al. (2008).

Therefore research on factors that help to objectify the purchase decision is essential for fact-based product management (Hofer and Mayerhofer, 2010). With a successful realization of the conceptual conversion of customer requirements into technical product characteristics, the customer's perceived product quality can be improved. Hence, the earlier the needs of customers are defined and converted into technical characteristics, the smaller the costs and duration of PDP are and the more the purchasing decision can be affected in a positive way, which leads to higher revenues and profits for the company (Beaujean et al., 2011).

Users' perception of any product or service can be obtained by analyzing their feelings. It is very common to use questionnaires in order to measure customer opinion. The methods based on questionnaires translate the answers of the customers into parameters related to the product. However, these methods use the subjective opinion of the customer, and the subjective opinion depends on individual circumstances. These methods have several shortcomings. It is possible that the customer will change their answer deliberately because they prefer to hide their feelings, and it is also possible for the costumer to get confused because the differences between products are very subtle (Laparra-Hernández et al., 2009). The experimenter can influence the customer and change their answer too (Czerwinski et al., 2001; Nielsen and Levy, 1994). Finally, it is important to consider that unconscious processes are involved when the customer makes a decision (Tversky and Kahneman, 1973). An example of unconscious process is the stress level of the customer. Stress is a common component when we make decisions and can influence the opinion (Regueiro and León, 2003).

2.3 Established methods and concepts for objectifying perceived quality

Quality function deployment (QFD), conjoint analysis, means–ends analysis and Kansei engineering are some established concepts that try to objectify perceived quality data for integration in emotional engineering. Basically, these concepts lack the ability to quantify emotions through measuring physiological signals. QFD hinders the quantification of qualitative information (Hawlitzky, 2002). Since conjoint analysis premises information about product features that determine people's preference valuation a priori, the definition of relevant product characteristics in particular is difficult with regard to innovative product ideas (Meffert, 2000; Sattler, 2006). Means–ends analysis only focuses on the cognitive view but disregards the significant activating and emotional aspects (Kroeber-Riel and Weinberg, 1996). Kansei engineering quantifies customers' emotions and perception extensively but in general only focuses on the description of customers' impressions by semantic concepts and disregards unconscious impressions (Schütte, 2002).

2.4 Measuring emotions

Emotions can be measured in different ways as either discrete categories or continuous dimensions. Desmet (2002) proposed seven negative emotions (anger, contempt, disgust, indignation, dissatisfaction, disagreement and boredom) and eight positive emotions (desire, pleasure, surprise, inspiration, amusement, admiration, satisfaction and fascination) related to product appearance. However, there are many possible classifications and the classification can be difficult. The relation between categories and physiological signals is not clear. Therefore, the dimensional approach was chosen. Emotions can be represented in a space of two dimensions: valence and arousal. Whether an emotional response is positive or negative, it is represented by the dimension valence. In the category of EMG (electromyography) measures, facial EMG measures are often applied to objectify emotion recognition (valence). Facial EMG can detect small changes in emotions even when the facial expression does not change (Cacioppo et al., 1990). Cacioppo et al. (2004) concludes that the muscles most involved in emotion detection are the corrugator supercilii (related to frowning), the zygomaticus major and the depressor anguli oris (both related to smiling). GSR (galvanic skin response) measures changes in the electrical skin conductance and is related to the level of arousal (Lang et al., 1993). Arousal characterizes elicitation of emotion and emotion intensity. Thus, for the understanding of product perception, facial EMG and GSR have a potential use (Laparra-Hernández et al., 2009). For these reasons, facial EMG and GSR are the most suitable bio-signals for applying the dimensional appraisal in order to distinguish between product design alternatives.

3 Objectifying user attention and emotion

This paper presents results of a comprehensive approach for objectifying user evaluation of product alternatives and product components, especially regarding user attention and the emotional behavior via analyzing bio-signals. The methodology investigates which products and which product attributes are important to the customer and how an arrangement of parts impresses the customer the most and fits best with the aligned product strategy and the customer's emotional feelings. The general objective is the improvement of perceived quality. In the process of product design and development it is important to gather educated information for the comparison between the global impression of product alternatives as well as for the decision in favor of one specific alternative. Furthermore it is important to focus on the most relevant quality attributes since the resources of time and costs are limited. Additionally, the relation between semantic concepts and relevant product components should be pointed out. In order to not only have relied on subjective data derived from questionnaires, the objectification of emotions is of significant importance. Therefore the differences in emotions while changing design parameters should be investigated and the relation between the semantic concepts and the emotions analyzed.

3.1 Research questions and general methodology

Derived from the presented state of the art and from the challenges of perceived quality and of emotional (Beaujean et al., 2011; Köhler et al., 2013) or affective engineering (Nagamachi, 2011), the following research questions are of importance:

First, concerning user attention, it is important to investigate the following questions:

1. Which are the most relevant components according to the visual cognitive impression of products?

2. How can eye tracking be applied in order to measure the relevant components in an objective way?

3. Do semantic concepts, or product strategies, have any influence on the attention of the user toward the most important product components?

Secondly, regarding the emotional evaluation during the process of product design, it is crucial to investigate

1. whether bio-signals are able to allow for distinguishing between product alternatives with regard to the overall product impression,

2. whether semantic concepts have any influence on emotions concerning special design alternatives.

The following methods are applied in order to investigate which product attributes customers subjectively perceive and what the customer really feels, as well as how this feeling influences the product evaluation:

– methodical collection of product strategies,

– gathering conscious product evaluation and decisions between design alternatives through use of questionnaires,

- identification of the most important components of a special product design with regard to user attention,

- applying semantic concepts to differentiate between design alternatives on global product impression as well as on a more detailed level of product attributes,

- applying eye tracking to investigate the user attention related to the aforementioned methods,

- measuring bio-signals with a two-dimensional approach (GSR, EMG) to objectify the emotional response

3.2 Experimental design and analysis procedure

The presented methodology was applied in several corresponding studies during the research project CONEMO (e.g., Köhler et al., 2013, 2014a, b). In this paper, results are presented with consideration of a case study on an optically refined SUV (sport utility vehicle). Since the refined car was modified with a body kit, it differs a lot with regard to the detailed level of quality attributes from its production model (Köhler and Schmitt, 2012).

Previous work: collection and reduction of semantic concepts

A holistic study of semantic concepts was conducted in advance. Several methods were applied in order to generate an overview of words that are related to the emotional description of the studied SUV. Afterwards, the company's point of view and the customers' perspective were aligned and the words systematically reduced. Four of the most important semantic concepts describing the product strategy are used for illustration: dynamic, powerful, sporty and modern.

For reasons of study design, and to reduce systematic influences on the results of eye tracking, the products were presented at the same angle and size and with the same background. In preliminary studies an investigation found that the length of time each picture should be presented is dependent on the type of product and the product complexity. For examining the exterior design of cars, the presentation of pictures for between 10 and 15 s was researched in previous studies of the presented approach in order to record data about the attention and the emotional response of the user. A time of 11 s for recording bio-signals was chosen in order to reduce the size of the files of bio-signals and the computing time. For both the serial model and the refined car, the exterior front of the car was subdivided into about 20 visible parts as well as clusters (e.g., the bumper, headlights, rims) by means of the structured approach towards quality perception (see Sect. 2 and Köhler and Schmitt, 2012).

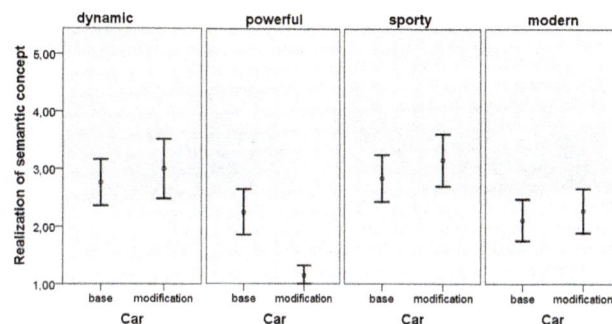

Figure 2. 95 % confidence interval – articulated evaluation of the realization of semantic concepts.

3.2.1 Capturing user attention with regard to product components

To deduce the attention provoked by perceived product components, a study with a screen-based eye-tracking system was performed. In order to link the product strategy with the product components, the theory that the gaze track is affected by current thoughts was used and applied (Yarbus, 1967). In this study, 25 automobile enthusiasts participated (18 to 27 years; mean: 21.4 years). Eye tracking was combined with ordinary questionnaires in order to contrast results of explicit and implicit user opinion.

A five-point Likert scale (1: strongly agree; 2: agree; 3: don't know; 4: disagree; 5: strongly disagree) was used to ask how the semantic concepts had been realized in one of the two product alternatives on the level of overall impression (see Fig. 2). ANOVA revealed only a significant difference concerning "powerful" ($F = 28.209$, $P = 0.000$). A post hoc paired t test showed that the refined car is evaluated as significantly more powerful than the serial model. These results only rely on the articulated opinion of users about the global impression (subjective source of information). Since there is no relation to special product attributes, it is hard to draw conclusions with regard to product design about manipulative design possibilities.

Capturing the most relevant product attributes

The participants were explained the eye-tracking experiment and the procedure. The study was conducted in an appropriate surrounding that assured, for example, constant light conditions and reduction of noise. The system was a screen-based combined pupil/corneal reflection system with the advantage that slight head movements do not lead to loss of the gaze track. Furthermore, the subject does not have any direct contact to the system. The study started with calibration and it was ensured the participant's eyes were maintained at a distance of about 60 cm from the monitor. Then, each alternative was presented without any comment (free interaction) for a duration of 15 s. The aim was to capture the areas of interest from the customer's point of view that are linked to

Figure 3. Individual gaze track for one user (free interaction vs. dynamic).

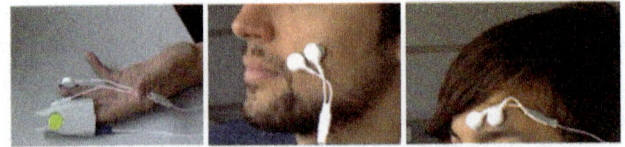

Figure 4. Placement of sensors for detecting bio-signals.

detailed product attributes or even more complex clusters of perception. All data were expressed in time ratios (dependent variables), since eye tracking is a relative measurement system and since the presented approach aims to compare the influence of different factors (independent variables) (Köhler et al., 2013). The time ratios were calculated by dividing the time a specific attribute was observed by the total time the product was observed (areas that do not belong to the car exterior such as the background of the picture were excluded). Pareto analysis was applied in order to focus on the most relevant product components.

Measuring the influence of semantic concepts on user's attention

Eye tracking was applied with the objective of relating bio-signals and responses with regard to quality attributes with the feelings and needs that are represented by semantic concepts. First, each participant was presented a slide with an introduction slide where *Think in dynamic* (that means the semantic concept dynamic) was written on it. Afterwards, a set of two pictures representing the refined and the serial model were shown. The duration of presenting the pictures was decreased after the free interaction because the subjects were already familiar with both product alternatives. Both the fixation time (expressed in time ratios) for each component (about 20 visible parts of the car exterior; see Sect. 3.2) and the gaze tracks representing the gaze while thinking about a certain product concept which was priorly introduced by the introduction slide were recorded (see Fig. 3). Besides these data, relevant information for user attention can also be deduced by analyzing the order of fixation and the frequency of fixating on an area of interest during a specific period of time.

Statistical methods

A within-subjects ANOVA was conducted in SPSS to compare the effect of semantic concepts on the time ratio of relevant quality attributes. For the significant effects, post hoc analysis of paired t tests was done and educated recommendations for product design were derived. The objective of the eye-tracking study was mainly to lead to valid information about which quality attribute the customers consciously and unconsciously relate to a certain semantic concept. To survey whether the customers' evaluation is positive or negative, the emotions also have to be investigated.

3.2.2 Bio-signals experimental design

Once the relevance of product components is known, the emotional response related to specific design alternatives should be objectified. Fourteen male undergraduate students (age from 18 to 30 years) took part in the study, with a single session lasting about half an hour. Again, the participants were explained the experiment (e.g., the procedure and the duration). In order for the participant's eyes to be maintained at a distance of 61 cm from the 17 in. monitor, adjustments were made (Laparra-Hernández et al., 2009). The EMG sensors were attached to the left corrugator and zygomatic muscles of the face, and the GSR sensors were attached to the middle and ring fingers of the left hand (see Fig. 4).

Two sets of pictures were used: the first set was used to adapt the participants to the experiment environment and to reduce the effect of surprise. The pictures used correspond to three different products: a fan, a lamp and a living room scene. The second set of pictures was the two car design alternatives. The subjects were instructed to examine the three pictures (6 s each). Then, the two car pictures were examined for 11 s (see duration concerning eye-tracking study). The aim of this step is to record the emotional responses without evaluating a particular semantic concept. After having been shown a question (semantic concept), the subject again examined the pictures and answered the question (using the Likert scale). This procedure allowed for identifying whether the bio-signals are influenced by a semantic concept. The order of questions and pictures was randomized in order to avoid potential ordering effects. Immediately before a picture was displayed to the user, a black slide was displayed for 10 s. For one thing, this additional slide separates the responses between pictures and, consequently, minimizes the impact of signals elicited by consecutive pictures. For another, the time span was applied to normalize the subject's response by inducing a state of relaxation (Laparra-Hernández et al., 2009).

Signal processing

EMG and GSR signals were processed offline; the processing technique is based on that used in other studies (Laparra-Hernández et al., 2009; Heino et al., 1990): on the one hand, the signals of EMG were fully rectified and filtered (2 Hz low-pass filter) to acquire the envelope and to eliminate fast changes. On the other hand, the GSR signal is characterized by two components: a fast fluctuation (phasic) and a slow component (tonic). The tonic component is the baseline level of skin conductance in the absence of any particular environmental event. The phasic component is more event-related, with fast changes (peaks) interpreted as responses to a specific stimuli, and is related to the arousal level. Furthermore, the GSR signal was filtered using a fifth Butterworth low-pass filter (cut-off frequency of 0.05 Hz) (Laparra-Hernández et al., 2009; Heino et al., 1990).

As done in the study for capturing the attention (see Sect. 3.2.1), each picture type (products and product associated with a semantic concept) is analyzed individually. To avoid transitory effects, the beginnings of all signals were discarded. First, EMG and GSR values are normalized to compare the values of each type of product for each subject. Two-step normalization process is used, with different possibilities. The first step is represented in Eq. (1), where S can be the 75th percentile of the signal during the entire exposure time of each image. The variables calculated with this criterion have the suffix **_MV**. Otherwise S can be the 75th percentile of the signal during a fixed time (6 s) for all the products. The variables calculated with this criterion have the suffix **_MF** (Laparra-Hernández et al., 2009).

$$Ns_{nj} = \frac{\hat{S} - \hat{R}}{\hat{R}}, \tag{1}$$

where n is the subject and j is the picture.

R can be the 75th percentile of the signal in the 10 s (black slide) before exposure to each stimulus. The variables calculated with this criterion have the suffix _ni_. Otherwise R can be the mean of the first three relaxations. At the beginning of the test there are three relaxations (three black slides) with three stimuli different to the evaluated products. These relaxations are done before asking the subject about any concept and are not affected by the previous history. Therefore R is the mean of the three 75th percentiles of the signals in the 10 s (black slide) before exposure to each stimulus. The variables calculated with this criterion have the suffix **_net**.

Secondly, for each user the signals are normalized again with respect to the maximum value recorded for the user in question. The second normalization (N_{nj}) is carried out according to Eq. (2), where N_{Snj} is the normalized value (with the step) of the signal that is elicited by picture j from subject n. N_{Sn} is a vector with the normalized values of the signals, elicited by all pictures, from subject n (Laparra-Hernández et al., 2009):

$$N_{nj} = \frac{N_{Snj} - \min(N_{Sn})}{\max(N_{Sn}) - \min(N_{Sn})}, \tag{2}$$

where n is the subject and j is the picture.

Table 1. Time ratio for the most relevant product attributes.

Item	Mean	SD
Name	0.069	0.0722
Rims	0.069	0.0427
Headlight_left	0.067	0.0562
Engine hood	0.115	0.0742
Windshield	0.095	0.0696
Side_door	0.089	0.0704
Grill	0.131	0.0823
Bumper	0.179	0.0897

Statistical methods

The analysis of the signal processing was performed with Octave and the statistical analysis with the statistics software R. To detect differences in the bio-signals (EMG and GSR), the univariate ANOVA was applied (Laparra-Hernández et al., 2009). The model applied is bio-signal $\sim$ CAR_TYPE + CAR_TYPE:CONCEPT, where bio-signal is the independent variable, CAR_TYPE is a factor and CAR_TYPE:CONCEPT is an interaction. Our model does not include CONCEPT as a factor because CONCEPT is a question about the product (CAR_TYPE). We consider that any influence of the question should be related to the product.

3.3 Results

As described before, differentiation between the investigation of attention and the objectification of emotional evaluation of visually perceived product attributes should be done.

3.3.1 Capturing user attention related to semantic concepts

For all subjects and for the two product alternatives the ratio of fixation duration is captured for each of the 20 areas of interest which were defined before (see Sect. 3.2). The Pareto analysis of the free examination of the products revealed that the subjects observed the following attributes more than 80 % of the whole mean time: the bumper (a cluster of perception formed by number plate and bumper bar), the grill, the engine hood, the windshield, the side door, the signature with the name/logo, the rims and the left headlight (see Table 1). Because of the special presentation of the product in the picture, almost all relevant product attributes are located in the front part of the car exterior.

Furthermore, a MANOVA, using the time ratios as dependent variables and the different defined areas of interest as independent variables, shows that there is only a slight significant effect of the car type according to the attention of the most important product attributes (Pillai's trace $= 0.580$; F value $= 1.936$; $p = 0.053$). Hence, there is no significant difference in the importance of the most relevant product

Table 2. Results of within-subject repeated-measures ANOVA.

Factor semantic concept (five levels)			
Car	Pillai's trace	F value	p level
Base car	0.535	$F(20, 284) = 2.194$	0.003
Modification	0.657	$F(20, 252) = 2.475$	0.001

Table 3. Results for the selected quality attributes.

Car 1: base car (serial model)		
Quality attribute	F value	p level
Name/signature	$F(4, 72) = 5.177$	0.009 (Greenhouse–Geisser)
Headlight left	$F(4, 72) = 5.055$	0.007 (Greenhouse–Geisser)
Bumper	$F(4, 72) = 0.874$	0.484
Car 2: modification		
Quality attribute	F value	p level
Name/signature	$F(4, 64) = 2.483$	0.081 (Greenhouse–Geisser)
Headlight left	$F(4, 64) = 3.355$	0.015
Bumper	$F(4, 64) = 5.481$	0.004 (Greenhouse–Geisser)

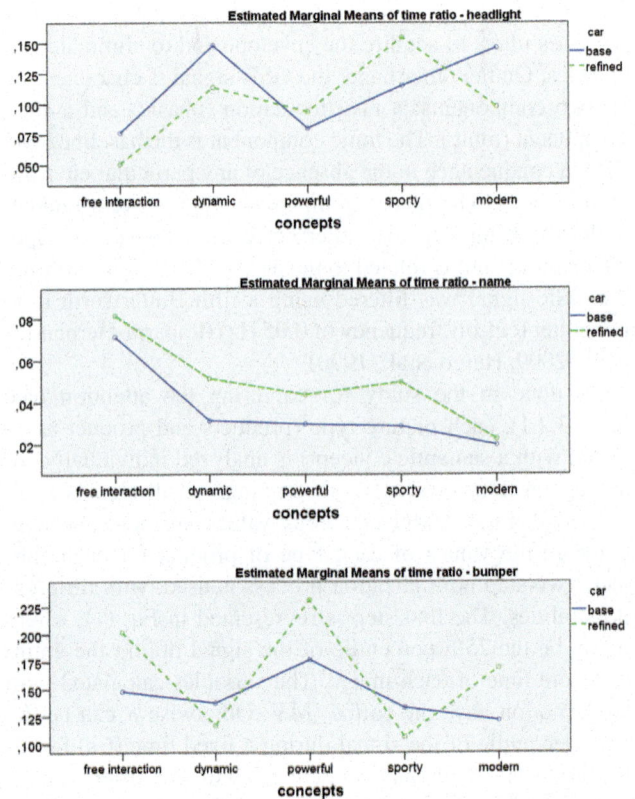

Figure 5. Means of time ratio for left headlight, name and bumper.

attributes, or, in other words, the most important attributes in the design of the refined car were also of similar importance in the design of the serial model. Analyzing the correlations between the time ratios for the different relevant product attributes revealed some weak correlations (Pearson correlation coefficient > 0.20). Obviously there are weak negative correlations between product attributes that are not directly next to each other (e.g., signature/name and bumper: $r_{Pearson} = -0.411$, $p = 0.003$).

To analyze the effect of semantic concepts on the user's attention, the 25 subjects were shown a set of stimuli: the factor "car" with two levels and the semantic concepts with five levels. The time ratio for each product attribute was recorded with specific interest on the most relevant attributes.

A repeated-measures ANOVA (Table 2), this time using the semantic concepts as independent variables, was conducted to test differences between means of fixation time ratios for significance. From analysis of the cars separately, there was a significant effect of the semantic concept on the time ratio for several relevant product attributes while thinking in semantic concepts (dynamic, powerful, sporty, modern in comparison to free interaction).

For purposes of illustration, the bumper, the signature with the name and the left headlights are selected for reporting results. Since the Mauchly's test of sphericity was not fulfilled ($p < 0.05$) for name and headlight in the case of the serial model car and for name and bumper in the case of the modified car, the Greenhouse–Geisser correction was applied for the p level of univariate tests (see Table 3).

The diagrams for the estimated marginal means for the time ratio of the selected product attributes (Fig. 5) give a first idea of the effect of semantic concepts. For both car

alternatives, post hoc pairwise comparisons were used to indicate which semantic concept differs significantly ($p < 0.1$) for both cars. That way, for each product attribute, design recommendations concerning the attention that is evoked by the semantic concepts could be derived.

For instance, it could be concluded that the left headlight of the refined car (green line in Fig. 5) was significantly more focused while thinking in the semantic concepts (e.g., dynamic ($T(24) = 3.45$, $p = 0.001$) or sporty ($T(23) = 3.76$, $p = 0.0001$)) in comparison to the free interaction. Similar results could be found in the case of the serial model car (blue line), except there was no significant difference between the free interaction and "powerful" ($T(18) = 0.915$, $p = 0.186$).

The diagram for the name reveals that the means for each semantic concept are higher in the case of the modified car than for the serial model. Furthermore, considering also slight significant influences in the case of the refined car ($p = 0.081$), there were significant differences between all semantic concepts and the free interaction.

Since, in the case of the serial model car, there was no significant effect of the semantic concepts regarding the time ratio of the bumper ($p = 0.484$), the interpretation of results is presented for the refined car. Specifically, the bumper was related more to "powerful" than to "dynamic" ($T(18) = 2.633$, $p = 0.008$), "sporty" ($T(17) = 3.315$, $p = 0.002$) or "modern" ($T(16) = 1.47$, $p = 0.08$). In combination with the

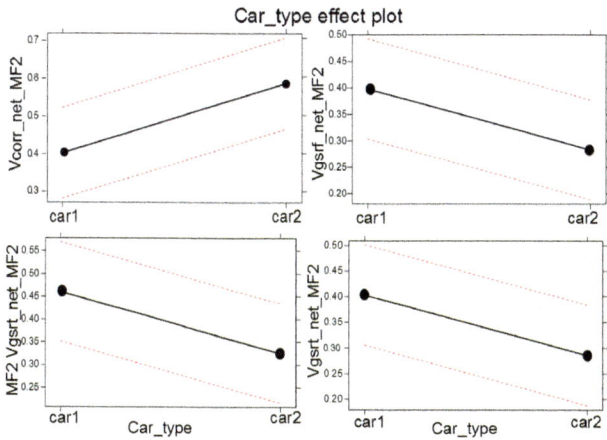

Figure 6. Results of EMG and GSR measures.

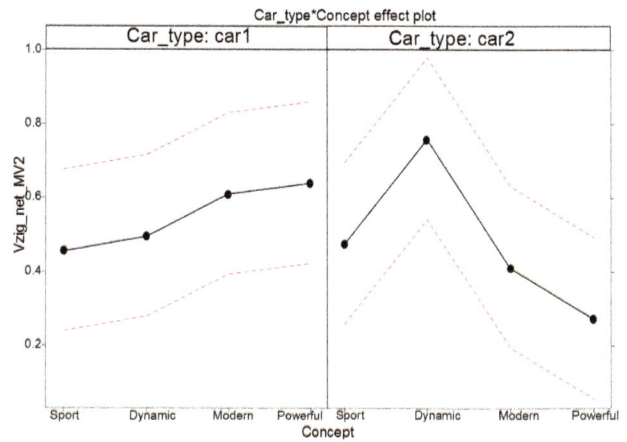

Figure 7. Interaction effects plot.

Table 4. Bio-signals.

Variable	Factor	p level
Vzig_net_MV2	Car_type:Concept	0.0709
Vgsrt_net_MV2	Car_type	0.09366
Vcorr_net_MF2	Car_type	0.03467
Vgsrf_net_MF2	Car_type	0.09067
Vgsrt_net_MF2	Car_type	0.08477

self-reporting questionnaire, a possible conclusion could be that the bumper is one important attribute responsible for the refined car to be evaluated as more powerful than the serial model car.

3.3.2　Measuring bio-signals

Considering the bio-signals, the significant ($p < 0.1$) effects are shown in Table 4, where V_{zig} is the zygomaticus EMG, V_{corr} is the corrugator EMG, V_{gsrt} is the tonic component of GSR, and V_{gsrf} is the phasic component of GSR.

The variables contain the indicator 2 in order to show that they have been normalized according to the two steps described above. No significant differences were found with variables normalized with the first step only, nor when using a normalization with the 75th percentile of the signal in the 10 s (black slide) before exposure to each stimulus.

Figure 6 shows that car 1 (serial model) better evokes emotions and more intensely (corrugators low and GSR high) than car 2 (modified model). Considering the interaction effect with the concepts (Fig. 7), it seems that "powerful" is the concept that provokes the highest positive opinion for car 1, and "dynamic" for car 2.

4　Discussion

Findings of a methodology are presented relating changes in bio-signals on the aesthetic design with the global product impression as well as with emotions and, subsequently, linking changes in bio-signals to the evaluation of semantic concepts and design parameters. It was shown that eye tracking and measuring bio-signals can be applied to indicate relevant product attributes and to measure their impact on customer's emotional response. Moreover, the results suggest that semantic concepts affect the individual gaze track and, subsequently, the time ratio for the attention evoked by relevant product attributes. Furthermore, results should be interpreted on a very detailed level of quality attributes in order to link feelings expressed through semantic concepts with the attention measured by means of the time ratio. To match customers' needs and feelings, this awareness is of particular importance in early steps of PDP and can be applied, for example, in product advertisements to attract and to point out special design alternatives or product components. Deeper analysis of the recorded eye-tracking data should be done in order to interpret the significance of why subjects really focus on one component longer than another.

According to the technical analysis of measuring bio-signals, it was shown that two-step normalization is more effective than using only one step. This kind of normalization considers for the maximum value recorded for each user and all the images that the subject examines in one session (see also Laparra-Hernández et al., 2009). Significant differences have been found with the normalization with the relaxations (black slides) done before asking the subject about any concept and not affected by the previous history (_net_), whereas no differences were found with the use of the relaxation before stimulus (_ni_). It is assumed that subjects could be thinking about the questions or semantic concepts during the relaxing time, which could be considered as a stimulus. For this reason the relaxing periods before the images had some noise. Therefore, it is better to use a method of relaxation that is not affected by any stimulus, i.e., not an image or concept (mental process).

Further suggestions according to the experimental design of the preliminary work for conducting the studies of eye tracking and measuring bio-signals can be derived. For one thing, the influence of the order of the presented pictures must be considered as done in further studies (Köhler et al., 2013, 2014a, b). It seems to be important whether or not a random order for each subject leads to data with higher quality regarding the eye-tracking experiment. For another, semantic concepts should be carefully selected in a way that users have a good understanding of the concept. In addition, it is crucial to apply an appropriate level of detail with regard to the product structure, which could also affect the results presented in the methodology. Moreover, it has to be considered that the results of the conducted studies only indicate influences regarding semantic concepts and possibilities concerning eye tracking and EMG/GSR. One specific sample of proper respondents for both kinds of emotional studies should be used in order to apply the same profile of respondents and to derive more significant and valid data. In addition, the number of subjects should be increased to improve the value of results of statistical analysis.

5 Conclusions and future directions

The presented approach provides conclusions and valid information about products as well as product components that provoke certain emotions and about product components linked to a certain product strategy. Consequently, the objective is to use that information for enhancing the perceived quality and to derive design rules and recommendations. The results shown are an excerpt of a comprehensive methodology for measuring and objectifying customers' attention and emotions evoked by products and product components as well as the systematic integration of the results into the early product development and design process. This also includes the comparison with self-reporting methods to identify similarities and differences between explicit expressed and implicit recorded customer requirements (Köhler and Schmitt, 2012; Köhler et al., 2013, 2014a, b). The possibility to generalize the methodology for other products on other levels of detail was proved.

Acknowledgements. The paper presents results from the cooperative research project CONEMO (Consumer Evaluation Measurement for Objectified Industrial Use – funded by the COR-NET program) of the Laboratory for Machine Tools and Production Engineering (WZL), RWTH Aachen University, Germany, together with the Institute of Biomechanics of Valencia (IBV), Spain. The funding agencies are IMPIVA (operational program FEDER of the Comunidad Valenciana, Spain) and AiF (Germany). The CORNET promotion plan 47EN of the Research Community for Quality (FQS), August-Schanz-Str. 21A, 60433 Frankfurt/Main, was funded by the AiF within the program for sponsorship by Industrial Joint Research and Development (IGF) of the German Federal Ministry of Economic Affairs and Technologies based on an enactment of the German parliament. The authors would like to express their gratitude to all parties involved.

Edited by: R. Tutsch
Reviewed by: two anonymous referees

References

Beaujean, P., Grob, R., Häfen, K., Köbler, E., Köhler, M., Quattelbaum, B., Schmitt, R., Seitz, R., Wagner, M., and Willach, A.: Emotionale Produktgestaltung – Wert der wahrgenommenen Qualität, in: Proceedings: Tagungsband zum AWK, edited by: Brecher, C., Klocke, F., Schmitt, R., and Schuh, G., Shaker, 2011.

Berghaus, N.: Eye-Tracking im stationären Einzelhandel. Eine empirische Analyse der Wahrnehmung von Kunden am Point of Purchase, dissertation, University Duisburg-Essen, 2005.

Cacioppo, J. T., Tassinary, L. G., and Fridlund, A. J.: The skeletomotor system, in: Principles of Psychophysiology: Physical Social, and Inferential elements, edited by: Caccioppo, J. T. Tassinary, L. G., Cambridge University Press, New York, 325–384, 1990.

Cacioppo, J. T., Bernston, G. G, Larsen, J. T., Poehlmann, K. M., and Ito, T. A.: The psychophysiology of emotion, in: Handbook of Emotions, edited by: Lewis, M. and Haviland-Jones, J. M., The Guilford Press, New York, 173–191, 2004.

Czerwinski, M., Horvitz, E., and Cutrell, E.: Subjective duration assessment: an implicit probe for software usability, in: Proceedings of the IHM-HCI 2001 Conference, Vol. 2, 167–170, 2001.

Desmet, P.: Designing Emotions. Delft University of Technology, Department of Industrial Design, 2002.

Duchowski, A. T.: Eye Tracking Methodology: Theory and Practice, Vol. 2., Springer, London, 2007.

Falk, B., Schmitt, R., and Quattelbaum, B.: Product Quality from the Customers' Perspective – Systematic Elicitation and Deployment of Perceived Quality Information, in: Proceedings of the 6th CIRP-Sponsored International Conference on Digital Enterprise Technology, 216 pp., 2008.

Hawlitzky, N.: Integriertes Qualitätscontrolling von Unternehmensprozessen, Gestaltung eines Quality-Gate-Konzeptes, dissertation, Technical University München, 2002.

Heino, A., Van der Molen, H. H., and Wilde, G. J. S.: Risk-homeostatic processes in car following behaviour: electrodermal responses and verbal risk estimates as indicators of the perceived level of risk during a car-driving task, Report VK 90-22, Traffic Research Centre, University of Groningen, Haren, 1990.

Hofer, N. and Mayerhofer, W.: Die Blickregistrierung in der Werbewirkungsforschung. Grundlagen und Ergebnisse, in: Der Markt, No. 49, 149 pp., 2010.

Jiao, R., Zhang, Y., and Helander, M.: A Kansei mining system for affective design, Journal of Expert Systems with Applications, 30, 658–673, 2006.

Köhler, M. and Schmitt, R.: Systematic Consumer Evaluation Measurement for Objectified Integration into the Product Development Process, in: Advances in Affective and Pleasurable Design, edited by: Ji, Y. G., CRC Press, 503–512, 2012.

Köhler, M., Falk, B., and Schmitt, R.: Objectifying user attention caused by visually perceived product components, in: Proceedings of the 16th International Congress of Metrology, doi:10.1051/metrology/201314002, 2013.

Köhler, M., Falk, B., and Schmitt, R.: Applying Eye-Tracking in Kansei Engineering Methodology for Design Evaluations in Product Development, KEER conference, Linköping, available at: http://dqi.id.tue.nl/keer2014/papers/KEER2014_125 (last access: 30 September 2014), 2014a.

Köhler, M., Falk, B., and Schmitt, R.: Integrating User Attention for Design Evaluations in Customer-Oriented Product Development, in: Advances in Affective and Pleasurable Design, edited by: Ji, Y. G. and Choi, S., Proceedings of the 5th International Conference on Applied Human Factors and Ergonomics AHFE 2014, 19–23 July 2014, CRC Press, 428–439, 2014b.

Kroeber-Riel, W. and Weinberg, P.: Konsumentenverhalten, Vol. 6, Vahlen, 1996.

Lang, P. J., Greenwald, M. K., Bradley, M. M., and Hamm, A. O.: Looking at pictures: affective, facial, visceral, and behavioral reactions, Psychophysiology, 30, 261–273, 1993.

Laparra-Hernández, J., Belda-Lois, J. M., Medina, E., Campos, N., and Poveda, R.: EMG and GSR signals for evaluating user's perception of different types of ceramic flooring, Journal of Industrial Ergonomics, 39, 326–332, 2009.

Meffert, H.: Marketing. Grundlagen marktorientierter Unternehmensführung, Vol. 9, Gabler, 2000.

Nagamachi, M.: Kansei/Affective Engineering, CRC Press, Florida, 2011.

Nielsen, J. and Levy, J.: Measuring usability: preference vs. performance, Communications of the ACM 37, No. 4, 66–75, 1994.

Regueiro, R. and León, O.: Estrés en desiciones cotidianas, Journal of Psicothema, 15, 533–538, 2003.

Sattler, H.: Methoden zur Messung von Präferenzen für Innovationen, Zeitschrift für betriebswirtschaftliche Forschung (zfbf), Vol. 54/06, 154–176, 2006.

Schütte, S.: Designing Feelings into Products. Integrating Kansei Engineering Methodology in Product Development, dissertation, University Linköping, 2002.

Tversky, A. and Kahneman, D.: 1 Availability: a heuristic for judging frequency and probability, Cognitive Psychol., 5, 207–232, 1973.

Yarbus, A. L.: Eye Movements and Vision, Plenum Press, 1967.

Permissions

All chapters in this book were first published in JSSS, by Copernicus Publications; hereby published with permission under the Creative Commons Attribution License or equivalent. Every chapter published in this book has been scrutinized by our experts. Their significance has been extensively debated. The topics covered herein carry significant findings which will fuel the growth of the discipline. They may even be implemented as practical applications or may be referred to as a beginning point for another development.

The contributors of this book come from diverse backgrounds, making this book a truly international effort. This book will bring forth new frontiers with its revolutionizing research information and detailed analysis of the nascent developments around the world.

We would like to thank all the contributing authors for lending their expertise to make the book truly unique. They have played a crucial role in the development of this book. Without their invaluable contributions this book wouldn't have been possible. They have made vital efforts to compile up to date information on the varied aspects of this subject to make this book a valuable addition to the collection of many professionals and students.

This book was conceptualized with the vision of imparting up-to-date information and advanced data in this field. To ensure the same, a matchless editorial board was set up. Every individual on the board went through rigorous rounds of assessment to prove their worth. After which they invested a large part of their time researching and compiling the most relevant data for our readers.

The editorial board has been involved in producing this book since its inception. They have spent rigorous hours researching and exploring the diverse topics which have resulted in the successful publishing of this book. They have passed on their knowledge of decades through this book. To expedite this challenging task, the publisher supported the team at every step. A small team of assistant editors was also appointed to further simplify the editing procedure and attain best results for the readers.

Apart from the editorial board, the designing team has also invested a significant amount of their time in understanding the subject and creating the most relevant covers. They scrutinized every image to scout for the most suitable representation of the subject and create an appropriate cover for the book.

The publishing team has been an ardent support to the editorial, designing and production team. Their endless efforts to recruit the best for this project, has resulted in the accomplishment of this book. They are a veteran in the field of academics and their pool of knowledge is as vast as their experience in printing. Their expertise and guidance has proved useful at every step. Their uncompromising quality standards have made this book an exceptional effort. Their encouragement from time to time has been an inspiration for everyone.

The publisher and the editorial board hope that this book will prove to be a valuable piece of knowledge for researchers, students, practitioners and scholars across the globe.

List of Contributors

J. Berthold
VDI/VDE GMA within VDI e.V., Dusseldorf, Germany

D. Imkamp
Carl Zeiss Industrielle Messtechnik GmbH, Oberkochen, Germany

A. Hosoya
Department of Applied Chemistry, Faculty of Engineering, Osaka University, 2-1 Yamadaoka, Suita, Osaka 565-0871, Japan

S. Tamura
Department of Applied Chemistry, Faculty of Engineering, Osaka University, 2-1 Yamadaoka, Suita, Osaka 565-0871, Japan

N. Imanaka
Department of Applied Chemistry, Faculty of Engineering, Osaka University, 2-1 Yamadaoka, Suita, Osaka 565-0871, Japan

T. Waber
Munich University of Applied Sciences, Lothstraße 64, 80335 Munich, Germany

M. Sax
Munich University of Applied Sciences, Lothstraße 64, 80335 Munich, Germany

W. Pahl
EPCOS AG, a member of TDK-EPC Corporation, Anzingerstraße 13, 81617 Munich, Germany

S. Stufler
EPCOS AG, a member of TDK-EPC Corporation, Anzingerstraße 13, 81617 Munich, Germany

A. Leidl
EPCOS AG, a member of TDK-EPC Corporation, Anzingerstraße 13, 81617 Munich, Germany

M. Günther
Technical University of Dresden, Helmholtzstraße 18, 01062 Dresden, Germany

G. Feiertag
Munich University of Applied Sciences, Lothstraße 64, 80335 Munich, Germany

R. Gruden
Seuffer GmbH & Co. KG, Bärental 26, 75365 Calw-Hirsau, Germany

A. Buchholz
Seuffer GmbH & Co. KG, Bärental 26, 75365 Calw-Hirsau, Germany

O. Kanoun
Technische Universität Chemnitz, Reichenhainer Strasse 70, 09126 Chemnitz, Germany

F. Fedi
CNR-Institute for Composite and Biomedical Materials, Portici (Naples), Italy
Faculty of Physics, University of Vienna, Strudlhofgasse 4, 1090 Vienna, Austria

F. Ricciardella
ENEA UTTP-MDB Laboratory, R. C. Portici (Naples), Italy
University of Naples "Federico II", Department of Physics, Naples, Italy

M. L. Miglietta
ENEA UTTP-MDB Laboratory, R. C. Portici (Naples), Italy

T. Polichetti
ENEA UTTP-MDB Laboratory, R. C. Portici (Naples), Italy

E. Massera
ENEA UTTP-MDB Laboratory, R. C. Portici (Naples), Italy

G. Di Francia
ENEA UTTP-MDB Laboratory, R. C. Portici (Naples), Italy

M. Bektas
Department of Functional Materials, University of Bayreuth, 95447 Bayreuth, Germany

D. Hanft
Department of Functional Materials, University of Bayreuth, 95447 Bayreuth, Germany

D. Schönauer-Kamin
Department of Functional Materials, University of Bayreuth, 95447 Bayreuth, Germany

T. Stöcker
Department of Functional Materials, University of Bayreuth, 95447 Bayreuth, Germany

G. Hagen
Department of Functional Materials, University of Bayreuth, 95447 Bayreuth, Germany

R. Moos
Department of Functional Materials, University of Bayreuth, 95447 Bayreuth, Germany

M. Schüler
Laboratory for Measurement Technology, Department of Mechatronics, Saarland University, Saarbrücken, Germany

T. Sauerwald
Laboratory for Measurement Technology, Department of Mechatronics, Saarland University, Saarbrücken, Germany

A. Schütze
Laboratory for Measurement Technology, Department of Mechatronics, Saarland University, Saarbrücken, Germany

A.Lorek
German Aerospace Center (DLR), Berlin, Germany

E. Dilonardo
Department of Chemistry, Università degli Studi di Bari Aldo Moro, Bari, Italy

M. Penza
ENEA, Italian National Agency for New Technologies, Energy and Sustainable Economic Development, Technical Unit for Materials Technologies – Brindisi Research Center, Brindisi, Italy

M. Alvisi
ENEA, Italian National Agency for New Technologies, Energy and Sustainable Economic Development, Technical Unit for Materials Technologies – Brindisi Research Center, Brindisi, Italy

C. Di Franco
CNR-IFN Bari, Bari, Italy

D. Suriano
ENEA, Italian National Agency for New Technologies, Energy and Sustainable Economic Development, Technical Unit for Materials Technologies – Brindisi Research Center, Brindisi, Italy

R. Rossi
ENEA, Italian National Agency for New Technologies, Energy and Sustainable Economic Development, Technical Unit for Materials Technologies – Brindisi Research Center, Brindisi, Italy

F. Palmisano
Department of Chemistry, Università degli Studi di Bari Aldo Moro, Bari, Italy

L. Torsi
Department of Chemistry, Università degli Studi di Bari Aldo Moro, Bari, Italy

N. Cioffi
Department of Chemistry, Università degli Studi di Bari Aldo Moro, Bari, Italy

S. Nakagomi
Faculty of Science and Engineering, Ishinomaki Senshu University Minamisakai, Ishinomaki, Miyagi 986-8580, Japan

K. Yokoyama
Faculty of Science and Engineering, Ishinomaki Senshu University Minamisakai, Ishinomaki, Miyagi 986-8580, Japan

Y. Kokubun
Faculty of Science and Engineering, Ishinomaki Senshu University Minamisakai, Ishinomaki, Miyagi 986-8580, Japan

F. P. Pentaris
Dept. of Electronic and Computer Engineering, Brunel University, London, UK

U. Kienitz
Optris GmbH, Berlin, Germany

B. Komander
TU Braunschweig, Institut für Analysis und Algebra, Braunschweig, Germany

D. Lorenz
TU Braunschweig, Institut für Analysis und Algebra, Braunschweig, Germany

M. Fischer
TU Braunschweig, Institut für Produktionsmesstechnik, Braunschweig, Germany

M. Petz
TU Braunschweig, Institut für Produktionsmesstechnik, Braunschweig, Germany

R. Tutsch
TU Braunschweig, Institut für Produktionsmesstechnik, Braunschweig, Germany

M. Leidinger
Saarland University, Lab for Measurement Technology, Saarbrücken, Germany

T. Sauerwald
Saarland University, Lab for Measurement Technology, Saarbrücken, Germany

W. Reimringer
S GmbH, Saarbrücken, Germany

G. Ventura
IDMEC – Institute of Mechanical Engineering, Porto, Portugal

A. Schütze
Saarland University, Lab for Measurement Technology, Saarbrücken, Germany

T. Mazingue
Univ. Savoie, SYMME, 74000 Annecy, France

M. Lomello-Tafin
Univ. Savoie, SYMME, 74000 Annecy, France

M. Passard
Univ. Savoie, SYMME, 74000 Annecy, France

C. Hernandez-Rodriguez
Univ. Savoie, SYMME, 74000 Annecy, France

L. Goujon
Univ. Savoie, SYMME, 74000 Annecy, France

J.-L. Rousset
Institut de Recherches sur la Catalyse et l'Environnement de Lyon (IRCELYON, CNRS – University of Lyon), 2 avenue Albert Einstein, 69626 Villeurbanne CEDEX, France

F. Morfin
Institut de Recherches sur la Catalyse et l'Environnement de Lyon (IRCELYON, CNRS – University of Lyon), 2 avenue Albert Einstein, 69626 Villeurbanne CEDEX, France

J.-F. Laithier
Comelec SA, Rue de la Paix 129 – 2301 La Chaux-de-Fonds, Switzerland

H. Madokoro
Faculty of Systems Science and Technology, Akita Prefectural University, 84-4 Tsuchiya Aza Ebinokuchi, Yurihonjo City, Akita, 015-0055, Japan

N. Shimoi
Faculty of Systems Science and Technology, Akita Prefectural University, 84-4 Tsuchiya Aza Ebinokuchi, Yurihonjo City, Akita, 015-0055, Japan

K. Sato
Faculty of Systems Science and Technology, Akita Prefectural University, 84-4 Tsuchiya Aza Ebinokuchi, Yurihonjo City, Akita, 015-0055, Japan

F. Vita
Istituto Nazionale di Geofisica e Vulcanologia – Sezione di Palermo – Via Ugo La Malfa, 153, 90146 Palermo, Italy

C. Kern
U.S. Geological Survey, Cascades Volcano Observatory, 1300 SE Cardinal Ct S100, Vancouver, Washington 98683, USA

S. Inguaggiato
Istituto Nazionale di Geofisica e Vulcanologia – Sezione di Palermo – Via Ugo La Malfa, 153, 90146 Palermo, Italy

F. Di Maggio
Department of Chemistry, University College London, 20 Gordon Street, London WC1H 0AJ, UK

M. Ling
Department of Chemistry, University College London, 20 Gordon Street, London WC1H 0AJ, UK

A. Tsang
Department of Chemistry, University College London, 20 Gordon Street, London WC1H 0AJ, UK

J. Covington
School of Engineering, University of Warwick, Coventry CV4 7AL, UK

J. Saffell
Alphasense Ltd, 300 Avenue West, Skyline 120, Great Notley, Essex CM77 7AA, UK

C. Blackman
Department of Chemistry, University College London, 20 Gordon Street, London WC1H 0AJ, UK

P. Gembaczka
Fraunhofer Institute for Microelectronic Circuits and Systems, Duisburg, Germany

M. Görtz
Fraunhofer Institute for Microelectronic Circuits and Systems, Duisburg, Germany

Y. Celik
Fraunhofer Institute for Microelectronic Circuits and Systems, Duisburg, Germany

A. Jupe
Fraunhofer Institute for Microelectronic Circuits and Systems, Duisburg, Germany

M. Stühlmeyer
Fraunhofer Institute for Microelectronic Circuits and Systems, Duisburg, Germany

A. Goehlich
Fraunhofer Institute for Microelectronic Circuits and Systems, Duisburg, Germany

H. Vogt
Fraunhofer Institute for Microelectronic Circuits and Systems, Duisburg, Germany

W. Mokwa
Institute of Materials in Electrical Engineering, RWTH Aachen University, Aachen, Germany

M. Kraft
Fraunhofer Institute for Microelectronic Circuits and Systems, Duisburg, Germany

C. Bur
Lab for Measurement Technology, Saarland University, Saarbrücken, Germany
Div. of Applied Sensor Science, Linköping University, Linköping, Sweden

M. Bastuck
Lab for Measurement Technology, Saarland University, Saarbrücken, Germany

A. Schütze
Lab for Measurement Technology, Saarland University, Saarbrücken, Germany

J. Juuti
Microelectronics and Material Physics Laboratories, University of Oulu, Oulu, Finland

A.Lloyd Spetz
Div. of Applied Sensor Science, Linköping University, Linköping, Sweden
Microelectronics and Material Physics Laboratories, University of Oulu, Oulu, Finland

M. Andersson
Div. of Applied Sensor Science, Linköping University, Linköping, Sweden
Microelectronics and Material Physics Laboratories, University of Oulu, Oulu, Finland

S. Kishimoto
National Institute of Technology, Kochi College, Nankoku-shi, Japan

S. Akamatsu
National Institute of Technology, Kochi College, Nankoku-shi, Japan

H. Song
Kochi University of Technology, Kami-shi, Japan

J. Nomoto
Kochi University of Technology, Kami-shi, Japan

H. Makino
Kochi University of Technology, Kami-shi, Japan

T. Yamamoto
Kochi University of Technology, Kami-shi, Japan

A. Hannon
NASA Ames Research Center, Moffett Field, CA 94035, USA
ERC at NASA Ames Research Center, Moffett Field, CA 94035, USA

Y. Lu
NASA Ames Research Center, Moffett Field, CA 94035, USA
ELORET Corporation at NASA Ames Research Center, Moffett Field, CA 94035, USA

J. Li
NASA Ames Research Center, Moffett Field, CA 94035, USA

M. Meyyappan
NASA Ames Research Center, Moffett Field, CA 94035, USA

R. Schmitt
Laboratory for Machine Tools and Production Engineering WZL, Chair of Metrology and Quality Management, RWTH Aachen University, Steinbachstr. 19, 52074 Aachen, Germany

M. Köhler
Laboratory for Machine Tools and Production Engineering WZL, Chair of Metrology and Quality Management, RWTH Aachen University, Steinbachstr. 19, 52074 Aachen, Germany

J. V. Durá
Institute of Biomechanics of Valencia (IBV), Universitat Politècnica de València, Camino de Vera s/n, 46022 Valencia, Spain

J. Diaz-Pineda
Institute of Biomechanics of Valencia (IBV), Universitat Politècnica de València, Camino de Vera s/n, 46022 Valencia, Spain

CPSIA information can be obtained
at www.ICGtesting.com
Printed in the USA
BVOW10*0908010716

454236BV00001B/24/P